PHASE DIAGRAMS

Materials Science and Technology

VOLUME II

The Use of Phase Diagrams in Metal, Refractory, Ceramic, and Cement Technology

REFRACTORY MATERIALS

A SERIES OF MONOGRAPHS

John L. Margrave, *Editor*
DEPARTMENT OF CHEMISTRY
RICE UNIVERSITY, HOUSTON, TEXAS

VOLUME 1. L. R. McCreight, H. W. Rauch, Sr., and W. H. Sutton
Ceramic and Graphite Fibers and Whiskers
A Survey of the Technology

VOLUME 2. Edmund K. Storms
The Refractory Carbides

VOLUME 3. H. W. Rauch, Sr., W. H. Sutton, and L. R. McCreight
Ceramic Fibers and Fibrous Composite Materials

VOLUME 4. Larry Kaufman and Harold Bernstein
Computer Calculation of Phase Diagrams
With Special Reference to Refractory Metals

VOLUME 5. Allen M. Alper, Editor
High Temperature Oxides (In Four Parts)

VOLUME 6. Allen M. Alper, Editor
Phase Diagrams: Materials Science and Technology (In Three Volumes)

In Preparation

Louis E. Toth
Transition Metal Carbides and Nitrides

PHASE DIAGRAMS

MATERIALS SCIENCE AND TECHNOLOGY

Edited by ALLEN M. ALPER

Chemical and Metallurgical Division
Sylvania Electric Products Inc.
Subsidiary of General Telephone and Electronics
Towanda, Pennsylvania

VOLUME II

The Use of Phase Diagrams in
Metal, Refractory, Ceramic,
and Cement Technology

1970

ACADEMIC PRESS New York and London

ACADEMIC PRESS, INC.
111 Fifth Avenue, New York, New York 10003

United Kingdom Edition published by
ACADEMIC PRESS, INC. (LONDON) LTD.
Berkeley Square House, London W1X 6BA

LIBRARY OF CONGRESS CATALOG CARD NUMBER: 78-97487

PRINTED IN THE UNITED STATES OF AMERICA

DEDICATED TO

Professor Wilbur T. Valentine

for his guidance and instruction in geoscience while I was a student at Brooklyn College.

Contents

List of Contributors

Numbers in parentheses indicate the pages on which the authors' contributions begin.

A. M. ALPER* (117), Research and Development Laboratories, Corning Glass Works, Corning, New York

IVAN B. CUTLER (265), Division of Materials Science and Engineering, University of Utah, Salt Lake City, Utah

R. C. DOMAN (117), Research and Development Laboratories, Corning Glass Works, Corning, New York

F. P. GLASSER (147), Department of Chemistry, University of Aberdeen, Old Aberdeen, Scotland

D. LYNN JOHNSON (265), Department of Materials Science, Northwestern University, Evanston, Illinois

HOBART M. KRANER (67), Consultant, Bethlehem, Pennsylvania

GEORGE KRAUSS (293), Department of Metallurgy and Materials Science, Lehigh University, Bethlehem, Pennsylvania

JOSEPH F. LIBSCH (293), Department of Metallurgy and Materials Science, Lehigh University, Bethlehem, Pennsylvania

R. N. McNALLY (117), Research and Development Laboratories, Corning Glass Works, Corning, New York

T. B. MASSALSKI (221), Mellon Institute of Carnegie-Mellon University, Pittsburgh, Pennsylvania

ARNULF MUAN (1), The Pennsylvania State University, University Park, Pennsylvania

HORACE POPS (221), Mellon Institute of Carnegie-Mellon University, Pittsburgh, Pennsylvania

A. PRINCE (319), The General Electric Company Limited, Central Research Laboratories, Hirst Research Center, Wembley, England

J. TAYLOR (191), Department of Extraction Metallurgy, University of Strathclyde, Glasgow, Scotland

JAMES WHITE (21), Department of Ceramics with Refractories Technology, University of Sheffield, Sheffield, England

H. C. YEH† (117), Research and Development Laboratories, Corning Glass Works, Corning, New York

* Present address: Chemical and Metallurgical Division, Sylvania Electric Products Inc., Subsidiary of General Telephone and Electronics, Towanda, Pennsylvania.

† Present address: Department of Metallurgical Engineering, Cleveland State University, Cleveland, Ohio.

Foreword

Perhaps no area of science is regarded as basic in so many disciplines as that concerned with phase transitions, phase diagrams, and the phase rule. Geologists, ceramists, physicists, metallurgists, material scientists, chemical engineers, and chemists all make wide use of phase separations and phase diagrams in developing and interpreting their fields. New techniques, new theories, computer methods, and an infinity of new materials have created many problems and opportunities which were not at all obvious to early researchers. Paradoxically, formal courses and modern, authoritative books have not been available to meet their needs.

Since it is the aim of this series to provide a set of modern reference volumes for various aspects of materials technology, and especially for refractory materials, it was logical for Dr. Allen Alper to undertake this new coverage of "Phase Diagrams: Materials Science and Technology" by bringing together research ideas and innovative approaches from diverse fields as presented by active contributors to the research literature. It is my feeling that this extensive and intensive treatment of phase diagrams and related phenomena will call attention to the many techniques and ideas which are available for use in the many materials-oriented disciplines.

John L. Margrave

Preface

Many recent advances in materials science and technology have been made by scientists, engineers, and technologists who have used phase diagrams to solve materials problems. Several books have been written on the use of phase diagrams in the heat treatment of metals; however, the use of phase diagrams in other areas of materials science and technology has not been previously compiled. Also, there have been numerous advances in this field in the last five years which have not appeared elsewhere than in this volume.

This volume covers the use of phase diagrams in metals, refractories, ceramics, and cements. Each article has been written by an authority in the field. The contents should be extremely useful to all scientists and engineers who are investigating and developing materials and to those who are using materials. It should also help in the education of materials science students.

The editor wishes to thank Professor John L. Margrave of Rice University and Mrs. Thyrza C. Hanson and Dr. John H. Munier of Corning Glass Works; thanks are also due to Corning Glass Works and Sylvania Electric Products Inc. for their assistance. Special thanks are given to all the authors who have contributed articles to this volume.

Contents of Other Volumes

Volume I: Theory, Principles, and Techniques of Phase Diagrams

Volume III: The Use of Phase Diagrams in Electronic Materials and Glass Technology

I. The Use of Phase Diagrams in Crystal Growth

J. W. Nielsen and R. R. Monchamp

II. The Use of the Phase Diagram in Investigations of the Properties of Compound Semiconductors

M. B. Panish

III. Superconductivity and Phase Diagrams

V. F. Zackay, M. F. Merriam, and K. M. Ralls

IV. Rapidly Quenched (Splat-Cooled) Metastable Alloy Phases: Their Phase-Diagram Representation, Preparation Methods, Occurrence and Properties

B. C. Giessen and R. H. Willens

V. Liquid Immiscibility in Oxide Systems

Ernest M. Levin

VI. The Use of Phase Diagrams in Dissolution Studies

Alfred R. Cooper

VII. Relationships between Phase Diagrams and the Structure of Glass-Forming Oxide Melts

E. F. Riebling

VIII. Phase Relations and Dilute Molten Salt Solutions—The Cryoscopic Approach

T. R. Kozlowski

PHASE DIAGRAMS

Materials Science and Technology

VOLUME II

The Use of Phase Diagrams in Metal, Refractory, Ceramic, and Cement Technology

I

The Effect of Oxygen Pressure on Phase Relations in Oxide Systems

ARNULF MUAN

THE PENNSYLVANIA STATE UNIVERSITY
UNIVERSITY PARK, PENNSYLVANIA

I. INTRODUCTION

The representation and interpretation of phase relations in systems containing oxides of transition-metal oxides as components is more complicated than in systems of noble-metal oxides, because the former oxides are relatively unstable and because the elements commonly occur in several states of oxidation. The stability range of the various phase assemblages, therefore, are functions not only of temperature but also of the oxygen pressure* of the system. In order to describe these systems adequately, provisions must be made in the phase diagrams to include the oxygen pressure as a parameter in addition to those usually represented in phase diagrams, i.e., temperature and composition of the condensed phases. For binary systems, this

* The term *oxygen pressure* will be used throughout this chapter to represent the oxygen potential of the system. This "oxygen pressure" may be established by using oxygen gas, other oxygen-containing gaseous species (e.g., CO_2–CO mixtures), or it may be defined in a condensed system by temperature and composition of the phases present.

is easily accomplished by superimposing on the temperature–composition diagram a family of curves of equal oxygen pressures (i.e., oxygen isobars). For ternary and quaternary systems, however, this approach would result in diagrams that are practically illegible because of excessive cluttering of lines. In the latter case, therefore, it is found more practical to present supplementary diagrams to represent the oxygen isobars.

In the following we will derive some of the main features of phase diagrams representing systems in which the oxygen pressure is an important parameter, starting with binary systems and proceeding toward the more complicated ternary and quaternary systems. The principles involved will be derived for hypothetical systems. Experimental data will then be presented for a few systems of technological importance.

II. BINARY SYSTEMS

Consider a binary system A–O, where A represents a transition metal. Assume that this element occurs in two different oxidation states, viz., A^+ and A^{2+}, in addition to the neutral species A. A phase diagram representing

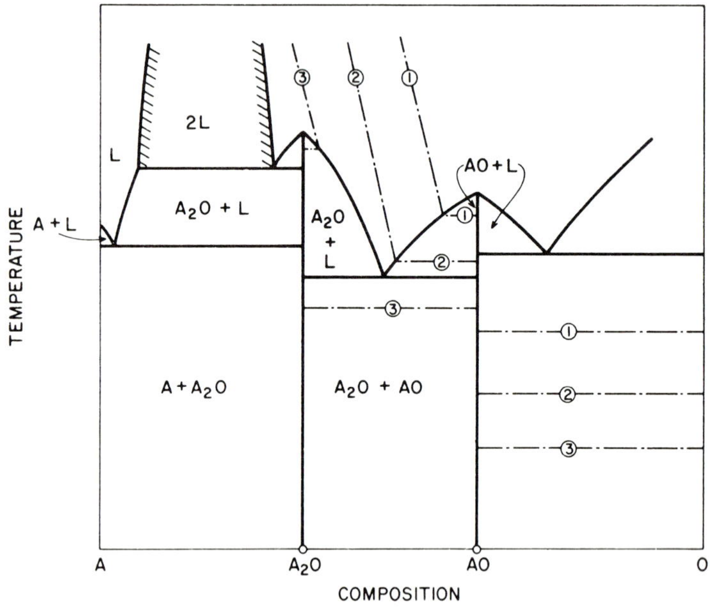

Fig. 1. Sketch of a hypothetical binary system A–O, showing boundary curves as heavy solid lines and oxygen isobars as light dash–dot lines. Lines with stippling on one side outline an area of two immiscible liquids.

this system is shown in Fig. 1. In order to provide a sound basis for an understanding of the more complex ternary and quaternary diagrams to be presented in subsequent sections, we will discuss in considerable detail some main factors involved in the use of this diagram.

Consider first equilibria between pairs of solid phases and a gas phase according to the equations

$$2\,AO_{(s)} = A_2O_{(s)} + \tfrac{1}{2}\,O_{2(g)} \tag{1}$$

and

$$A_2O_{(s)} = 2\,A_{(s)} + \tfrac{1}{2}\,O_{2(g)} \tag{2}$$

The two-component system has only one degree of freedom ($F = C + 2 - P = 2 + 3 - 3 = 1$) when three phases are present (two solids and a gas). Hence, choice of temperature for each of the univariant equilibria above fixes the oxygen pressure, and vice versa. The same situation prevails if one (or both) condensed phase(s) is (are) liquid. In a temperature–composition diagram, therefore, the oxygen isobars will be straight lines parallel to the composition axis in an area of the diagram where two condensed phases coexist in equilibrium.

If, on the other hand, only one condensed phase, either solid or liquid, is present in addition to the gas phase, the system has two degrees of freedom. Hence, temperature and oxygen pressure can vary simultaneously in an area of the diagram where only one condensed phase is present. The oxygen isobars therefore cross such an area diagonally rather than horizontally. These features are illustrated in the sketch in Fig. 1.

Consider next the problem of determining the highest temperature to which a mixture of composition AO can be heated without formation of a liquid phase, i.e., the solidus temperature. The answer depends on the conditions under which the heating is carried out (see Fig. 2). If conditions are such that the composition of the condensed phase remains constant, i.e., if there is no reduction or oxidation of A^{2+}, the phase changes can be read off the diagram by following a vertical line (i.e., isopleth) from low to high temperature (line *a–a′* in Fig. 2). Under these conditions, *a′* is the solidus temperature. Such constancy of total composition is usually achieved in systems of the type discussed here only if precautions are taken to avoid reaction between the condensed phase and the oxygen of the gas phase. This can be done, for instance, by sealing the condensed phase(s) into an inert, gas-tight container.

If such precautions are not taken, there will usually be sufficient reaction between the condensed phase and the gas to change significantly the composition of the former. However, under given sets of experimental conditions, it may be possible to control and graphically represent these compositional changes, and hence to show the phase relations prevailing under equilibrium

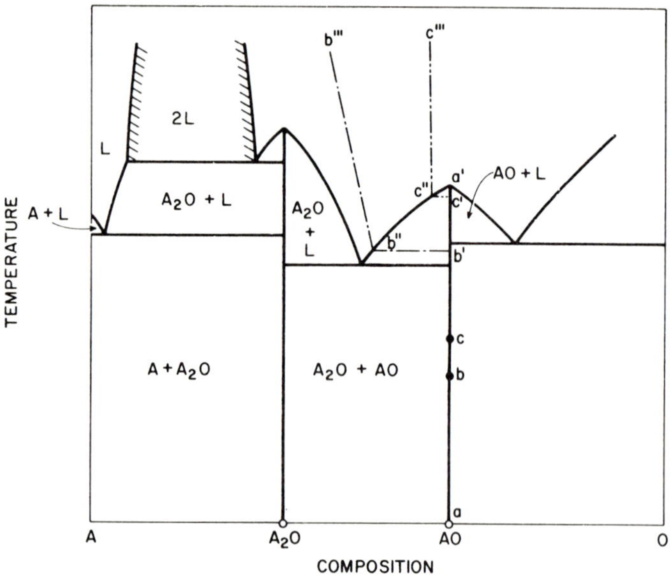

Fig. 2. Sketch illustrating use of the phase diagram A–O. In addition to the line symbols explained in the legend to Fig. 1 the light dash–double dot lines represents constant CO_2/CO ratio of the gas phase.

conditions. One such experimental condition consists in keeping the oxygen pressure constant during the heating. In this case the composition of the condensed phase(s) is represented by points on the oxygen isobar corresponding to the oxygen pressure in question. Curve b–b'–b''–b''' in Fig. 2 is an example of such a case. The sequence of phase changes occurring in a mixture AO as temperature increases at constant oxygen pressure are read off the diagram by following the oxygen isobar b–b'–b''–b''' from low to high temperature.

A third method of heating a mixture under controlled atmospheric conditions is to use a gas phase of constant CO_2/CO ratio. Under these conditions, the sequence of phase changes taking place is derived by following the appropriate line of constant CO_2/CO ratios superimposed on the diagram. An example of this is shown schematically by curve c–c'–c''–c''' in Fig. 2. The relation between oxygen pressure and CO_2/CO ratios is readily derived from thermodynamic data that are available for these species [see, for instance, the recent book of Muan and Osborn (1965) for a graphical illustration of these relations].

One important difference between melting at constant total composition of condensed phases and melting at constant oxygen pressure or constant CO_2/CO ratios of the gas phase should be noted in particular. Whereas in the former case solid and liquid can coexist over a temperature range, these two phases can coexist only at one temperature when the oxygen pressure or the CO_2/CO ratio of the gas phase is maintained constant. This is readily understood from the phase rule, which is concerned with the chemical potentials and hence chemical compositions of the individual phases present, not the total composition of a mixture. Changes of the latter, within certain limits, only cause changes in the relative amounts of the phases present, not their compositions. Hence, fixing the composition of the gas phase, for instance, the O_2 pressure or the CO_2/CO ratio of the gas, expends one degree of freedom, whereas fixing the total composition of the mixture does not impose a restriction on the system in terms of the phase rule. Hence, liquid and solid

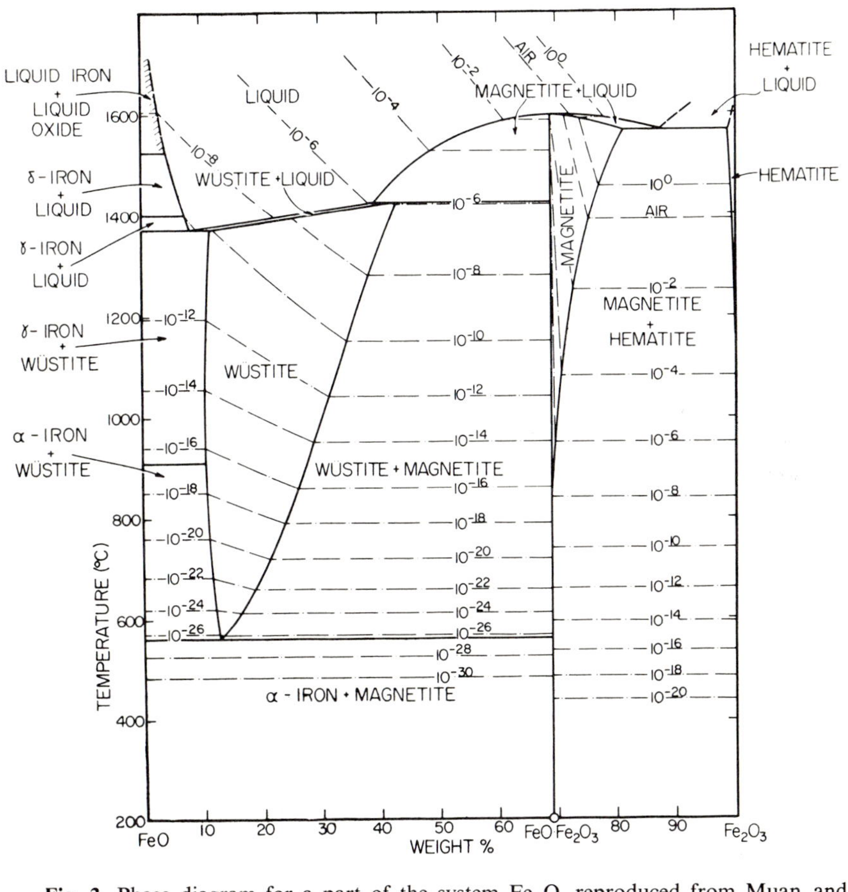

Fig. 3. Phase diagram for a part of the system Fe–O, reproduced from Muan and Osborn (1965). Heavy lines are boundary curves, lines with stippling on one side outline the two-liquid region, and light dash–dot lines are oxygen isobars.

can coexist over a temperature range when the total composition is kept constant, but not when the O_2 pressure or the CO_2/CO ratio of the gas phase is kept constant.

An example of a technologically important binary system of the type discussed above is Fe–O, as illustrated in Figs. 3 and 4. Detailed discussions of equilibria in this system have been presented in previous literature (Darken and Gurry, 1945, 1946; Muan, 1958; Muan and Osborn, 1965). Suffice it here to stress that phase changes at constant total composition of condensed phases are read off the diagram by following vertical lines, those during heating at constant oxygen pressures by following oxygen isobars (Fig. 3),

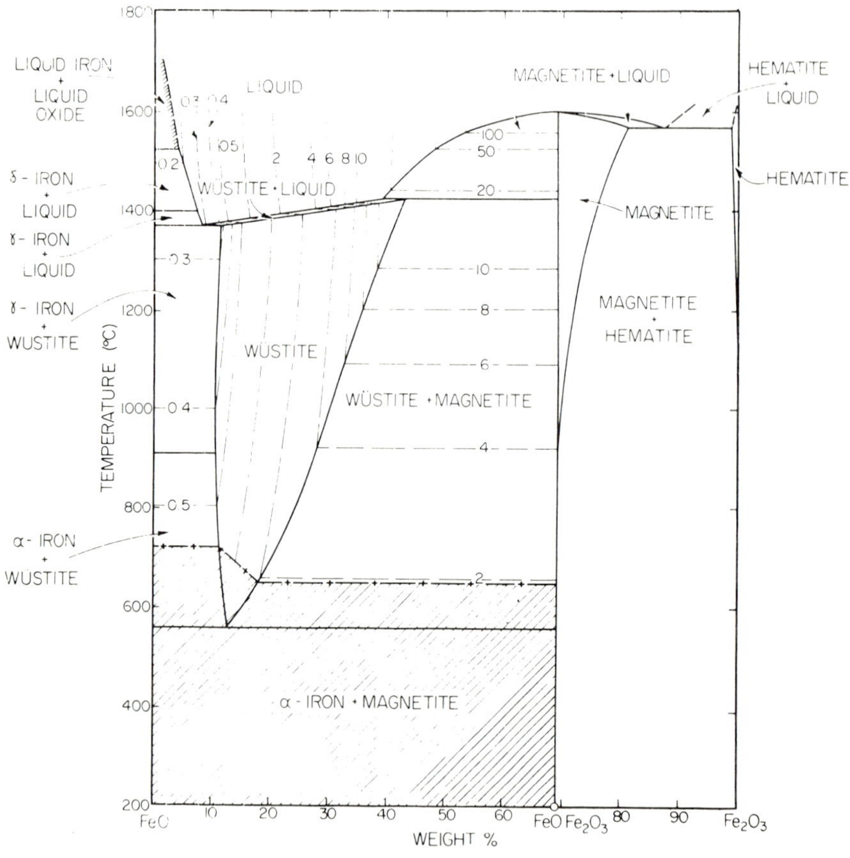

Fig. 4. Phase diagram for a part of the system Fe–O, reproduced from Muan and Osborn (1965). Light dash–dot lines represent constant CO_2/CO ratios of the gas phase. The shaded area below the dash–cross line in the lower part of the diagram represents an area within which carbon is present as a phase under equilibrium conditions.

and those during heating at constant CO_2/CO mixing ratios by following curves of constant CO_2/CO ratios (Fig. 4).

III. TERNARY SYSTEMS

The relations discussed above for the hypothetical binary system A–O will now be extended into a ternary system by adding B as a component. For sake of simplicity it will be assumed that B forms only one oxide phase, BO. Furthermore, it will be assumed that a compound A_2BO_2 is formed between A_2O and BO, that there is no other compound formation in the system, and that there is no appreciable mutual solubility among the components in the solid state. The general relations are shown in Fig. 5.

In the use of a diagram of the type shown in Fig. 5, the so-called oxygen reaction lines play a particularly important role. These are lines describing the changes in composition of condensed phases taking place as a result of

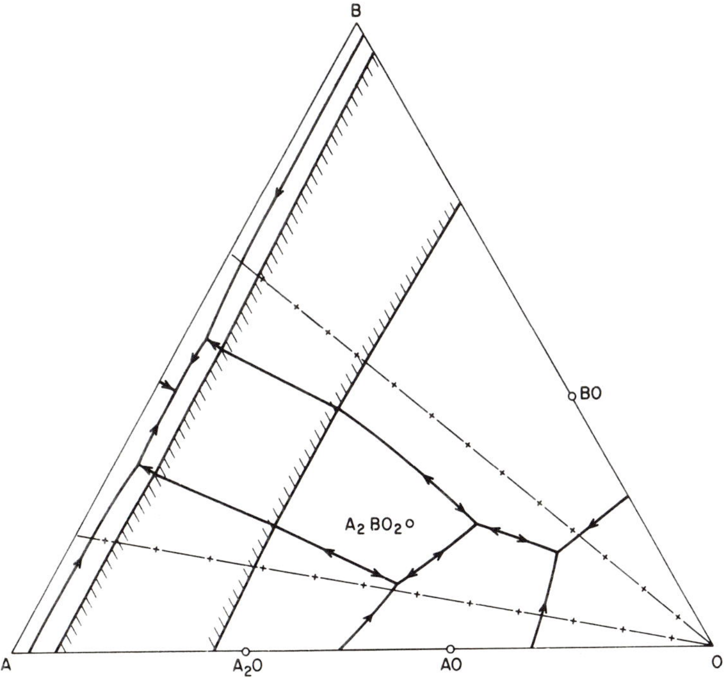

Fig. 5. Sketch of the liquidus surface of a hypothetical ternary system A–B–O. Heavy lines are boundary curves between the various primary-phase areas, lines with stippling on one side outline an area of two immiscible liquids, and light dash–cross lines are oxygen reaction lines.

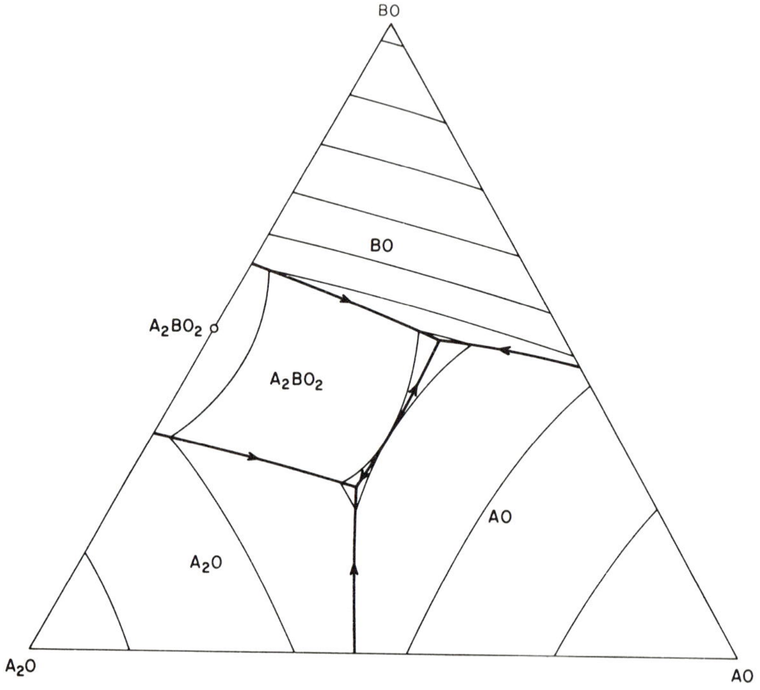

Fig. 6. Diagram showing the liquidus surface of the hypothetical system A_2O–AO–BO. Heavy lines are boundary curves, and light lines are liquidus isotherms.

their reactions with oxygen of the gas phase. The oxygen-reaction lines are straight lines radiating from the oxygen apex of the triangle representing the system. A number of these oxygen-reaction lines are shown as dash–cross lines in Fig. 5.

In the present discussion, attention will be focused on equilibria existing among the oxide phases. Hence, the composition area A_2O–AO–BO is of particular interest. This area is shown as an equilateral triangle in Fig. 6. Included in this diagram are boundary curves between adjacent primary-phase areas and liquidus isotherms.

A supplementary diagram of the system A_2O–AO–BO is presented in Fig. 7, showing the liquidus surface with boundary curves and oxygen isobars along the liquidus surface. The significance of these oxygen isobars is probably best understood by using the binary system A_2O–AO as a starting point and considering the effect of adding BO as another component. In the binary system (see Figs. 1 and 2), the temperature along the liquidus curve is fixed if the oxygen pressure is fixed, and vice versa. However, with

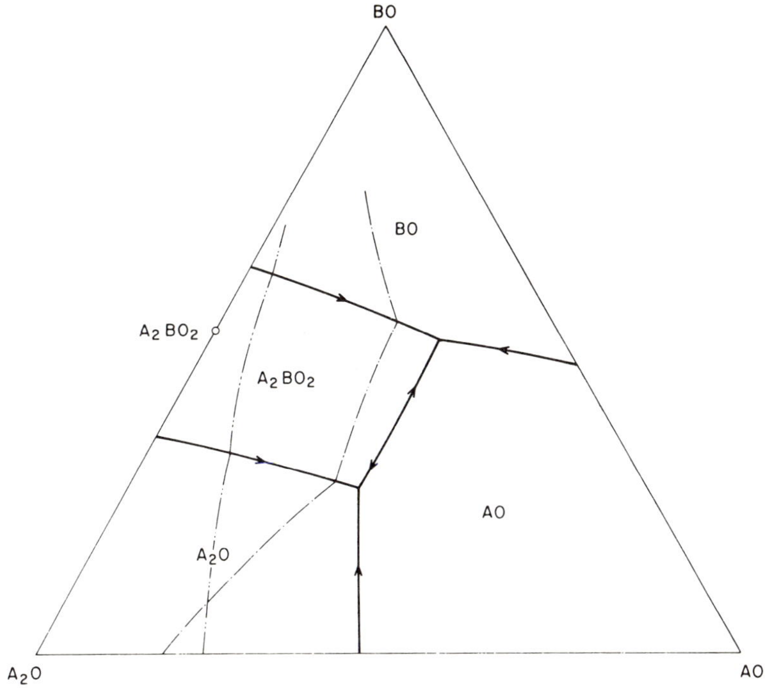

Fig. 7. Diagram showing liquidus surface of the hypothetical system A_2O–AO–BO. Heavy lines are boundary curves, light dash-dot line is an oxygen isobar, and light dash-double-dot line is a line of constant CO_2/CO ratio of the gas phase.

the addition of another component (BO), the system has one additional degree of freedom when the same phase assemblage is present (one crystalline phase and liquid). Hence, temperatures may change at constant oxygen pressure along the liquidus surface of the three-component system, and oxygen isobars are traced as univariant curves along this surface in the ternary system.

We will consider in some detail the phase relations existing in the temperature range between liquidus and solidus temperatures. These relations will be illustrated by a derivation of paths of equilibrium crystallization or equilibrium melting of selected mixtures. As in the case of the binary system A_2O–AO, we will distinguish between a number of idealized experimental conditions under which the crystallization or melting process is carried out.

First, it will be assumed that the total composition of condensed phases remains constant. This situation is approximated for instance when the oxide mixture is sealed in an inert, gas-tight container. Under these conditions, the derivation of path of crystallization or path of melting is no more complex than in "ordinary" oxide or silicate systems containing only oxides of

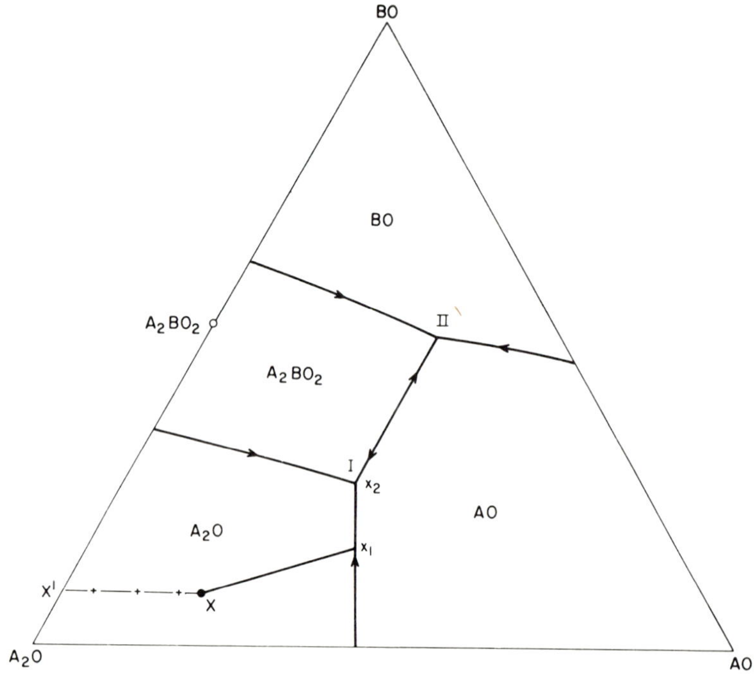

Fig. 8. Diagram showing path of crystallization or path of melting of a mixture X in the ternary system A_2O–AO–BO at constant total composition of condensed phases. The very heavy line X–x_1–x_2 shows the course of change in liquid composition, and the light line between the composition points AO and A_2BO_2 divides the triangle into two composition triangles each with an invariant point (I and II).

noble-gas-type ions as components. Suffice it here to present one example as shown in the sketch of Fig. 8. The liquid composition during crystallization of mixture X changes along the path X–x_1–x_2, and during melting along the same path in the opposite direction, i.e., x_2–x_1–X. Liquids between X and x_1 are in equilibrium with A_2O, liquids between x_1 and x_2 in equilibrium with AO and A_2O, and liquid x_2 is in equilibrium with AO, A_2O, and A_2BO_2. The temperature of the latter point is the solidus temperature of mixture X, i.e., the temperature at which the last trace of liquid disappears during equilibrium cooling, or the temperature at which the first trace of liquid appears during equilibrium heating of the mixture under the specified conditions. It is to be noted that this temperature is that of the liquidus invariant point I at which the liquid is in equilibrium with the three crystalline phases A_2O, AO, A_2BO_2 whose composition points constitute a composition triangle within which point X is located. The temperature of this invariant point is the solidus temperature for all mixtures located within this composition triangle when the heating is carried out at constant total compositions of the condensed phases. Similarly, the solidus temperature of mixtures within the composition triangle AO–BO–A_2BO_2 is that of invariant point II.

If the same mixture X is cooled or heated at constant oxygen pressure, however, the phase relations as a function of temperature are quite different from those described above. At constant oxygen pressure, the composition of the liquid coexisting with a crystalline phase must change along the appropriate oxygen isobar along the liquidus surface. The path of crystallization or melting under these conditions is shown in the sketch of Fig. 9. During equilibrium cooling, the liquid composition changes along the path X-x_3, as increasing amounts of A_2O crystallize out. Liquid of composition x_3 is in equilibrium with both A_2O and A_2BO_2. With oxygen pressure fixed, the phase assemblage

$$A_2O + A_2BO_2 + \text{liquid} + \text{gas}$$

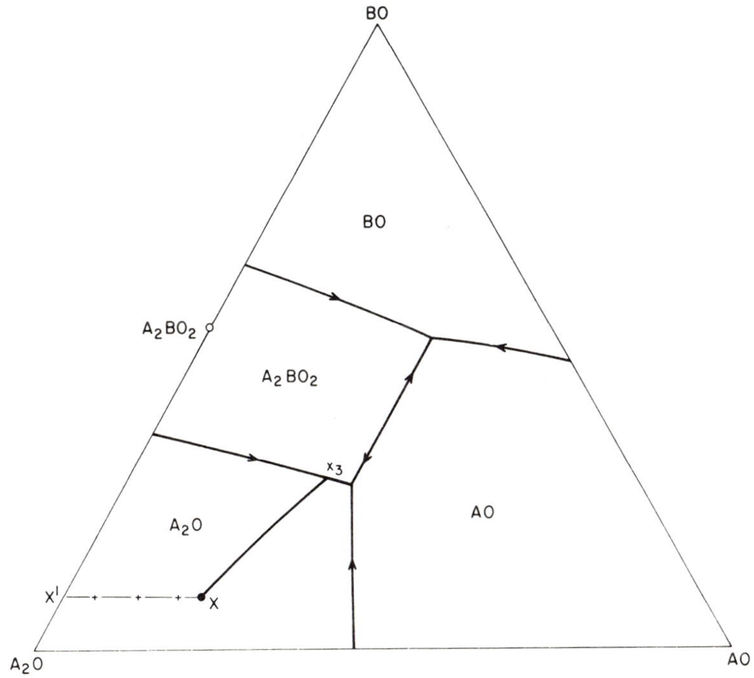

Fig. 9. Path of crystallization or path of melting in the hypothetical ternary system A_2O–AO–BO at constant oxygen pressure. The very heavy line X–x_3 shows the course of change in liquid composition, and the light dash–cross line is the oxygen reaction line through point X.

represents an invariant situation, and the liquid phase must disappear before the temperature can decrease further. Hence, the temperature of point x_3 is the solidus temperature when the oxygen pressure of the system remains constant. Under equilibrium heating of the same mixture, at constant oxygen pressure, the same sequence of phase changes in reverse direction take place. The first liquid to develop (at temperature t_{x_3}) has composition x_3. Upon further heating, the liquid composition changes along the path x_3–X, and the last trace of crystals disappears at the temperature of the latter point. In general, at constant oxygen pressure the solidus temperature in a ternary system involving changes in oxidation states is that of the intersection between the oxygen isobar in question and a liquidus-univariant curve.

Melting or freezing relations of the same mixture under conditions of constant CO_2/CO ratios of the gas phase are similar to those just described for constant oxygen pressure. However, the curves of constant CO_2/CO ratios along the liquidus surface have a different course than those of the oxygen isobars. The approximate relations are indicated in the sketch of Fig. 10. During equilibrium cooling of mixture X, the composition of the liquid

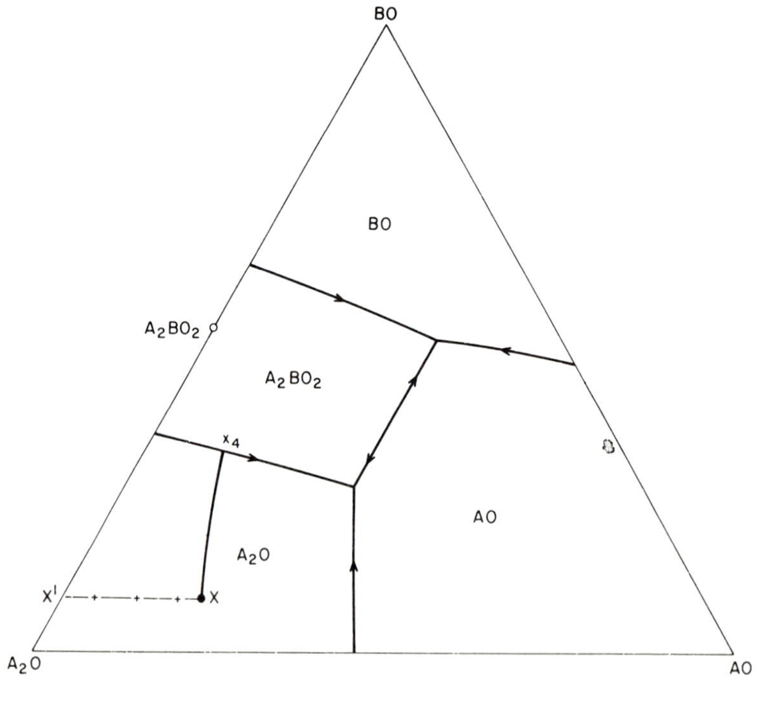

Fig. 10. Path of crystallization or path of melting in the hypothetical ternary system A_2O–AO–BO at constant CO_2/CO ratio of the gas phase.

follows the path X–x_4, and reverse during equilibrium heating of the mixture. The solidus temperature is that of the point of intersection between the liquidus univariant curve and the curve of constant CO_2/CO ratio. This applies as a general rule for deriving solidus temperatures for all mixtures under conditions of constant CO_2/CO ratios.

During cooling or heating at constant oxygen pressure or constant CO_2/CO ratio, the total composition of the mixture changes by reaction with oxygen of the gas phase. In both cases the total composition changes from X at the liquidus temperature to X′ at the solidus temperature.

An example of a technologically important ternary system with features similar to those described above is $FeO–Fe_2O_3–SiO_2$ (Muan, 1956). Projections of the liquidus surface of this system are shown in Figs. 11 and 12. These, as well as other aspects of phase relations in this system, have been discussed in detail elsewhere (Muan, 1956, 1958; Muan and Osborn, 1965).

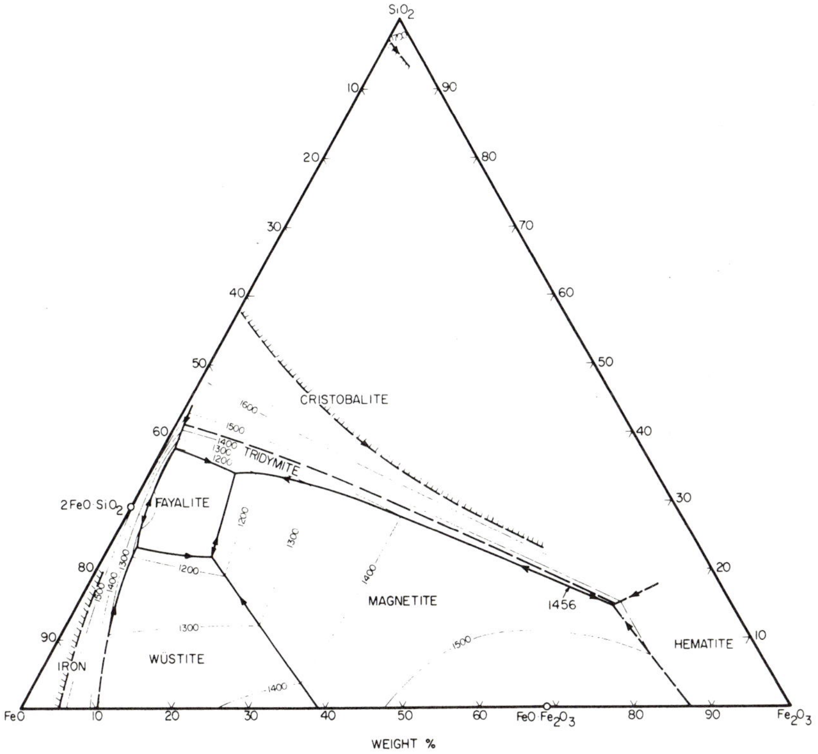

Fig. 11. Projection of the liquidus surface of the system $FeO–Fe_2O_3–SiO_2$, reproduced from Muan and Osborn (1965). Heavy lines are boundary curves, heavy lines with stippling on one side outline an area of coexistence of two immiscible liquids, and light lines are liquidus isotherms.

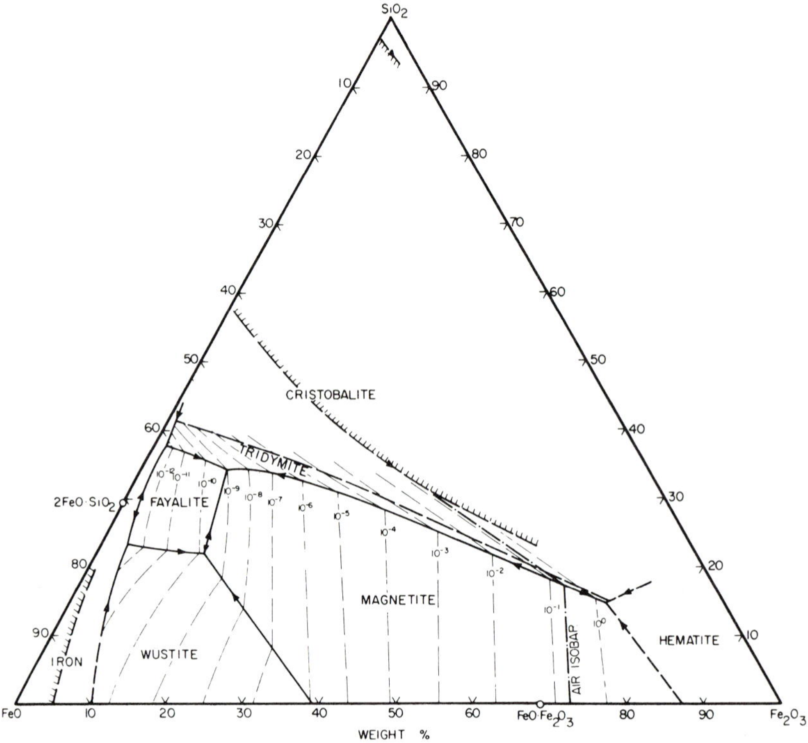

Fig. 12. Projection of the liquidus surface of the system FeO–Fe_2O_3–SiO_2, reproduced from Muan and Osborn (1965). In addition to lines explained in legend to Fig. 11, light dash–dot lines are oxygen isobars along the liquidus surface.

IV. QUATERNARY SYSTEMS

The reasoning involved in representing and interpreting phase relations in binary and ternary systems, as discussed in the foregoing, applies also to quaternary systems. In the latter systems, however, it is more difficult to present the relations in a simple form because of the many variables present. If detailed knowledge of data for a quaternary system is available, it is usually advantageous to build a tetrahedral model of the system in order to portray satisfactorily phase relations in the system in sufficient detail for practical use. In the present discussion, however, we will confine ourselves to only a brief consideration of the main features and the principles involved in extending into a quaternary system the relations just discussed for binary and ternary systems involving changes in oxidation states.

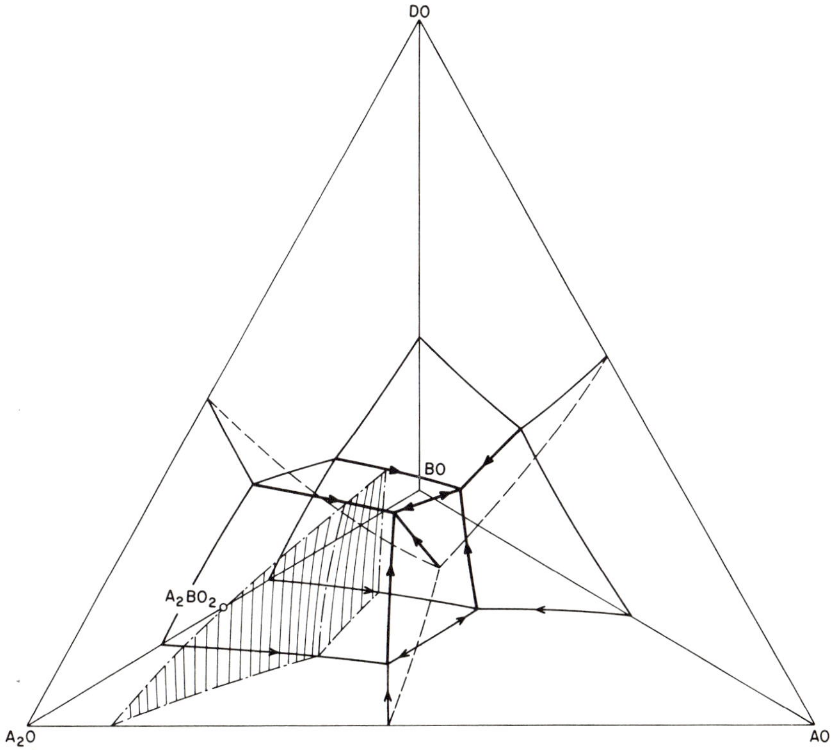

Fig. 13. Sketch showing phase relations at liquidus temperatures in a hypothetical quaternary system A_2O–AO–BO–DO. Medium lines represent boundary curves in the bounding ternary systems, heavy lines are univariant liquidus curves within the tetrahedron representing the quaternary system, and the shaded surface is an oxygen isobaric surface.

Assume that we add another component DO to the ternary system A_2O-AO-BO to form the quaternary system A_2O–AO–BO–DO, as sketched in Fig. 13. It is assumed that there is no compound formation or solid solubility between DO and the other components, and there is complete mutual solubility among all components in the liquid state. The primary phase areas in the ternary system become primary phase volumes in the quaternary system. The surfaces between two adjacent phase volumes are divariant surfaces representing compositions of liquids in equilibrium with two crystalline phases, the lines of intersection between three adjacent phase volumes are univariant lines representing compositions of liquids in equilibrium with three crystalline phases, and the point where four adjacent phase volumes touch each other represents the composition of a liquid in invariant equilibrium with four crystalline phases.

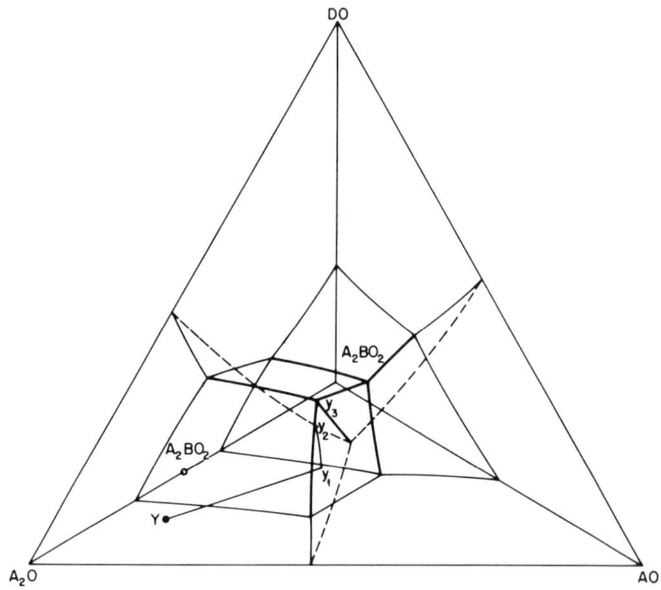

Fig. 14. Sketch showing path of equilibrium crystallization or path of equilibrium melting of a mixture Y in the quaternary system A_2O–AO–BO–DO at constant total composition of condensed phases. The curve Y–y_1–y_2–y_3 is the course of change of liquid composition. The other lines have the same meaning as explained in the legend to Fig. 11.

Similarly, the oxygen isobars in the ternary system become oxygen isobaric surfaces in the quaternary system.

During equilibrium cooling of a mixture Y in the quaternary system under conditions of constant total composition, the sequence of phase changes may be described briefly as follows (see Fig. 14). At the liquidus temperature, crystals of A_2O start to separate out, and the composition of the liquid changes along the straight line Y–y_1. The latter point is on the divariant surface between the primary-phase volumes of A_2O and AO. From here on, crystalline A_2O and AO coexist with liquid of composition changing from y_1 to y_2 along the divariant surface. A third crystalline phase, A_2BO_2, appears when the liquid composition reaches the univariant curve along which $A_2O + AO + A_2BO_2$ are in equilibrium with the liquid. With these three crystalline phases coexisting with liquid over a temperature range, the liquid composition follows the univariant curve until the quaternary invariant point $A_2O + AO + A_2BO_2 + DO +$ liquid is reached. At this point (y_3) the liquid disappears, giving rise to a final product consisting of the four crystalline phases $A_2O + AO + A_2BO_2 + DO$.

In general, if no solid solutions are involved, the solidus temperature in a quaternary system at constant total composition of condensed phases is the temperature of one of the quaternary liquidus-invariant points. Which among the liquidus-invariant points is the solidus temperature is determined by the location of the point representing the constant total composition, relative to the composition volumes in the system. The solidus temperature is that of the invariant point representing the composition of a liquid that is in equilibrium with the four crystalline phases forming the composition volume within which the point representing the total composition of the mixture is located.

Under conditions of constant oxygen pressure, the situation is quite different from that described above. As can be derived by the same reasoning as was used in extending binary-phase relations into ternary-phase relations, the liquid composition during cooling or heating at constant oxygen pressure must stay within the oxygen isobaric surface corresponding to the oxygen pressure in question. By analogy with the ternary systems, it follows that the solidus temperature at constant oxygen pressure in a quaternary oxide system involving components occurring in different oxidation states usually is the temperature of the point of intersection between a quaternary liquidus-univariant curve and the oxygen isobaric surface corresponding to the chosen oxygen pressure. For the hypothetical system under consideration, the relations are indicated in the sketch in Fig. 15. During equilibrium crystallization, the liquid first changes composition along the path Y–y_4, as crystals of A_2O separate out. When the divariant surface between the primary phase volumes of A_2O and A_2BO_2 is reached at point y_4, the liquid starts to change its composition along the path y_4–y_5, with A_2O and A_2BO_2 coexisting with the liquid. At point y_5, the univariant curve representing the compositions of liquids existing in equilibrium with the three crystalline phases A_2O, A_2BO_2, and DO is reached, and the liquid phase disappears at the temperature of this point, leaving as end product of the crystallization a mixture of the three crystallines phase A_2O, A_2BO_2, and DO. The total composition of the mixture during the crystallization process has changed along the oxygen reaction line from Y to Y′.

Under conditions of constant CO_2/CO ratios of the atmosphere, the relations will be similar to those just described, with surfaces of equal CO_2/CO ratios playing a role similar to that of the oxygen isobaric surfaces, as discussed above. During equilibrium cooling or heating, the liquid composition changes along the pertinent surface of constant CO_2/CO ratios

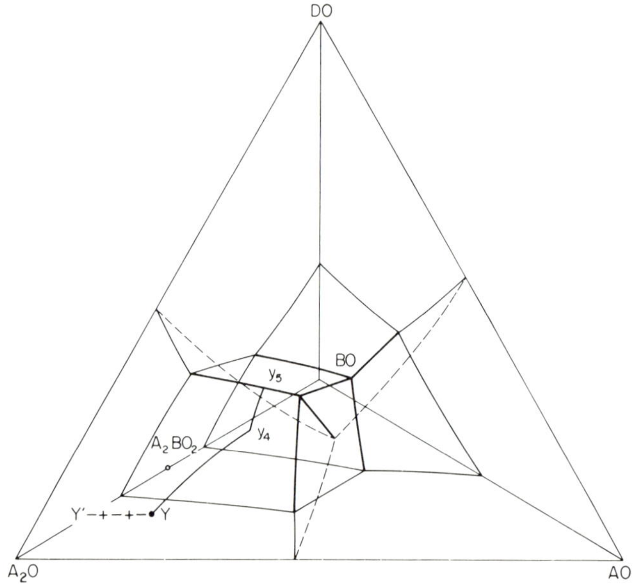

Fig. 15. Sketch showing path of equilibrium crystallization or path of equilibrium melting of a mixture Y in the quaternary system A_2O–AO–BO–DO at constant oxygen pressure. The curve Y–y_4–y_5 is the course of change of liquid composition. The other lines have the same meanings as explained in the legend to Fig. 14.

until a quaternary liquidus-univariant curve is reached. The solidus temperature at constant CO_2/CO ratio, therefore, usually is the point of intersection between a quaternary liquidus-univariant curve and the surface of constant CO_2/CO ratio.

V. GENERAL CONSIDERATIONS

We have simplified the presentation of the material herein by assuming that the crystalline phases have essentially constant compositions rather than being solid solutions of variable compositions. Systems in which solid-solution phases appear, in conjunction with varying oxygen pressures, have been discussed elsewhere [see, for instance, the recent book of Muan and Osborn (1965)].

The main differences from the present discussion resulting from the variability of composition of the crystalline phases are encountered under conditions of constant total composition of condensed phases. In this case, the liquid composition changes along a curved path rather than along a straight line when one crystalline phase and liquid coexist in equilibrium. Furthermore, the solidus temperature may be that of a point along a univariant curve rather than that of an invariant point.

During crystallization or melting at constant oxygen pressures or at constant CO_2/CO ratios, however, the pattern of change of liquid composition is similar, whether the crystalline phases have constant composition or are members of solid solutions with variable composition. In either case, the composition of the liquid phase at liquidus temperatures (a) is fixed in a binary system, (b) follows the oxygen isobar in a ternary system, (c) follows the oxygen isobaric surface in a quaternary system. The solidus temperature in all cases is that of the intersection between a liquidus univariant curve and an oxygen isobar or an oxygen isobaric surface.

References

Darken, L. S., and Gurry, R. W. (1945) *J. Am. Chem. Soc.* **67**, 1398.

Darken, L. S., and Gurry, R. W. (1946). *J. Am. Chem. Soc.* **68**, 798.

Muan, A. (1956). *J. Am. Ceram. Soc.* **39**, 121.

Muan, A. (1958). *Am. J. Sci.* **256**, 171.

Muan, A., and Osborn, E. F. (1965). "Phase Equilibria among Oxides in Steelmaking." Addison-Wesley, Reading, Massachusetts.

II

The Relationship of Phase Diagrams to Constitution and Microstructure in Ceramic and Ceramic–Metal Systems

JAMES WHITE

DEPARTMENT OF CERAMICS WITH REFRACTORIES TECHNOLOGY
UNIVERSITY OF SHEFFIELD, SHEFFIELD, ENGLAND

I. INTRODUCTION

The reason why phase-equilibrium diagrams are still investigated some 70 years after the first diagrams were constructed, and in spite of changes in fashion in research, is simply that they are still essential to an understanding of multiphase systems, which means, in turn, that they are essential to an understanding of most manufactured constructional materials and of many industrial processes.

This is particularly true of metals and ceramics and the processes by which they are made. Thus, to understand the relationship between the chemical composition and the constitution (i.e., the phases occurring) in a metallic alloy or a ceramic material, a knowledge of the relevant phase diagrams is necessary. Similarly, to understand the changes taking place during the heat treatment of an alloy or during the firing of a ceramic body, we must know what the equilibrium phases in it at each temperature should be.

It is also true, whether the material after cooling from the heat-treatment process is actually in an equilibrium condition or not. The heat treatment of steels, for example, is based on the iron–carbon phase diagram (which incidentally as normally used is a diagram of metastable equilibrium) although heat-treated steels are almost invariably in a nonequilibrium state. Similarly the fact that say a firebrick contains a glassy phase, and is, therefore not in equilibrium, does not prevent us from relating its constitution to the Al_2O_3–SiO_2 phase diagram.

In such cases we are usually dealing with arrested equilibria and, if metastable changes that may occur during cooling are allowed for, such as the austenite-to-martensite change during the quenching of steel or the β to α cristobalite change during the cooling of silica refractories, the condition that is found to exist at room temperature can be regarded as a frozen-in condition representing equilibrium at some higher temperature.

Phase equilibrium also plays an important part in many metal-extraction processes. For example, in steelmaking processes, while it is usually necessary that the slag be completely liquid at the operating temperature, it is also necessary that it be not too far removed from equilibrium with the refractory if excessive attack is to be avoided. In acid steelmaking processes this requirement largely determines the type of iron to be used, since it is necessary that it should contain sufficient silicon to form a silica-rich slag.

In basic processes, where lime must be added to the slag to promote phosphorus removal, the use of a basic refractory is necessary for similar reasons. In general, therefore, slags, like the refractories used, are either strongly acid or strongly basic, and processes that call for slags falling between these extremes usually earn a reputation for being troublesome.

While thermal equilibrium relationships determine the nature and the amounts of the phases occurring in high-temperature systems, there are many cases where the geometrical distribution of the phases is important. For example, in a metallic alloy containing a brittle phase, or a furnace refractory containing a liquid phase at high temperatures, the effect of these phases will usually be much less if they occur as discrete globules than if they occur as continuous films round grain boundaries.

The practical importance of these considerations was first demonstrated in the field of metallurgy by such examples as the action of FeS, which forms grain-boundary films, in causing "hot shortness" in steel and of Bi, which embrittles high-purity copper for the same reason. With the advent of "direct-bonded" basic refractories in which the normal silicate bond between the grains is replaced by direct grain-to-grain contact, they have become equally important in the field of ceramics. This aspect is discussed in Section IV.

II. LIQUID IMMISCIBILITY IN SILICATE MELTS

A classical example of the value of phase-equilibrium studies in the field of refractories is the development of the so-called superduty silica brick of low alumina and alkali contents. This took place in 1943 (USP, 1944) although the phase diagrams on which it was based had been published in 1927. In that year Greig (1927) showed that, in many basic oxide–silica systems, compositions high in silica form two immiscible liquid phases on melting. As a result the liquidus curve is displaced toward lower SiO_2 contents so that the solubility of SiO_2 in the melt is decreased.

The nature of this effect is shown in Fig. 1 where the liquidus curves of a number of binary basic oxide–silica systems have been drawn. The existence of immiscibility is shown by the occurrence on the curve of a monotectic horizontal, the termination of which indicates the limit of the miscibility gap at the monotectic temperature. (In each case the other limit, which lies

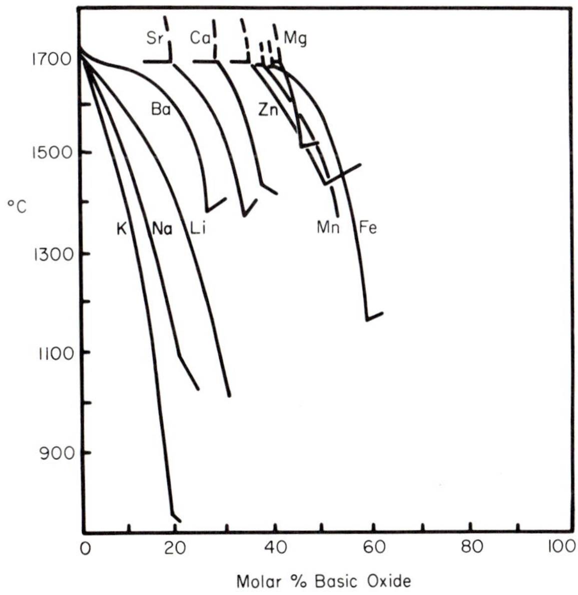

Fig. 1. Liquidus curves of binary basic oxide–silica systems.

close to the composition of silica, has been omitted to avoid overlap.) It will be seen that immiscibility occurs with all the oxides of the divalent cations except Ba^{2+}, which is the largest, and that, with these oxides, the width of the gap increases as the size of the cation decreases, showing that the positive deviation from Raoult's ideal-solution law, which must be large to give immiscibility, is increasing. No immiscibility occurs in the alkali–metal oxide–silica systems, but the steepness of the liquidus curves decreases as the size of the cation decreases showing that the positive deviation from Raoult's law is again increasing as the cation size decreases.

In the system Al_2O_3–SiO_2 similarly, no liquid immiscibility occurs and Greig showed that, when alumina or alkali oxides are added to silicate systems in which immiscibility occurs, their effect is to make the miscibility gap narrower and ultimately to eliminate it entirely. From 0.3 to 0.5% alkali or 2.5–3.0% Al_2O_3 is sufficient to do this in most cases. As a consequence, the presence of these oxides, even in small amounts, causes a marked decrease in the resistance of silica bricks to attack by steelmaking slags. (It also causes increased formation of liquid phase at high temperatures in unused bricks, which contain up to 2.0% CaO added as a "mineralizer," so

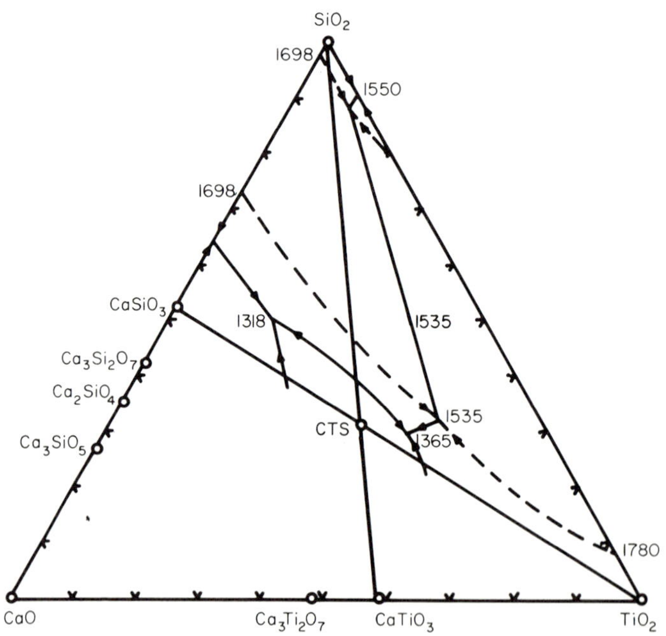

Fig. 2. Liquidus surface of ternary system CaO–TiO_2–SiO_2 showing extensive area of liquid immiscibility crossed by TiO_2–SiO_2 eutectic boundary at 1535°C. Immiscibility is normal on silica side of boundary line and inverse on TiO_2 side.

that the refractoriness under load is lowered.) On the basis of these relationships, a specification for a superduty brick low in alkalis, alumina, and titania was patented in the United States in 1943.

The conclusion that TiO_2 was harmful was based on the then accepted phase diagram of the system TiO_2–SiO_2, which indicated that a liquid miscibility gap did not occur in this system. It was, however, found that bricks containing up to 2.0% TiO_2 performed well in service and it appeared that the behavior of TiO_2 might be anomalous until a revision of the TiO_2–SiO_2 diagram showed that the earlier diagram had been in error, since immiscibility did occur, although on the TiO_2 side of the eutectic (De Vries *et al.*, 1954). In the ternary system CaO–TiO_2–SiO_2, this had the effect of producing a broad band of immiscibility across the diagram as shown in Fig. 2 (De Vries *et al.*, 1955). Consequently, in this system the solubility of solid silica in the liquid phase actually decreases slightly when TiO_2 is present. As shown by the diagram, some lowering of the temperature of complete melting is also to be expected but at the TiO_2 contents encountered this will be very small. An investigation of the system CaO–Al_2O_3–TiO_2–SiO_2 showed that a low Al_2O_3 content should be as desirable in TiO_2-containing bricks as in the more normal kind (Agamawi and White, 1952, 1953, 1954).

Figure 3 shows an attempt by Glasser *et al.* (1960) to correlate the width of the immiscibility gap in binary silicate systems with the electrostatic

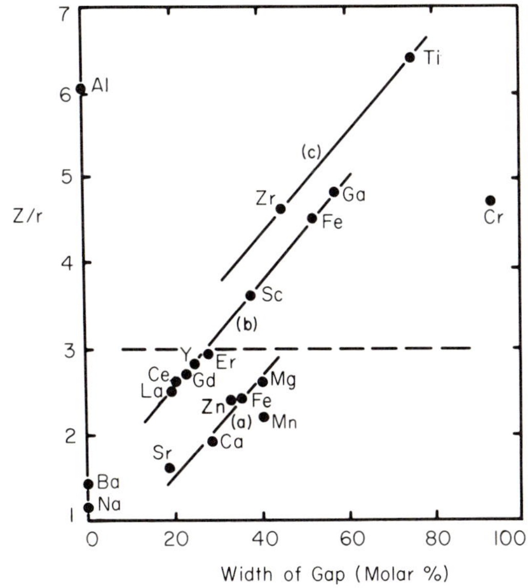

Fig. 3. Plots showing variation of width of miscibility gap in binary silicate systems with Z/r of cations. (a) Divalent cations. (b) Trivalent cations. (c) Quadrivalent cations.

potential of the cation as measured by the ratio Z/r, where Z is the charge on the cation in electron units and r its (Goldschmidt) radius. The plots for the divalent and trivalent cations lie along two straight lines that are approximately parallel, the width of the gap increasing with increasing Z/r. A straight line of similar slope was drawn for systems containing quadrivalent cations but only one of the two points shown on it, that for Ti^{4+}, was experimental, that for Zr^{4+} being inserted to show the width of gap predicted in the system ZrO_2–SiO_2. No miscibility gap has, of course, yet been found in this system.

Another discrepancy was noted with Cr_2O_3 since the experimentally observed gap was much wider than would have been predicted from its value of Z/r. Finally, the fact that Al_2O_3 and B_2O_3 ,which would be expected to give wide miscibility gaps from their values of Z/r do not give immiscibility at all was attributed to the fact that they could accept fourfold coordination by oxygen and thus replace Si^{4+} in SiO_4 tetrahedra.

It was also found that, where the value of Z/r was less than 3.0 (the value indicated by the broken horizontal line), the gap occurred over the primary crystallization field of silica (normal immiscibility), whereas, when it was greater than 3.0, it occurred over the primary crystallization field of the other oxide (inverse immiscibility).

In ionic terms these effects can be broadly correlated with the ability of the cations to fit into the interstices between the SiO_4 tetrahedra in the silicate structure. This requires that the cations should be able to surround themselves (1) with a coordination shell having the correct number of O^{2-} ions to satisfy radius-ratio requirements and (2) with the correct number of nonbridging oxygens to neutralize their charge. At high SiO_2 contents, when the content of bridging oxygens is high, (1) will be satisfied more readily by large cations having a large coordination number for O^{2-} ions, since the SiO_4 tetrahedra will tend to arrange themselves in "networks" having large interstices similar to those observed in solid silica, while (2) will be most easily satisfied by cations of small charge, since they will allow maximum randomness in the arrangement of the nonbridging oxygens.

Analogous effects are observed in silicate systems in the solid state in that (1) the number of silicates formed in a system, and in particular the ability to form silicates of high SiO_2 content and (2) the heats of formation of silicates of the same type, e.g., orthosilicates or metasilicates, in different systems both increase as Z/r decreases. Further, large (negative) heats of formation of the solid silicates in a system are in general associated with a decreased positive deviation or a negative deviation from Raoult's law in the liquid state or both. (In most silicate systems in which compounds occur, there is a positive deviation at high silica contents and a negative deviation within the composition range in which the compounds occur. As the negative deviation increases, the positive deviation decreases so that the width of the miscibility gap decreases and finally disappears.)

These effects can be correlated in terms of a treatment by Ramberg (1954), which, though chemically naive, is useful in indicating the nature of the factors involved. Ramberg considered that the heat of formation of the silicates was associated with polarization of the oxygen ions in the structure, which would increase as the difference between the polarizing powers of the cations and the Si^{4+} ion increased, and he showed that, for the orthosilicates of ideal-gas-type divalent cations, the calculated polarization energies (i.e., energy release on polarization) of the O^{2-} ions with increasing cation size (decreasing Z/r) was roughly proportional to the observed increase in ΔH. (Owing to the simplifying assumptions made in his treatment, Z canceled out from his equations so that he could not predict the effect of changing the charge on the cations.)

On this basis it is possible to suggest explanations for some of the discrepancies observed by Glasser *et al.* Thus the fact that no immiscibility has been found to occur with ZrO_2 can possibly be correlated with the fact that it forms a silicate. Similarly, the fact that values of Z/r less than 3.0 are associated with normal immiscibility and values greater than 3.0 with inverse can possibly be related to the fact that, while the former form silicates, the latter in general do not. Consequently, whereas in the former case the positive deviation from Raoult's law is restricted to a range of high-silica compositions, in the latter it is likely to extend over the whole range of compositions.

On the other hand, the abnormally wide miscibility gap in the case of Cr^{3+}, which Glasser *et al.* suggest is in error, is probably associated with the energy of *d*-level splitting in this ion when it is octahedrally coordinated by oxygen ions. This energy is a maximum in the first transition series at Cr^{3+}. When the oxygen ions surrounding the Cr^{3+} ions are simultaneously bonded to Si^{4+}, it would be expected that polarization would decrease their electronegativity toward the Cr^{3+}, thus lowering the energy associated with splitting so that it would tend to separate out in a phase low in silica. There is some evidence in support of this suggestion (Curtis, 1963). Thus, the solubility of Cr_2O_3 appears to be generally low in silicate melts (other than those rich in CaO or alkalis) from which it separates as Cr_2O_3 or as mixed oxide phases such as spinels, in each of which Cr^{3+} will be surrounded by relatively unpolarized O^{2-} ions. Again Cr^{3+} does not form any silicates, whereas Be^{2+} and Al^{3+}, which have higher values of Z/r do. (Although Ti and Zr are transition elements, no effects of this kind are to be expected with Ti^{4+} and Zr^{4+} since they contain no *d*-level electrons.)

The fact that with ideal-gas-type cations the stabilities of silicates of the same type increase as Z/r for the cation decreases, and hence, with cations of constant charge, as the size of the cation increases has important consequences in the technology of basic refractories and cements. Thus if SiO_2 is added in limited amounts to a mixture of MgO and CaO, 3 CaO·SiO_2 and 2 CaO·SiO_2 are formed preferentially, depending on the CaO/SiO_2 ratio, while the MgO remains free. Only when the molar ratio falls below 2.0 does MgO begin to react with the silica and then as a partial replacement for CaO in 3 CaO·MgO·2 SiO_2 (merwinite) and CaO·MgO·SiO_2 (monticellite).

III. PHASE EQUILIBRIA IN BASIC REFRACTORIES

Phase-equilibrium studies have also played an essential part in the development of the modern range of basic refractories. These are of three main types, dolomite and magnesite, both of which have been in use as refractories since the introduction of the basic steelmaking processes, and chrome–magnesite, the use of which dates from the early 1930's.

Chromite (chrome ore) had been used as a neutral refractory for many years before that, but its range of uses had been limited by its relatively low softening point under load. The addition of magnesite fines was found to give a material that was not only superior to chrome in this respect but also to the magnesite refractories then being produced and led to the development of the all-basic open-hearth furnace in which most of the upper structure of the furnace was of chrome–magnesite instead of silica.

It was early recognized that the superiority of chrome–magnesite relative to chrome refractories could be explained on the basis of the phase equilibrium diagram of the binary system MgO–SiO_2, which showed that the effect of adding magnesia would be to convert the relatively low-melting magnesium silicates, that occur in the ore in association with the highly refractory grains of chrome spine, to forsterite (2 MgO·SiO_2), which has a melting point of 1910°C. This explanation, however, could not account for the influence of the other constituents in the brick, some of which, e.g., lime, were known to be harmful, nor did it account for the fact that chrome–magnesite was generally superior to magnesite in load-bearing properties and in resistance to lime and iron-oxide attack.

The answer to these questions turned out to be a matter partly of phase relationships and partly of structure. The first step in this direction was the identification of the solid phase assemblages coexisting with the magnesiowustite phase in the seven-component system CaO–MgO–FeO–Fe_2O_3–Al_2O_3–Cr_2O_3–SiO_2. This problem was first tackled in the Metallurgy Department of the Royal Technical College, Glasgow, in the 1930's, and the work led to the formulation of a system of phase assemblages that defined the

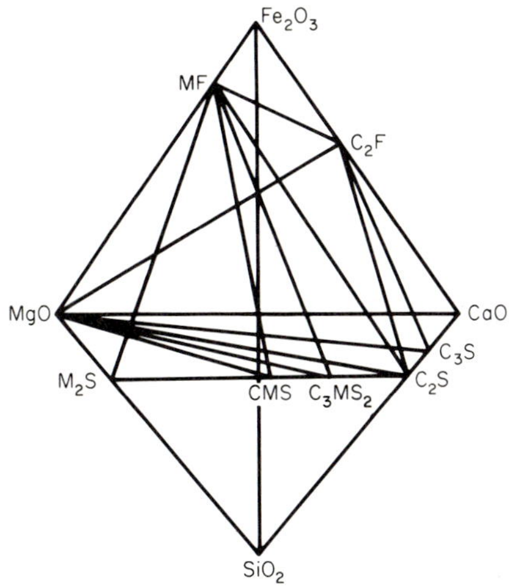

Fig. 4. Phase relationships in the solid state in the system CaO–MgO–Fe_2O_3–SiO_2. Phase volumes shown are those in which MgO occurs as a solid phase.

combinations of phases coexisting with magnesia in the solid state in the seven-component system.

Fortunately, the relationships were found to be fairly simple owing to the prevalence of solid solution and the principal features can be understood from the solid phase relationships in the corresponding part of the system CaO–MgO–Fe_2O_3–SiO_2 that are shown in Fig. 4 (White, 1962).

The latter shows the various phase combinations or assemblages that can occur in the presence of free MgO (periclase). Each assemblage consists of four phases whose compositions lie at the corners of a tetrahedron within the composition tetrahedron of the complete system. These smaller tetrahedra also define the composition limits within which the various phase assemblages exist.

An important feature is the critical nature of the CaO/SiO_2 ratio. When it is 2/1 molecularly, the composition lies on the plane MgO–2 CaO·SiO_2–MgO·Fe_2O_3. When it is greater than 2/1, all or part of the Fe_2O_3 occurs in combination with CaO as low-melting dicalcium ferrite. When it is less than 2/1, all the Fe_2O_3 occurs in combination with MgO as spinel.

Essentially similar relationships hold when FeO, Al_2O_3, and Cr_2O_3 are present. This is illustrated in Table I, which omits certain complications that arise at CaO/SiO_2 molar ratios greater than 2.0 (when CaO occurs in combination with Fe_2O_3, Al_2O_3, and Cr_2O_3). Up to a ratio of 2.0—i.e., so long as the CaO occurs entirely in the silicates—the phase combination is determined uniquely by the value of the ratio. With ratios greater than 2.0, this is no longer rigorously true, although if the sesquioxide content is kept constant,

TABLE I

PHASE COMBINATIONS IN BASIC REFRACTORIES

Molar ratio CaO/SiO_2	0–1.0 (1)	1.0–1.5 (2)	1.5–2.0 (3)	(4)	Greater than 2.0 (5)	(6)
	Magnesio-wustite	Magnesio-wustite	Magnesio-wustite	Magnesio-wustite	Magnesio-wustite	Magnesio-wustite
	Spinel	Spinel	Spinel	Spinel	Dicalcium silicate	Lime
	Forsterite	Monticellite	Merwinite	Dicalcium silicate	Tricalcium silicate	Tricalcium silicate
	Monticellite	Merwinite	Dicalcium silicate	Ca ferrites, aluminates and chromites	Ca ferrites, aluminates and chromites	Ca ferrites, aluminates and chromites

the tendency is still to move toward the right-hand side of the table as the ratio increases. Another point to note is that, up to a ratio of 2.0, each of the three phase combinations occurring contains only four phases: magnesio-wustite, spinel, and two silicates. For a more complete treatment, reference can be made to Rait (1950) or White (1962).

The phase assemblages have been found to account for the phases found both in unused fired bricks and in used bricks after CaO and iron oxides have been picked up in service. Thus the constitution of dead-burned dolomite is represented by assemblage (6) and that of dolomite stabilized against hydration by addition of serpentine by assemblage (5), while the constitution of magnesites usually corresponds to assemblages (2) or (3), or if the CaO/SiO_2 ratio is low enough to assemblage (1).

The constitution of fired chrome–magnesite and magnesite–chrome bricks, which normally have CaO/SiO_2 ratios less than 1.0, usually corresponds to phase assemblage (1). Hence all the CaO is present as monticellite, which melts at 1502°C. It is for this reason that a high CaO content has usually been considered undesirable in chrome-magnesite refractories. During service in basic furnaces, CaO is, of course, picked up by the bricks, and monticellite forms at the expense of forsterite and the liquid migrates back into the bricks. Consequently, in used bricks, the CaO/SiO_2 ratio usually increases behind the hot face, and phase combinations corresponding to (2) and (3) may occur. In some cases ratios greater than 2.0 can be reached and dicalcium ferrite, dicalcium silicate, and tricalcium silicate have been reported.

While the phase assemblages have proved useful in predicting the solid phases to be expected in basic refractories, some latitude must be allowed in applying the CaO/SiO_2 ratios given in Table I. The reason is that there is appreciable solubility between certain of the phases at high temperatures.

Thus monticellite and merwinite exist over appreciable ranges of CaO/MgO ratios at high temperatures, while part of the MgO in forsterite can be replaced by CaO and part of the CaO in Ca_2SiO_4 by MgO (Ricker and Osborn, 1954; Schlaudt and Roy, 1966). In both cases the solubility decreases with falling temperature and is small below 1000°C.

Again it is now known that, in the system CaO–MgO, the lime and magnesia are appreciably soluble in each other at high temperatures (though probably negligibly so at low), the solubility of CaO in MgO rising from approximately 0.9% by weight at 1620°C to 7% at 2370°C and that of MgO in CaO from 2.5% at 1620°C to 17% at 2370°C (Doman *et al.*, 1963).

Further it has been reported that SiO_2 in the system $MgO–SiO_2$, where it is combined as $2\,MgO{\cdot}SiO_2$ is appreciably soluble in MgO at high temperatures (Schlaudt and Roy, 1965). The solubilities proposed are unexpectedly high, being of the order of 8% at 1590°C and 12% at 1780°C, and it seems unlikely that solubilities of this order occur in commercial refractories.

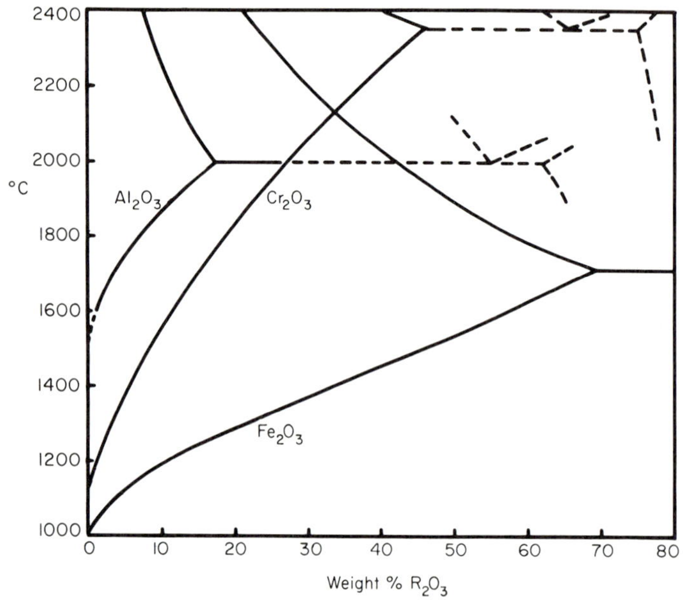

Fig. 5. Solid solubilities of sesquioxides in periclase at high temperatures.

When CaO and SiO_2 occur in combination with each other, their solubilities in MgO, particularly that of SiO_2, are probably considerably lower (2 $CaO{\cdot}SiO_2$, for example, is appreciably more stable than 2 $MgO{\cdot}SiO_2$) and there is evidence that, when MgO is in contact with Ca_2SiO_4 at 1700°C some CaO, but very little SiO_2, is dissolved (Melford, 1967). Such selective solution could appreciably increase the CaO/SiO_2 ratio required to give Ca_2SiO_4 at low SiO_2 contents.

Al_2O_3, Cr_2O_3, and iron oxide are also partially soluble in MgO at high temperatures. Figure 5 shows the solubilities of these oxides in the systems MgO–$MgAl_2O_4$ (Alper *et al.*, 1962), MgO–$MgCr_2O_4$ (Hayhurst, 1961; Alper *et al.*, 1964), and MgO–$MgFe_2O_4$ (Woodhouse and White, 1955; Phillips *et al.*, 1961). In the last system loss of oxygen occurs at high temperatures to an extent that depends on the oxygen pressure, so that the magnesiowustite formed contains both Fe^{2+} and Fe^{3+} cations. (See Section V,A.) The solubility curve shown was determined in air.

Since all three solubilities increase with rising temperature and are very low below 1000°C, it follows that, when chrome–magnesite refractories are heated to high temperatures, the periclase grains will dissolve Al_2O_3, Cr_2O_3, Fe_2O_3, and FeO from the grains of chrome spinel. Since the only other constituent of the spinel is MgO, this means that, in effect, solution of the spinel in the MgO will occur. On subsequent cooling the dissolved oxides precipitate as particles of a "mixed" spinel within the periclase grains. Because solution is selective, owing to the different solubilities of the sesquioxide, the particles of precipitated spinel will have a different composition from the original spinel. Figure 8 shows the spinel precipitate in the periclase grains of a MgO–$CaMgSiO_4$–Cr_2O_3 mixture that was cooled slowly from 1700° to 1500°C and then quenched.

At CaO/SiO_2 ratios greater than 2.0, when the sesquioxides occur in combination with CaO, their solubilities in the periclase should be considerably lower than those shown in Fig. 5 owing to the greater stabilities of the calcium compounds. Stephenson and White (1967) have shown that at 1550°C in air, MgO in contact with a melt rich in dicalcium ferrite dissolved only about 2% iron oxide.

The phase assemblages also make no provision for the occurrence of a new orthosilicate, which occurs between dicalcium silicate and merwinite, and has been called phase *T* (Johnson, 1967; Gutt, 1961, 1965; Schlaudt and Roy, 1966). The existence of this silicate would mean that phase assemblage (3) in Table I should be subdivided into two phase assemblages containing $2\,CaO{\cdot}SiO_2 + T$ and $T +$ merwinite, respectively. T has, however, been found to dissociate into merwinite and $2\,CaO{\cdot}SiO_2$ in the solid state below the temperature of melt formation and seems, further, to be formed with some difficulty, so that it has been considered expedient to ignore it for the present purpose.

Although the phase assemblages indicate the solid phases to be expected in a basic refractory of given composition, and hence the conditions under which low melting phases like monticellite and dicalcium ferrite will be formed, they give no indication of the extent to which temperatures of melt formation may be lowered by the formation of eutectics. Such lowering may be considerable. Thus although dicalcium silicate and $MgO{\cdot}Al_2O_3$ melt individually at 2130° and 2135°C, they form a eutectic that melts at 1418°C. Similarly dicalcium silicate and $MgO{\cdot}Fe_2O_3$ form a eutectic that melts in air at 1415°C.

The problem of studying melting relationships in a seven-component system is, of course, a formidable one. A simplified approach was, therefore, adopted by El-Shahat and White (1964, 1966a, b). to establish trends in melting behavior with composition in assemblages (1), (2), and (3) of Table I. These authors pointed out that in each of these assemblages only four solid phases coexist and that if the system contained only four components, melting would begin in each at a quaternary invariant point at which the four phases would coexist with a liquid phase. In the quaternary system CaO–MgO–Al_2O_3–SiO_2, Solacolu (1960) had shown that all three points were peritectic points and that they occurred at 1388°, 1366°, and 1387°C, respectively. In each of these four-phase assemblages in the seven-component system, on the other hand, the four solid phases would coexist with the liquid phase over a range of temperatures while the temperature of initial melting would vary with the composition of the solid-solution phases, viz., spinel and magnesiowustite.

They further argued that if a change in the composition of one of the solid-solution phases was found to be accompanied by say an increase in the temperature of initial melting in one of the component three-phase or two-phase combinations, then the same change of composition would cause an increase in the temperature of initial melting in the three-phase assemblages in which this combination occurred. They, therefore, undertook an investigation of the influence of the composition of the spinel on melting behavior in the spinel-silicate combinations occurring in the four-phase assemblages.

Figure 6 shows the melting relationships established in the system $MgAl_2O_3$–$MgCr_2O_4$–Ca_2SiO_4. (Only the part of the composition triangle containing over 50% Ca_2SiO_4 is actually shown.) It will be seen that the temperature of the binary eutectic formed between the silicate and the continuous series of spinel solid solutions increases from 1418° to 1700°C as the Al_2O_3 of the spinel is replaced by Cr_2O_3. Another feature of the diagram is

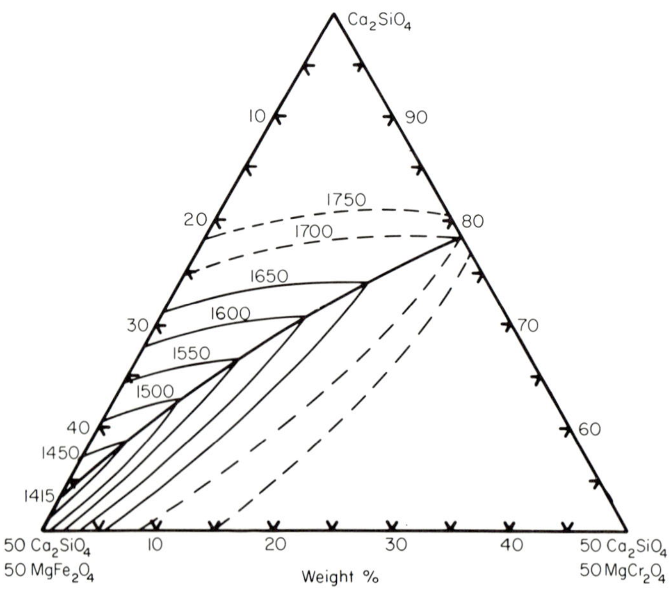

Fig. 6. Liquidus surface of the system $MgFe_2O_4$–$MgCr_2O_4$–Ca_2SiO_4.

that the isotherms on the liquidus surface in the primary field of the spinel slope away from the base of the triangle, so that the solubility of the spinel in the liquid silicate decreases, as the Al_2O_3 is replaced by Cr_2O_3. Consequently, in mixtures of spinel with minor amounts of $CaSi_2O_4$, the quantity of melt formed at a given temperature will decrease as the Cr_2O_3/Al_2O_3 ratio in the spinel increases.

Similar relationships were found to exist in the system $MgFe_2O_4$–$MgCr_2O_4$–Ca_2SiO_4, which was investigated in air, i.e., the temperature of the eutectic increased from 1415° to 1700°C as the iron oxide in the spinel was replaced by Cr_2O_3, while the solubility of the spinel in the melt decreased even more steeply than in Fig. 6. In the system $MgFe_2O_4$–$MgAl_2O_4$–Ca_2SiO_4, on the other hand, the temperature of the eutectic increased only slightly from 1415° to 1418°C as Fe_2O_3 was replaced by Al_2O_3. The solubility of the spinel in the melt, however, decreased appreciably so that the order of increasing solubility of the spinel sesquioxides in the melt is

$$Cr_2O_3 < Al_2O_3 < Fe_2O_3$$

Cr_2O_3 being considerably less soluble than the other two. This series may be

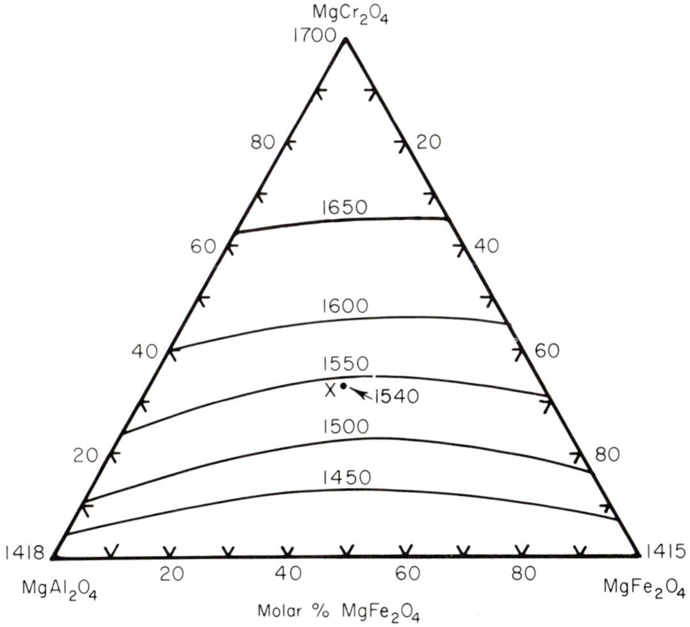

Fig. 7. Variation in temperature of melting of eutectic in the system $MgAl_2O_4$–$MgFe_2O_4$–$MgCr_2O_4$–Ca_2SiO_4 with variation in composition of spinel solid solution. Compositions giving the same melting temperature are joined by isotherms.

compared with the order of increasing solid solubility of the spinel sesquioxides in the magnesiowustite phase shown in Fig. 5, which is

$$Al_2O_3 < Cr_2O_3 < \text{“}Fe_2O_3\text{”}$$

where the “ Fe_2O_3 ” is considerably more soluble than the other two.

Figure 7 shows how the temperature of complete melting of the eutectic varies with the relative proportions of Al_2O_3, “ Fe_2O_3,” and Cr_2O_3 in the spinel. It will be evident that the melting temperature is determined mainly by the Cr_2O_3 content, the relative proportions of Al_2O_3 to Fe_2O_3 having little effect.

Similar trends with spinel composition were observed in the corresponding systems containing monticellite and merwinite, although the phase relationships were considerably more complex owing to the fact that both these compounds melt incongruently with separation of MgO. Extensive fields of primary crystallization of MgO, therefore, occurred in these systems. Another feature of these systems was that, because of the relatively low temperatures of peritectic melting of the silicates (1492°C with monticellite, 1575°C with merwinite), the influence of Cr_2O_3 on the temperature of initial melting was considerably less than with Ca_2SiO_4. Thus the effect of replacing all the Al_2O_3 in $MgAl_2O_4$–monticellite mixtures by Cr_2O_3 raised the temperature of initial melting by 80° only, viz., from 1410° to 1490°C.

Fig. 8. Photomicrograph of specimen containing 70% MgO, 15% $CaMgSiO_4$, 15% Cr_2O_3 by weight, fired at 1700°C for 2 h, then furnace cooled to 1500°C and quenched. (Reflected light. 320 ×.)

These temperatures are in good agreement with an early observation by Berry *et al.* (1950), who found that the addition of $MgCr_2O_4$ to a mixture of MgO, forsterite, and monticellite had negligible effect on the temperature of complete melting, whereas the addition of $MgAl_2O_4$ lowered it to 1420°C. They also found, in agreement with the relative solubilities of the sesquioxides in the liquid silicates given above that, in zones adjacent to the hot face of used chrome–magnesite roof blocks, the Al_2O_3 content of the spinel grains had decreased while that of the silicate matrix had increased.

Another consequence of the solubility of the spinel sesquioxides in liquid silicates is that, when chrome–magnesite refractories are fired at high temperatures and subsequently cooled, crystals of spinel can be seen to separate from the melt between the periclase grains, forming "bridges" between them. This effect has been used to produce what is sometimes referred to as "spinel bonding" in fired chrome-magnesite refractories (Hayhurst and Laming, 1963). Crystals of this so-called secondary spinel, which have crystallized from the liquid silicate during cooling, can be seen in Fig. 8.

IV. THE EQUILIBRIUM DISTRIBUTION OF THE PHASES IN MULTIPHASE BODIES

A. Nature of the Factors Controlling Phase Distribution

As already mentioned in Section I, the geometrical distribution of the phases has an important bearing on the properties of multiphase alloys and ceramics. The factors determining the distribution in equilibrium structures were first established by work on alloys, notably by Smith (1948), which showed that the distribution of the minor phase in a two-phase alloy is governed by the geometrical balance of the surface-tension forces, where interfaces between different phases and grain boundaries between like phases intersect.

In particular the condition that a phase β should be able to penetrate completely between the grains of a phase α, i.e., should form continuous films round the α grains, is that

$$\gamma_{\alpha\alpha} \geqslant 2\gamma_{\alpha\beta}$$

where $\gamma_{\alpha\alpha}$ is the surface tension of the α–α grain boundaries and $\gamma_{\alpha\beta}$ that of the α–β interfaces.

When $\gamma_{\alpha\alpha} < 2\gamma_{\alpha\beta}$ complete penetration will not occur but a balance of forces will be reached when

$$\gamma_{\alpha\alpha} = 2\gamma_{\alpha\beta}\cos(\phi/2), \tag{1}$$

where ϕ is the dihedral angle formed by the intersecting α–β interfaces. ϕ is measured at right angles to the edge along which the two α–β interfaces meet the α–α boundary, i.e., the edge along which the two α grains and the β grain meet.

Smith showed that, as ϕ increased from zero, penetration of β between the α grains would decrease, though up to a value of 60° it would still be capable of penetrating along such three-grain edges. The structure would thus consist of two continuous interpenetrating phases. When $\phi > 60°$ the second phase would occur as discrete inclusions along four-grain junctions.

It is necessary, of course, that at the temperature concerned, the grains can change their shape, however slowly, by diffusion or other processes, under the influence of the surface tension forces. It is also assumed that these forces are exerted tangentially to the surfaces and are independent of their direction in the surfaces and of the relative crystallographic orientation of the grains so that the surface tensions in the two α–β interfaces are equal. Smith showed that the equilibrium structures observed after annealing in the microstructures of two- and three-phase alloys could be accounted for on the basis of these assumptions.

In general, when the grains of the major phase are rounded, they seem to be justified. Where the grains are markedly anisometric and have well-marked crystallographic faces, they will no longer hold. It can be assumed that such faces will represent surface orientations of minimum energy and that any rotation of such a surface relative to the crystallographic axes will be accompanied by an increase in surface energy. The surface forces will then no longer be entirely tangential and isotropic, since there will be additional forces acting normal to the surfaces concerned, tending to oppose rotation (torque terms), and they will no longer be numerically and dimensionally equivalent to the specific surface free energies (Mykura, 1966), which was tacitly assumed in deriving Eq. (1). Because of the torque terms, it is not then very precise to use the term *surface tension*, and Eq. (1) should be replaced by an equation due to Herring (1951), which was derived by minimizing the energy and allows for the effect of the torque terms. This equation is not, however, usually very helpful, since it contains too many unknown variables. [The real criterion for equilibrium of the surface forces is that the surface energy should be a minimum and Eq. (1) defines this condition only when the surface tensions and specific surface free energies are equivalent.]

In practice the sizes of the dihedral angles observed in a polished section vary between 0° and 180°, even when ϕ is constant throughout the structure, because the three-grain edges intersected by the section are randomly oriented. It was, however, shown by Harker and Parker (1945) that, if a large enough number of angles is measured and a histogram constructed, the most frequently occurring angle should be the true angle. Alternatively, the median has been used as a measure of the true angle (Riegger and Van Vlack, 1960). Other structural parameters obtainable from such a section are the mean grain size (Fullman, 1953) and the ratio of the contact area between grains (grain-boundary area) to their total surface area (grain-boundary area + phase-boundary area) (Jackson *et al.*, 1963). The latter ratio is obtained by drawing straight lines across a micrograph and counting the numbers of intersections they make with grain boundaries ($N_{\alpha\alpha}$) and phase boundaries ($N_{\alpha\beta}$). Then the ratio is given by

$$N_{\alpha\alpha}/(N_{\alpha\alpha} + N_{\alpha\beta}) = N_{\alpha\alpha}/N \tag{2}$$

B. "Direct Bonding" in Refractories

The importance of controlling the distribution of the low-melting phases in furnace refractories was pointed out by Allison *et al.* (1959). Previous to this Parikh and Humenik (1957) had shown that in certain cermets, agglomeration and growth of the particles of the dispersed phase during firing were inhibited when they were completely wetted by the liquid metal so that a film of the latter was maintained between them. They assumed that the condition for complete penetration was that θ, the wetting angle formed by a drop of liquid metal lying on the ceramic (carbide) phase, should be zero, which is not necessarily a sufficient condition. White (1968), however, has pointed out that, under certain conditions, which may have been fulfilled by the systems investigated, ϕ would be zero when $\theta = 0$.

Subsequently, Van Vlack (1960) studied the influence of temperature and atmosphere on the value of ϕ in quenched bodies consisting of tridymite and cristobalite and a liquid iron silicate phase at the firing temperature and found that, in the bodies containing cristobalite, ϕ was about 55° at 1550°C and decreased only slowly with rising temperature so that a considerable degree of direct contact was maintained between the cristobalite grains up to the temperature of complete melting. He concluded that the high value of ϕ could explain the high refractoriness under load of silica refractories. Another contributory factor is, of course, the existence of liquid immiscibility discussed in Section II.

The influence of chemical composition and temperature on ϕ in basic refractories has been the subject of a number of papers from the writer's Department. The initial investigations were aimed at establishing the type of distribution to be expected in bodies consisting of periclase (magnesiowustite) with minor amounts of silicate and with examining the effects, on ϕ and the distribution, of the other oxides that may be present in commercial basic refractories.

For this purpose bodies consisting of 85% MgO and 15% $CaMgSiO_4$, which would consist of periclase and a liquid phase saturated with MgO at temperatures above 1502°C, were selected initially and additions of Fe_2O_3, Cr_2O_3, Al_2O_3, and TiO_2 were made to them as replacements for equal weights of MgO so that the monticellite content remained constant.

Figure 9 shows the effect of progressive additions of Fe_2O_3 and Cr_2O_3, singly and together, on bodies that had been fired at 1550°C in air for 2 h and quenched (Jackson *et al.*, 1963). ϕ is plotted against the total content of R_2O_3 so that the curve marked "No Fe_2O_3" shows the effect of adding Cr_2O_3 alone, and so on. In all cases the addition of Cr_2O_3 increased ϕ, whereas the addition of Fe_2O_3 lowered it.

The dashed line ab represents the saturation limit at which spinel crystals were present at the firing temperature. On crossing ab the curves become horizontal, which is to be expected since two solid phases then coexist with the liquid and the compositions of the phases would then remain constant, only their proportions varying, with increasing content of R_2O_3.

Figure 10 shows how the degree of contact between the periclase grains (measured in this case by $N_{\alpha\alpha}/N_{\alpha l}$) increased with increasing Cr_2O_3 and decreased with increasing Fe_2O_3, showing that it varied in the same sense as ϕ. Figure 11 shows the microstructures of two of the bodies examined. A greater incidence of large angles and an increased contact between the periclase grains can be seen in the body containing Cr_2O_3.

Further investigations (Jackson and Ford, 1966) at temperatures up to

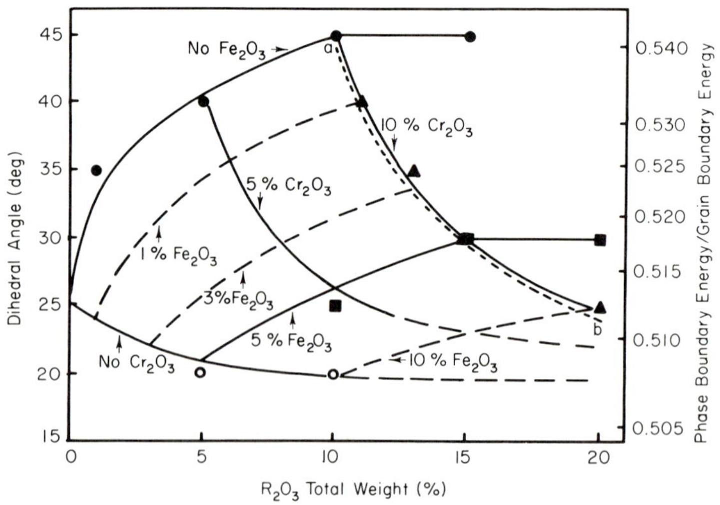

Fig. 9. Effect of additions of Cr_2O_3 and Fe_2O_3 on dihedral angle formed between periclase grains in periclase–monticellite mixtures at 1500°C.

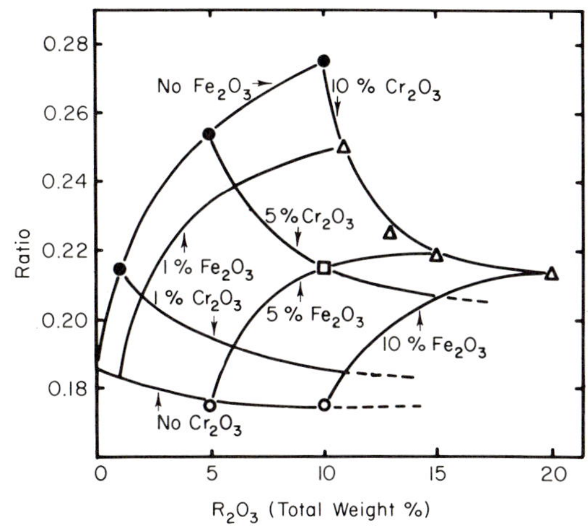

Fig. 10. Effect of additions of Cr_2O_3 and Fe_2O_3 on ratio of solid–solid boundary area to solid–liquid boundary area in periclase–monticellite mixtures at 1550°C.

1725°C showed that additions of Al_2O_3 and TiO_2 lower ϕ and decrease the degree of contact between the periclase grains to a greater extent than Fe_2O_3, while increasing the CaO/SiO_2 ratio in the silicate increases ϕ and the degree of contact, possibly because, as shown by the phase diagram of the system CaO–MgO–SiO_2, the MgO concentration in the saturated liquid phase decreases as the CaO/SiO_2 ratio decreases. Smith (1948) found similarly that, in alloys, ϕ increased as the concentration of the solid phase in the saturated liquid phase decreased, i.e., as the difference in composition between the solid and liquid phases increased, which, he considered, would increase $\gamma_{\alpha l}$. The action of Cr_2O_3 with increasing ϕ is not at present understood. It seems possible that it does so by lowering the energy of the periclase–periclase grain boundaries.

Two other significant observations were made in the course of these investigations, viz., that the grain size after a given firing time and the degree to which the samples densified both decreased as ϕ increased. The former effect, which is illustrated in Fig. 11, led to a study of grain growth in periclase–liquid, lime–liquid, and corundum–liquid bodies by Buist *et al.* (1965), who found that, when the grains were rounded, the cube of the mean grain diameter increased linearly with time, i.e.,

$$d^3 - d_0{}^3 = k(t - t_0)$$

The significance of this finding arises from the fact that it is the relationship predicted by Greenwood (1956) and later by Wagner (1961) and by

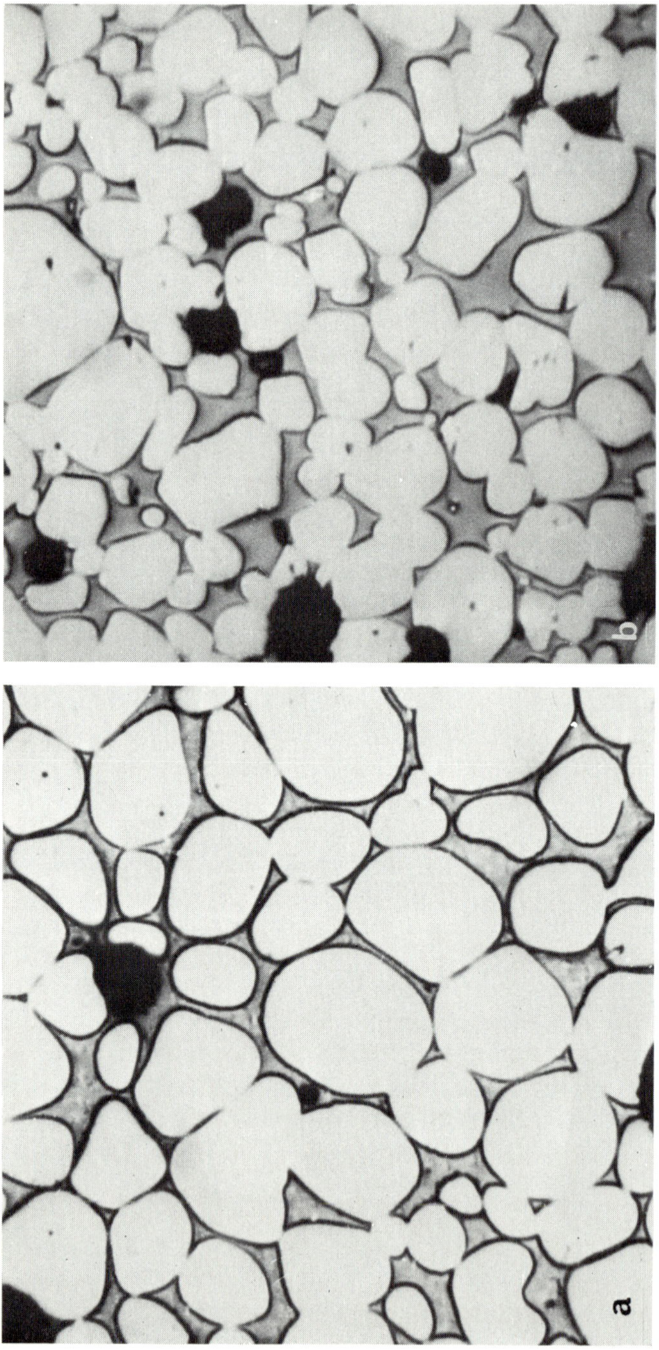

Fig. 11. Micrograph of mixtures containing (a) 80% MgO, 15% $CaMgSiO_4$, 5% Fe_2O_3, (b) 80% MgO, 15% $CaMgSiO_4$, 5% Cr_2O_3, after firing for 8 h at 1550°C in air. Round grains are periclase.

Lifschitz and Slezov (1961) for growth in dispersions of spherical particles in a liquid phase in which they are appreciably soluble. The basis of all three treatments was the well-known Thomson–Freundlich relation between particle radius and solubility in the liquid phase and the mechanism postulated involved solution of the smaller grains and growth of the larger, with transport of material by diffusion through the liquid phase.

The finding that the d^3 relationship might still apply when the particles were in contact had been foreshadowed by Greenwood, who found that his equation predicted growth rates of the right order in settled slurries of UPb_3 in liquid lead and, to test its applicability to the bodies studied by them, Buist *et al.* used it to calculated values of D, the diffusion coefficient for transport through the liquid phase, from the values they had obtained for k. The values obtained were found to be of the right order and they concluded that grain growth occurred by solution and deposition at the curved surfaces that were in contact with the liquid phase.

This finding at once raised the question as to how large grains growing by this mechanism while in contact across grain boundaries with other grains could remain isometric (cf. Fig. 11). To overcome this difficulty Buist *et al.* pointed out that, when grains of different sizes are in contact, deposition on the larger grain and solution (or slower deposition) at the surface of the smaller would displace the root of the neck between the grains, and hence the periphery of the grain boundary between them, towards the smaller grain. Hence grain boundary migration towards the smaller grain would occur. Evidence supporting this conclusion has been observed by Stephenson and White (1967) who observed that, in certain bodies containing two solid phases and a liquid, large grains of the faster growing phase do, in fact, tend to grow round small grains of the slower growing phase, forming a curved interface.

To explain the effect of ϕ on grain growth Buist *et al.* (1965) suggested that in an assemblage of grains with finite contact areas, the surface curvature would tend to decrease as ϕ increased, assuming that the diameters of the contact areas remained constant. The latter assumption was later rejected by Stephenson and White who suggest that ϕ may affect particularly the curvature and hence the solubility of small particles that are in contact with large ones. They argue that, when a small sphere is in contact with two or more larger spheres of given size and separation, there will be a critical size of the small sphere that meets the surfaces of the larger spheres at the equilibrium angle and still remains spherical. Below the critical size, which increases as ϕ increases, the mean surface curvature and solubility of the smaller sphere will decrease relative to those of a perfect sphere of the same volume.

An explanation for the effect of ϕ on densification was also suggested by Stephenson and White. These authors calculated the relative change in total boundary energy (grain-boundary energy plus solid–liquid-interfacial energy) that would occur with various values of $A = \gamma_{\alpha\alpha}/\gamma_{\alpha l}$ in an assemblage of uniform spherical grains as the grain centers are brought closer together causing flat grain boundaries to develop between neighboring grains. It was assumed that the grain volumes remained constant, and it was then found that at all values of A less than 2.0 the energy passed through a minimum that corresponded to the distance between the centers at which the surfaces of the spheres would meet at the equilibrium angle defined by Eq. (1). This is illustrated in Fig. 12 in which the change in energy per sphere, expressed as a fraction of the initial energy when the spheres were in point contact, is plotted against p, the decrease in center-to-center distance expressed as a fraction of $2r_0$, the initial distance.

It will be evident from the curves that if shrinkage depended on the development of necks between the grains, as in the sintering of single phase powders, it would increase as ϕ increased. The fact that the reverse is the case suggests that when the liquid phase is forming the shrinkage is determined by the ability of the solid particles to rearrange themselves. Any tendency for the particles to cohere on touching would thus oppose shrinkage. The existence of a force causing cohesion and proportional to the initial

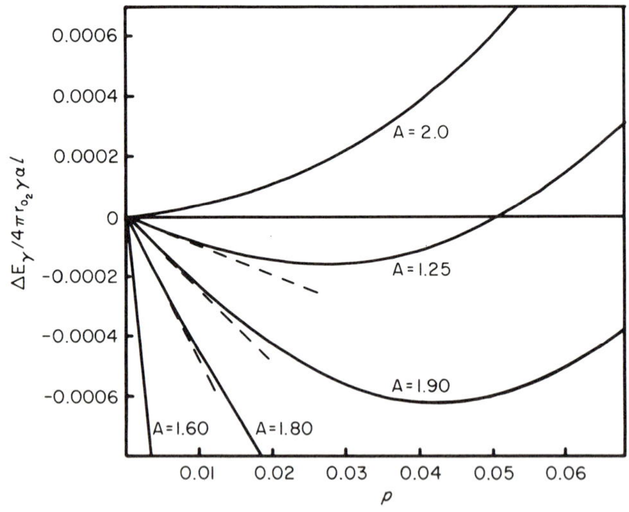

Fig. 12. Change in total interfacial energy per sphere at various values of $A = \gamma_{\alpha\alpha}/\gamma_{\alpha l}$ in a regular assemblage of uniform spherical grains as the grain centers are brought together causing flat contact areas (grain boundaries) to develop between them. The energy change is expressed as a fraction of the initial energy when the spheres are in point contact. p is the decrease in the center-to-center distance expressed as a fraction of $2r_0$, the initial distance.

slopes of the curves can be deduced from the shape of the latter and it will be evident that this force, and hence the tendency for the particles to cohere, increases as A decreases, i.e., as ϕ increases. Support for this mechanism was provided by the observation that the differences in shrinkage occurred during the initial rapid stage of densification.

Similar investigations have been carried out on bodies consisting, at the firing temperature, of two solid phases and a liquid. In the sections of these bodies, five different kinds of boundary appear but, by drawing lines on the micrographs and counting the number of intersections with each kind of boundary separately, we can estimate the fraction of the total boundary area contributed by each kind of boundary. Thus in the case of a body consisting of magnesia, spinel, and a liquid phase

$$\frac{\text{area of MgO–spinel interfaces}}{\text{total boundary area}} = \frac{N_{\text{M.Sp}}}{N}$$

where $N = N_{\text{M.M}} + N_{\text{Sp.Sp}} + N_{\text{M.Sp}} + N_{\text{M.}l} + N_{\text{Sp.}l}$, the suffixes M, Sp, and l denoting magnesia, spinel, and liquid, respectively.

Probably the most significant finding from this work has been that, in the systems so far examined, the extent to which the liquid phase penetrates between unlike grains is less than that between like grains. This behavior is

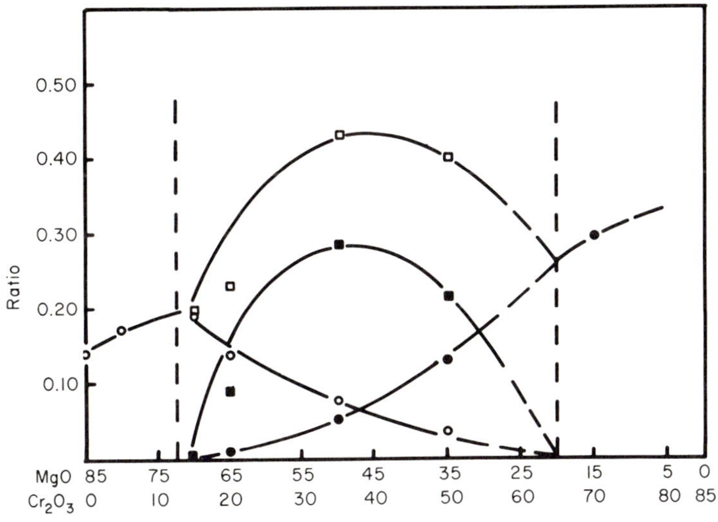

Fig. 13. Variation with MgO/Cr_2O_3 contents of total solid–solid contact ratio (N_{ss}/N), magnesite–spinel contact ratio ($N_{\text{M.Sp}}/N$), magnesia–magnesia contact ratio ($N_{\text{M.M}}/N$), and spinel–spinel contact ratio ($N_{\text{Sp.Sp}}/N$) in mixtures of $MgO–MgCr_2O_4–CaMgSiO_4$ after firing for 2 h at 1700°C. $CaMgSiO_4$ content constant at 15%.

illustrated in Fig. 13 for the case of a series of MgO–$MgCr_2O_4$–$CaMgSiO_4$ mixtures in which the ratio of $MgO/MgCr_2O_4$ was varied while the content of $CaMgSiO_4$ was kept constant. Between the dashed vertical lines, these mixtures would consist of solid MgO and $MgCr_2O_4$ with a constant amount of liquid of constant composition. To the left of this range, they would consist of solid MgO and liquid, and to the right of solid spinel and liquid.

It will be seen that N_{ss}/N, which represents the ratio of the total solid–solid contact area to the total boundary area, increases to a maximum within the three-phase range and that this increase is due to the high degree of contact between the MgO and spinel as indicated by the plot of $N_{M.Sp}/N$.

Similar results have been obtained in mixtures containing other spinels (Richmond, unpublished work), though the effect appears to be greatest with $MgCr_2O_4$, and in mixtures consisting of solid lime, solid magnesia, and a liquid phase (Stephenson and White, 1967). Further, in the system MgO–$MgAl_2O_4$–$CaMgSiO_4$, an increase in the solid–solid contact ratio within the three-phase range was observed even though Al_2O_3, unlike Cr_2O_3, when added to MgO–$CaMgSiO_4$ mixtures in amounts smaller than those required to give solid spinel at the firing temperature, caused the solid–solid contact area to decrease.

In all these cases, where it has been possible to measure the dihedral angles formed at the different grain-to-grain contacts, it has been found that the interfacial energy between unlike grains is smaller than that between like grains. The reason for this is not known, but it is significant that in two-phase alloys studied by Smith (1948), the interfacial energy was generally less than the grain-boundary energy in either phase as shown by the fact that ϕ was generally less than 120°. It was also observed by Riegger *et al.* (1963) that, in mixtures consisting of periclase, magnesioferrite, and a liquid phase, penetration of the latter between unlike grains was less than that between like grains, suggesting that in this system also, the interfacial energy between the periclase and spinel was less than the grain-boundary energies.

V. PHASE EQUILIBRIA IN SYSTEMS CONTAINING A GASEOUS COMPONENT

A. Oxide Systems

Studies of oxide systems containing elements of variable valency have found important application in the fields of chemical metallurgy, refractories, magnetic materials, and nuclear power. Thus, during firing and in service, furnace refractories containing iron oxides, such as chrome and chrome–magnesite refractories suffer changes in the state of oxidation of their iron content, as a result of changes in temperature and furnace atmosphere, and these can cause phase changes, accompanied in some cases by changes in volume that can have a detrimental effect on their performance. Similar changes occur during the firing of ferrites for use as magnets and may adversely affect their magnetic properties.

Because of their importance to an understanding of the reactions involved, studies have now been made of equilibrium relationships in a large number of systems containing iron oxides, nickel oxides, chromium oxides, copper oxides, and uranium oxides.

The nature of the composition and phase changes that occur in systems of this type when the temperature is varied at constant oxygen pressure is shown in Fig. 14. It shows the phase boundaries in the binary system Fe_3O_4–Fe_2O_3 (part of the binary system Fe–O) according to Phillips and Muan (1960) and composition isobars along which the composition will vary with temperature at the oxygen pressures indicated. Thus, when Fe_2O_3 is heated

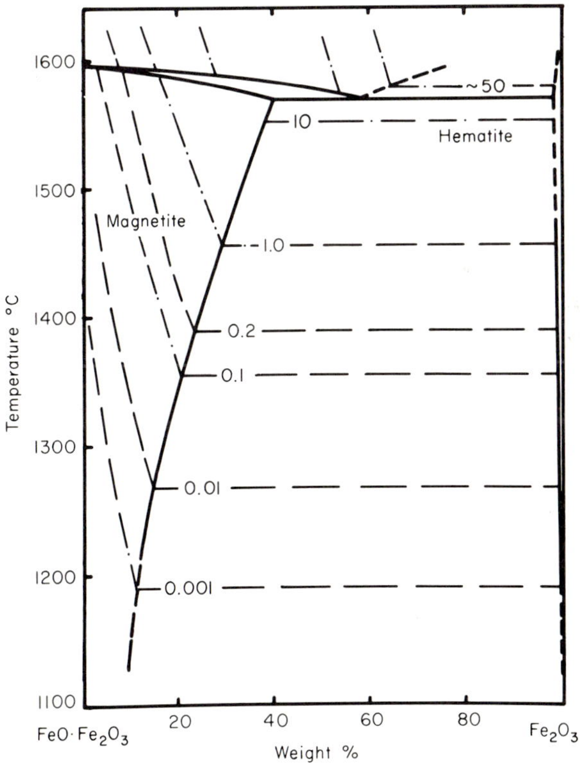

Fig. 14. Phase diagram of system Fe_3O_4–Fe_2O_3. Dashed lines are isobars at indicated oxygen pressures in atmospheres.

in air under such conditions that equilibrium with the atmosphere is maintained, loss of oxygen occurs at 1388°C until conversion to nonstoicheiometric (oxygen-excess) magnetite is complete. During this transition, since three phases coexist (two solids and the gas phase) the system is monovariant so that, at constant oxygen pressure, the transition occurs at constant temperature.

Once conversion to magnetite is complete, the system becomes bivariant and oxygen loss occurs progressively with rising temperature until the isobar reaches the solidus curve and melting begins. Since three phases again coexist (magnetite, liquid, and gas) the system is again bivariant and melting proceeds to completion at constant temperature with a pickup of oxygen. On cooling the reverse changes occur, oxygen being given off during freezing and picked up with oxidation of magnetite to Fe_2O_3 at 1388°C.

The isobars at 0.1, 1.0, and 10 atm are similar except that the temperature of the Fe_2O_3–magnetite transition increases and the melting temperature decreases with increasing oxygen pressure. At 50 atm oxygen pressure, on the other hand, Fe_2O_3 (with a slight deficiency of oxygen) remains stable to above the eutectic temperature and melting occurs with a loss of oxygen. Melting to form the eutectic liquid will thus occur at some pressure between 10 and 50 atm. Since four phases coexist at the eutectic, the latter represents a monovariant condition of the system, i.e., there exists only one pressure at which melting at the eutectic temperature will occur. At pressures above this, the melting temperature will increase with increasing oxygen pressure.

At 0.01 and 0.001 atm, the isobars pass right through the magnetite field to intersect the boundary on the Fe-rich side of the latter, which coincides with the stoichiometric composition. When this occurs, melting takes place with a loss of oxygen to form a liquid phase having a composition between magnetite and wustite, which is not shown on the diagram.

The principles governing the phase transformations in ternary systems are essentially similar except that allowance must be made, in applying the phase rule, for the presence of a third component. Thus the system will be invariant when five phases coexist, e.g., three solid phases, a liquid, and the gas phase as at a ternary eutectic point, monovariant when four phases coexist, e.g., three solids and the gas phase or two solids, a liquid, and the gas phase, and so on. Figure 15a shows the ternary isobaric diagram of the system MgO–FeO–Fe_2O_3 in air that is particularly important for an understanding of the phase changes that occur in magnesite and chrome–magnesite refractories when they are heated at constant oxygen pressure (Willshee and White, 1967).

In the diagram the compositions of fully oxidized mixtures, which represent the stable state of the system at room temperature, lie along the MgO–Fe_2O_3 edge. As the mixtures lose oxygen with rising temperature, the compositions move along lines of constant Mg/Fe ratio (reaction paths). Five of

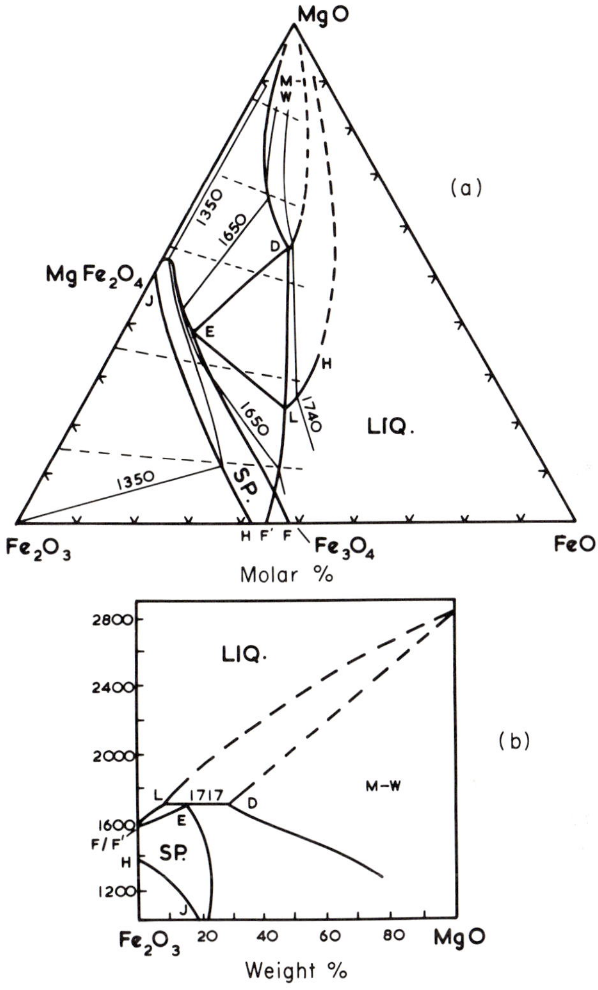

Fig. 15. (a) Isobaric ternary diagram of system MgO–FeO–Fe_2O_3 in air (see text). Dashed lines are reaction paths. (b) Quasibinary temperature-composition diagram of system MgO–FeO–Fe_2O_3 in air. (SP: spinel; M–W: magnesiowustite.)

these reaction paths are illustrated, the temperature variation along them being shown by their intersections with the three isotherms. (Only five reaction paths and three isotherms are shown to simplify the diagram.) The condensed phases occurring at any composition are determined by the phase field in which that composition lies.

Four single-phase fields occur, viz., those of Fe_2O_3 (which is very restricted), spinel, magnesiowustite, and the field of the liquid phase to the right of

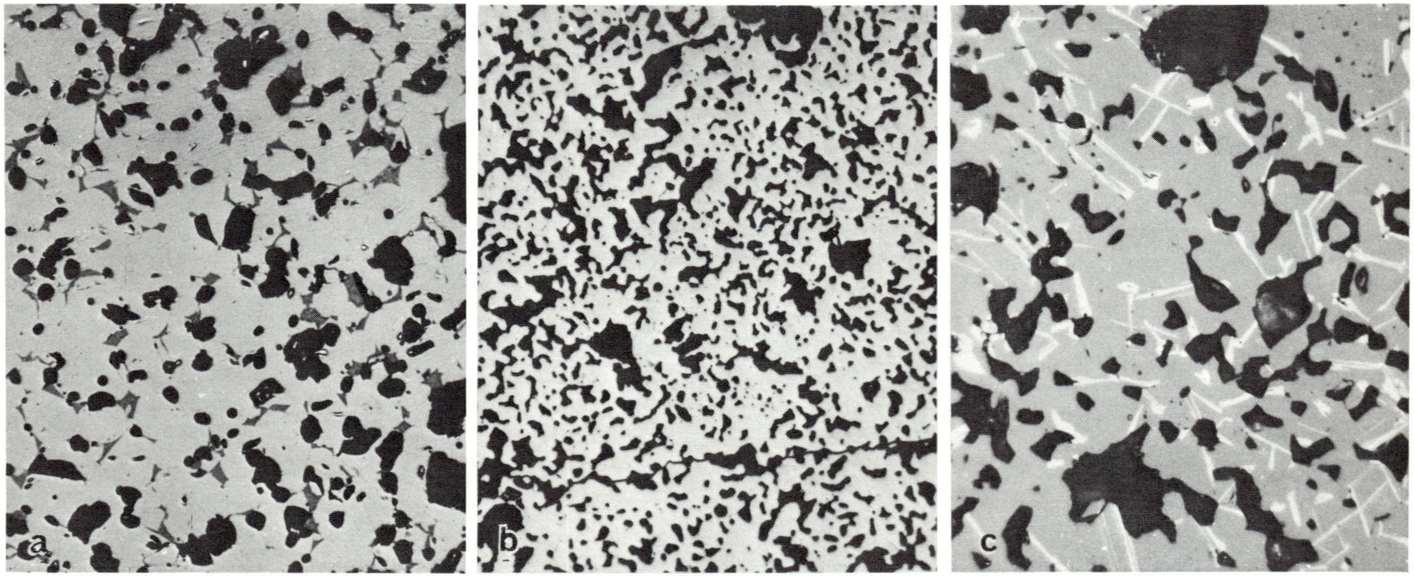

Fig. 16. Microstructures (reflected light) of mixture containing initially 55% Fe_2O_3, 45% MgO (molar) after quenching from (a) 1700°C, (b) 1150°C, (c) 1000°C. Phases present are (a) spinel + magnesiowustite (dark phase); (b) spinel; (c) spinel + Fe_2O_3 (light phase).

the line F′LH–MgO. Two phases occur in the areas Fe_2O_3–HJ (Fe_2O_3 + spinel), $MgFe_2O_4$–MgO–DE (spinel + magnesiowustite), FELF′ (spinel + liquid) and DL–MgO (magnesiowustite + liquid) and three phases within the tie triangle EDL (spinel E + magnesiowustite D + liquid L). Since the latter situation is a monovariant one, these three phases will coexist at only one temperature, if the oxygen pressure is fixed. In air the tie triangle occurs at 1717°C.

When the temperatures along the various phase boundaries are plotted against the initial composition of the fully oxidized mixtures, the temperature-composition diagram shown in Fig. 15b is obtained. In the latter diagram, which is lettered to correspond to Fig. 15a, the tie triangle EDL appears as a horizontal straight line. Figure 15b agrees well with an earlier diagram proposed by Phillips *et al.* (1961).

Figure 16 shows the microstructures observed in a sample containing initially 55% Fe_2O_3:45% MgO (molar) after it had been quenched from just above and just below the temperatures at which it would consist of a single spinel phase. A feature of the structures is that neither the magnesiowustite formed in the former case and the Fe_2O_3 formed in the latter occur at the free surfaces where oxygen loss and gain would be occurring, but in the interior of the sample. (The free surface probably included most of the pore surface, since most of the pores appeared to be open.)

To account for this distribution in the case of the magnesiowustite, the authors suggest that a process analogous to a scaling reaction in reverse had occurred, in which loss of oxygen from free surfaces was accompanied by the destruction of cation vacancies and formation of free electrons which diffused to the magnetite–wustite interface where more wustite was formed.

At the Magnetite–Gas Interface

$$4O^{2-} + V_c^{IV} + 2V_c^{VI} = 2O_2 + 8e^-$$

where V_c^{IV} and V_c^{VI} are vacant tetrahedral and octahedral sites in the spinel and e^- denotes an electron. Cations and electrons then diffuse inwards while vacancies diffuse outwards.

At the Magnetite–Wustite Interface

$$\underbrace{Fe^{2+} + 2Fe^{3+}}_{\text{spinel lattice}} + 2e^- = \underbrace{3Fe^{2+}}_{\text{wustite lattice}} + \underbrace{V_c^{IV} + 2V_c^{IV}}_{\text{spinel lattice}}$$

$$\underbrace{3Fe^{2+} + 6Fe^{3+} + 12O^{2-}}_{\text{spinel lattice}} + 6e^- = \underbrace{9Fe^{2+} + 12O^{2-}}_{\text{wustite lattice}}$$

Adjustment of the Mg/Fe ratios in the two phases would, of course, have to occur simultaneously to conform with the direction of the isotherms (conjugation lines) in Fig. 15a. A similar process involving oxidation can be formulated to account for the separation of Fe_2O_3 on cooling.

One of the original reasons for investigating this system was to throw light on the phase changes occuring when basic refractories containing iron oxides are subjected to changes of temperature or atmosphere during firing or in service. For this purpose similar studies were also made of the systems FeO–Fe_2O_3–Cr_2O_3 and FeO–Fe_2O_3–Al_2O_3 and the changes occurring in chrome and chrome–magnesite refractories when they are heated and cooled in an oxidizing atmosphere can be broadly understood in terms of the three diagrams. Thus when raw chrome ore, which contains most of its iron in the ferrous state, is fired in an oxidizing atmosphere, oxidation of Fe^{2+} to Fe^{3+} in the spinel phase will occur at low temperatures with separation of sesquioxide phase that is a solid solution of Fe_2O_3, Cr_2O_3, and Al_2O_3. At higher temperatures, assuming equilibrium has been reached at the lower temperatures, oxygen loss will occur with resolution of the sesquioxide. The latter process will be reversed on cooling with reprecipitation of sesquioxide.

In chrome–magnesite mixtures, on the other hand, the sesquioxide formed initially will tend to react with the added MgO to form more spinel. Subsequently, on progressive heating, loss of oxygen will occur while FeO, Fe_2O_3, Cr_2O_3, and Al_2O_3 dissolve in the unreacted magnesia. This process has already been mentioned in Section III, but the part played in it by conversion of Fe^{3+} to Fe^{2+} was not mentioned. Similarly, on subsequent cooling, reoxidation of the iron will contribute to the precipitation of spinel from the magnesiowustite formed.

A similar sequence of changes will occur at constant temperature when the partial pressure of oxygen in the atmosphere is reduced and then raised again. Harbach and Ford (1964) have shown that both the sesquioxide–spinel transition and the spinel–magnesiowustite transition cause test pieces to expand while the reverse changes cause them to contract and this behavior probably explains the disruptive effect of cyclic changes of temperature and atmosphere on chrome–magnesite refractories.

The stability relationships of the spinel phase shown in Fig. 15a have also a bearing on the firing of ceramic magnets based on magnesioferrite. These materials normally contain other cations, e.g., Mn^{2+} and Zn^{2+}, in addition to Mg^{2+} and Fe^{3+}, but the stability relationships are somewhat similar. To obtain optimum magnetic properties it is necessary that the fired material should consist of a single spinel phase, which means that the firing temperature and atmosphere must be such that the final composition lies within the spinel field. One way of achieving this is by firing in air at a temperature that will produce the spinel free from magnesiowustite and then changing to a CO/CO_2 atmosphere during cooling to prevent the separation of sesquioxide.

A feature of Fig. 15a, not shown in previous diagrams of this system, is the indication (shown by the bulge in the spinel phase boundary) that with starting compositions in the vicinity of $MgFe_2O_4$, spinels having an oxygen/cation ratio less than 1.33 can be obtained. Such spinels, though single phase, would have appreciable ferrous iron, which would probably be detrimental to their magnetic properties.

It has also been shown by Reijnen (1965) that, because of the shape of the isotherms in Fig. 15a, when Fe_2O_3 and MgO react to form spinel by cation diffusion in the solid state, a Kirkendall shift should be observed at the Fe_2O_3–spinel and spinel–magnesia interfaces. This can be seen from the form of the 1350°C isotherm, which shows that at temperatures approaching 1388°C in air the spinel in equilibrium with Fe_2O_3 has an O/Fe ratio approaching that of Fe_3O_4, i.e., 1.33, whereas the spinel in equilibrium with magnesiowustite has an O/Fe ratio approaching that of Fe_2O_3, i.e., 1.5. Hence, when a layer of spinel is forming by cation diffusion between Fe_2O_3 and MgO, oxygen will be lost to the gas phase at the Fe_2O_3–spinel interface and will be picked up from the gase phase at the spinel–MgO interface.

The interface reactions can be written formally:

At Fe_2O_3–Spinel Interface

$$3Fe_2O_3 + 2Mg^{2+} = 2MgFe_2O_4 + 2Fe^{2+} + \tfrac{1}{2}O_2$$

At Spinel–MgO Interface

$$2Fe^{2+} + 3MgO + \tfrac{1}{2}O_2 = MgFe_2O_4 + 2Mg^{2+}$$

Approximately twice as much spinel should therefore be formed at the former interface than at the latter. Further, since oxygen loss at the former is equivalent to removing part of the lattice from the Fe_2O_3 while addition of oxygen at the latter is equivalent to adding to the spinel lattice, a Kirkendall shift should be observed. Since one O^{2-} ion is transferred through the gas phase for every 12 combined in the spinel, the shift should have a value approaching $\frac{1}{12}$ of the thickness of the spinel layer.

Displacements observed during sintering experiments with spheres on planes (where the sphere was Fe_2O_3 and the plane MgO and vice versa) can be explained on the basis of this effect (Kooy, 1964: Kuczynski, 1965). Under these conditions, because counter diffusion of Mg and Fe ions is restricted to a narrow neck between the sphere and plane, the Fe_2O_3–spinel and spinel–MgO interfaces formed on each side of the neck are approximately hemispherical in shape. Further, since the volume of spinel formed on the Fe_2O_3 side of the neck is greater than that on the MgO side, the area of the former interface is greater than that of the latter. Consequently, removal of a given number of oxygen ions from the former will result in a smaller displacement than the addition of the same number of oxygen ions to the latter. Hence a displacement of the spinel and neck away from the MgO will occur.

An observation by Goodison and White (1961) that, during the sintering of MgO compacts having Fe_2O_3 particles embedded in them, expansions occurred while pores were formed where the Fe_2O_3 particles had been, can possibly be attributed to a similar mechanism. In this case the spinel layer formed round each Fe_2O_3 particle would have the form of a spherical shell and would be unable to shrink to compensate for removal of oxygen at its interface with the Fe_2O_3. Consequently, porosity would develop inside the shell while the addition of oxygen at the spinel–MgO interface would cause expansion of the compact. On the other hand, when particles of electrically fused magnesia were embedded in Fe_2O_3 compacts, the particles were found to be loose after sintering, presumably because removal of oxygen at the Fe_2O_3–spinel interface had caused the shell of the spinel to expand outwards.

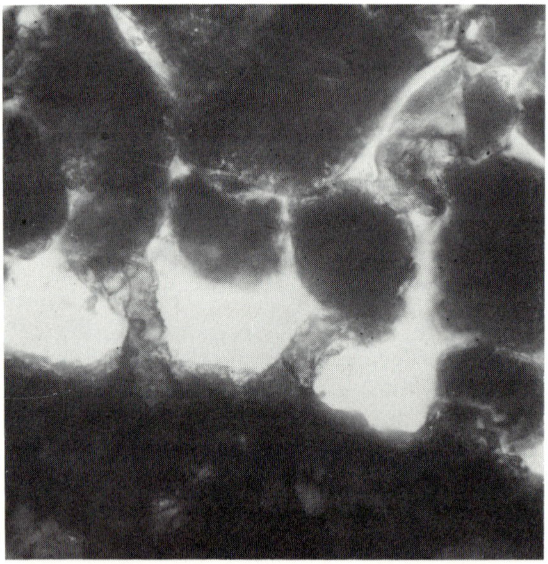

Fig. 17. Micrograph of chrome–magnesite refractory. Thin section viewed in natural light (300×) showing gap between large grain of chrome spinel and magnesia matrix.

An observation by Ford *et al.* (1961) that, in certain chrome–magnesite refractories that had been fired in the region of 1500°C, cavities had developed round the grains of chrome spinel (see Fig. 17) may have a similar explanation. Recently, Stubican and de Menezes (1966) have shown that, during the firing of chrome–magnesite bricks, as the ferrous oxide in the chrome grains oxidizes, the layer of sesquioxide formed initially on the surfaces of the grains consists mainly of Fe_2O_3. (They attribute this behavior to rapid diffusion of Fe ions in the chrome spinel relative to Al and Cr.) Reaction between this layer and the matrix magnesia in contact with it to form magnesioferrite-rich spinel, with loss of oxygen at the interface between this new spinel and the sesquioxide could possibly explain the formation of cavities.

Another significant conclusion from Fig. 15a, which follows from the fact that the liquidus isobar F′L and the spinel boundary FE cross each other, is that the addition of MgO to Fe_2O_3 eliminates oxygen pickup on melting, and consequently oxygen evolution on freezing. This finding is of potential practical importance since the liberation of oxygen during freezing is a potential cause of blow-hole formation in fused-cast magnesite–chrome refractories.

In the system Fe–O (Fig. 14), oxygen evolution on freezing, with crystallization of magnetite, occurs down to an oxygen pressure approaching 0.02 atm so that this mechanism of blow-hole formation could operate at oxygen potentials in the melt considerably lower than those prevailing under equilibrium conditions in air.

Figure 18 shows the results of a recent reinvestigation of the system FeO–Fe_2O_3–Al_2O_3 in air up to 1750°C by Willshee and White (1968). In Fig. 18a the method of representation is similar to that adopted in Fig. 15a, the solid phases being an Fe_2O_3-rich sesquioxide solid solution extending from Fe_2O_3 to A, a phase approximating to $(FeAl)O_3$ extending from B to C, an Al_2O_3-rich solid solution extending from Al_2O_3 to G, and the spinel phase bounded by the line HXYEF. The two solid-state tie triangles ABX and CGY were derived mainly from earlier work up to 1500°C in air by Muan and Gee (1956). FELF′ is a two-phase field in which spinel and liquid coexist, while DEL is a tie triangle within which sesquioxide of composition D, spinel of composition E and liquid L coexist. The line F′L–Al_2O_3 (shown dashed between L and Al_2O_3) is the liquidus isobar.

A feature of the melting relationships was that over the entire range of Al_2O_3 contents examined, melting occurred with a pickup of oxygen. This is shown by the fact that the liquidus isobar F′L in this system does not cross the spinel boundary FE and that the tie triangle DEL has its apex L to the left of DE.

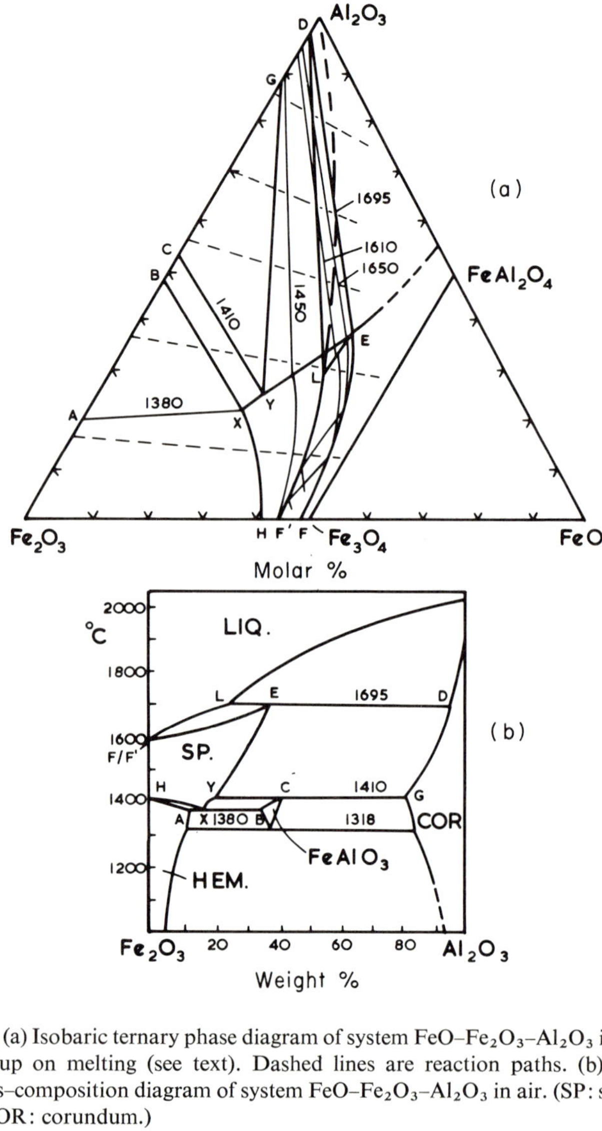

Fig. 18. (a) Isobaric ternary phase diagram of system $FeO–Fe_2O_3–Al_2O_3$ in air showing oxygen pickup on melting (see text). Dashed lines are reaction paths. (b) Quasibinary temperatures–composition diagram of system $FeO–Fe_2O_3–Al_2O_3$ in air. (SP: spinel; HEM: hematite; COR: corundum.)

Although difficulties were encountered due to the corrosive nature of the melts, indications were obtained that a similar situation exists in the system $FeO–Fe_2O_3–Cr_2O_3$. The presence of Al_2O_3 and Cr_2O_3 should, therefore, oppose the tendency of MgO to suppress oxygen evolution on freezing.

B. Oxide–Metal Systems

In constructing Figs. 15 and 18, the choice of Fe_2O_3, FeO, MgO, or Al_2O_3 as components was dictated mainly by convenience. We could, for example, have taken $\frac{1}{2}Fe_2O_3$ as a component, which is sometimes done to make the reaction paths parallel to the base of the composition triangle. We could also have taken one or both of the metals as components. Diagrams of the latter type are useful, since they often reveal features of the compatibility relationships between the metals and the oxide phases not indicated by

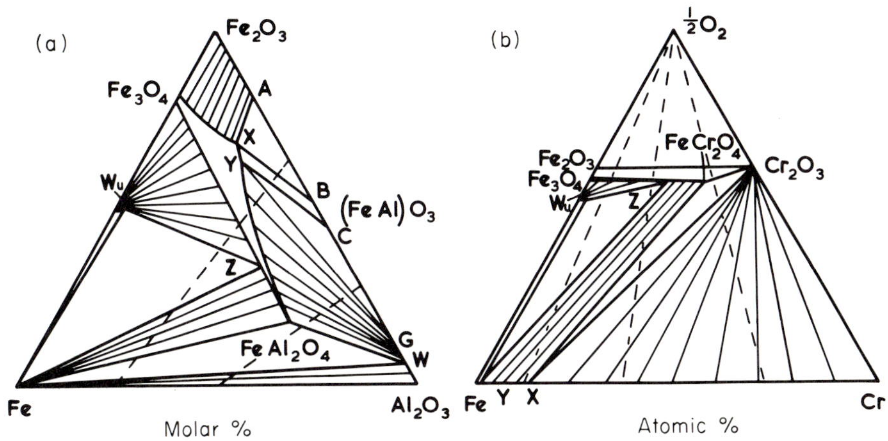

Fig. 19. Ternary diagrams showing solid phase relationships in systems (a) Fe–Al_2O_3–Fe_2O_3 at 900°C and (b) Fe–Cr–O at 1200°C. Thin straight lines show directions of isobars in two-phase regions. Dashed lines show directions of reaction paths. (Wu: wustite.)

the free-energy equations for the formation of the oxides from the pure metals. They can thus contribute to our understanding of the reactions taking place during the scaling of metals and of the conditions governing the formation of the different types of oxide inclusions formed during the deoxidation of metals.

Two diagrams that have proved useful in this way are shown in Figs. 19a and b. The former shows solid phase relationships in the system Fe–Fe_2O_3–Al_2O_3 at 900°C (i.e., it is an isothermal not an isobaric diagram like Figs. 15 and 18). The choice of Al_2O_3 as a component was possible because the concentration of aluminum in metallic iron in contact with Al_2O_3 could be considered negligible for the purposes of the diagram. The latter shows solid phase relationships in the system Fe–Cr–O in the region of 1200°C.

A feature of both diagrams is that the spinel phase in equilibrium with wustite and the metallic phase contains appreciable Fe_3O_4 in solid solution. Wustite is, therefore, incapable of coexisting with pure $FeAl_2O_4$ or $FeCr_2O_4$ and, when heated in contact with them, will react to form a spinel containing Fe^{3+} ions and a metallic phase. In the system Fe–Cr–O, the latter phase contains appreciable Cr (point Y) while $FeCr_2O_4$ and Cr_2O_3 coexist with an Fe–Cr solid solution of even higher Cr content (point X). In the Fe–Al–O system, the corresponding points occur at negligible Al contents (in terms of the composition scale used in the diagram). Consequently, iron containing more than a trace of Al cannot coexist with either wustite of $FeAl_2O_3$.

These relationships are of considerable practical significance. Thus in oxide scales formed on chromium steels the wustite layer normally present adjacent to the metal no longer occurs when the Cr content exceeds a critical amount. Spinel is then formed in direct contact with the metal. This can be a source of trouble during precision casting since the surface of the metal tends to be pitted by the formation of the spinel crystals. In Ni–Cr steels the critical Cr content is appreciably higher than that of point Y as would probably be expected since NiO will tend to increase the stability of the wustite.

The direction of the conjugation lines (isobars) in Fig. 19b also gives an indication of the extent to which the scale formed will be enriched in Cr relative to the metal. When the oxygen potentials corresponding to the various isobars are inserted, Fig. 19a also provides a guide to the kind of non-metallic inclusions to be expected in steels that have been deoxidized to various degrees with Al.

In both systems the existence of a point Z on the spinel phase boundary, at which the spinel would coexist with wustite and the metal, was originally deduced from the fact that the oxygen potential to reduce the pure end-member spinel was less than that required to reduce wustite to metal, and by making use of the principle that at constant temperature the oxygen pressure along a reaction path could never increase in the direction of decreasing oxygen content in the condensed phases. It could only stay constant, when crossing tie triangles, or decrease. The direction of the reaction paths in Figs. 19a and 19b are indicated by the dashed straight lines while the oxygen potentials corresponding to the various monovariant equilibria in the two systems so far as they were known, are shown in Fig. 20, which is based mainly on the data of Richardson and Jeffes (1948, 1949). The curve marked FeO is that for the wustite–Fe equilibrium in the system Fe–O while those marked $FeAl_2O_4$ and $FeCr_2O_4$ give the oxygen potentials at which the sesquioxide, and the end member spinel coexist with the metal. In each case the oxygen potential at which wustite and the metal coexist with spinel Z lie between the values indicated.

Subsequently the correctness of these predictions was confirmed by the

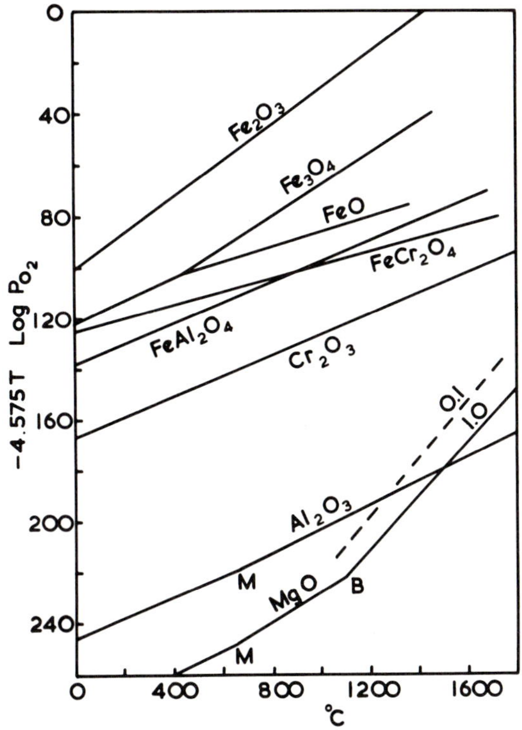

Fig. 20. Standard free energy curves showing oxygen potentials corresponding to monovariant equilibria in systems $Fe–Fe_2O_3–Al_2O_3$ and Fe–Cr–O. The curves marked Fe_2O_3, Fe_3O_4 and FeO correspond to the three monovariant equilibria in the system Fe–O. (The curves for MgO are not referred to in the text.)

work of Atlas and Sumida (1958) on the system Fe–Al–O and of Seybolt (1960) on the system Fe–Cr–O.

C. Systems Containing Oxycarbides

Diagrams of the same kind have also been constructed for the system U–C–O, using as a basis the thermodynamic data for the various monovariant reactions occurring in the system (Dutta and White, 1967).

An important feature of this system is that at high temperatures over the composition range of interest in nuclear technology, the compositions of the solid phases at equilibrium at any temperature are determined by the pressure of CO, not by the partial pressure of oxygen. The reasons are that, since the condensed phases contain both C and O, the equilibrium between them and the gas phase is determined by the product of the carbon and oxygen activities in them, which is proportional to the partial pressure of CO, and also because at temperatures over 1000°C, the CO_2/CO ratios in the equilibrium gas are extremely small over the part of the composition triangle of interest. Consequently the gas is nearly pure CO.

Figure 21 shows the compatibility relationships between the condensed phases at 1600°C as deduced mainly from the work of Piazza and Sinnott (1962), Brett *et al.* (1964), Magnier *et al.* (1964), Anselin *et al.* (1964), and Stoops and Hamme (1964). The condensed phases shown are UO_{2+x}, a

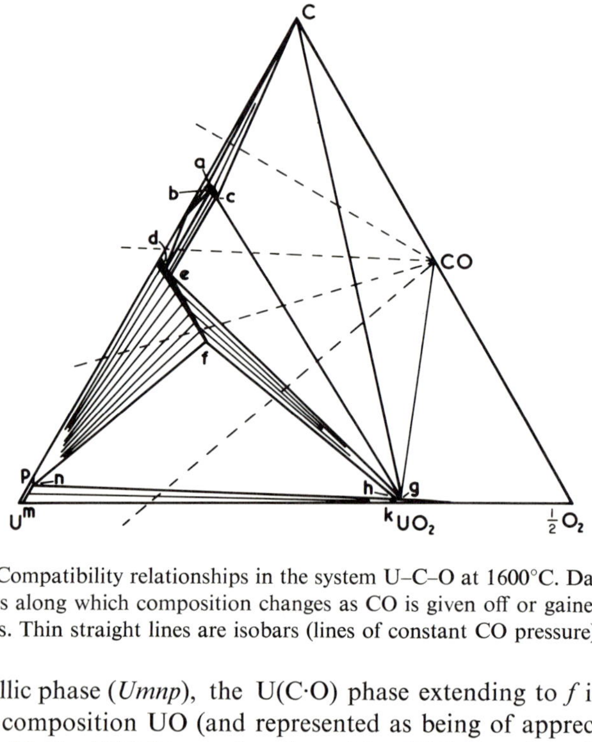

Fig. 21. Compatibility relationships in the system U–C–O at 1600°C. Dashed lines are reaction paths along which composition changes as CO is given off or gained by the condensed phases. Thin straight lines are isobars (lines of constant CO pressure).

liquid metallic phase (*Umnp*), the U(C·O) phase extending to *f* in the direction of the composition UO (and represented as being of appreciable width for clarity), U_2C_3, $U(C{\cdot}O)_2$, and carbon. The $U(C{\cdot}O)_2$ phase is shown to extend over a range of O/C ratios (which has been exaggerated) but has been considered, because of the effect of oxygen on the stability of UC_2, not to extend at this temperature to the U–C edge of the diagram. Recently, Henney (1966) has shown that at 1500°C a tie line exists between U_2C_3 and UO_2, showing these phases to be compatible at this temperature but that at 1700°C the U(C·O) phase and the $U(C{\cdot}O)_2$ phase are compatible. The latter situation has been assumed to exist at 1600°C. Since the compositions of the condensed phases are determined by the CO pressure, the reaction paths along which changes in composition occur in the condensed part of the system as equilibrium is approached at any pressure of CO must radiate from the composition of CO as shown by the dashed lines.

As pointed out earlier, at constant temperature, as the composition of the condensed part moves along a reaction path away from the composition of CO (CO being evolved) the equilibrium pressure can stay constant or decrease but it cannot increase. It will remain constant as the reaction path crosses a tie triangle and will decrease where the path crosses an area where fewer than three condensed phases coexist. One consequence of this requirement is that, in Fig. 21, along the boundary *def* of the U(C·O) phase, the

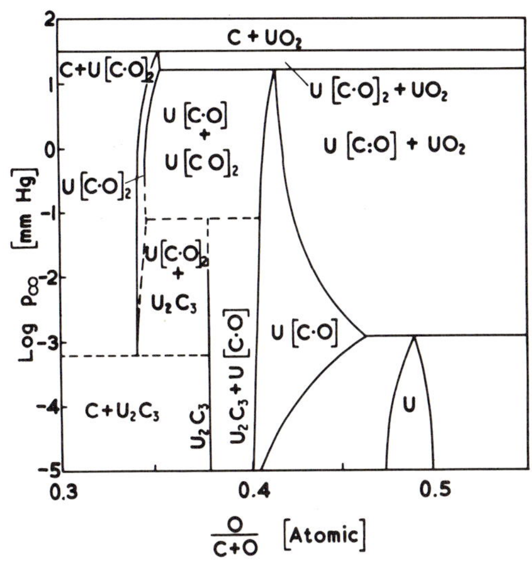

Fig. 22. Pressure–composition diagram of system U–C–O at 1600°C showing CO pressures at which various phase changes occur plotted against O/(C + O) ratio in original mixtures of C and UO_2.

CO pressure must increase from the edge of the diagram to *e* and decrease from *e* to *f*. (This follows because, for example, as the composition moves across an area such as *efh* with loss of CO, the CO pressure must drop.) Similarly, along the lower boundary of the U(C·O) phase the CO pressure must fall from *f* to the edge of the diagram.

Figure 22, which is partly qualitative, is the quasibinary pressure/composition diagram of the system at 1600°C, and shows the CO pressures at which the various phase changes will occur at this temperature in mixtures consisting initially of C and UO_2, pressure being plotted against the ratio O/(C + O) in the initial mixtures. The horizontal (constant pressure) lines in the diagram correspond to the tie triangles in Fig. 21, while the curved lines show how the pressure varies along the boundaries of the single-phase areas. Estimates of the pressures corresponding to the C–$U(C \cdot O)_2$–UO_{2+x} tie triangles were obtained from the data of Piazza and Sinnott and for the pressure corresponding to the $U(C \cdot O)$–U–UO_2 tie triangle from the data of Anselin *et al.* The pressure variations along the boundaries of the $U(C \cdot O)$ phase were also estimated from the data of the latter workers and, as was predicted above, the pressure is a maximum at the composition at which the $U(C \cdot O)$ phase is compatible with both $U(C \cdot O)_2$ and UO_{2+x}.

Figures 21 and 22 give considerable insight into the nature of the factors that control the composition of the uranium oxycarbides. Thus if any comsition lying to the left of the join C–UO_2 in Fig. 21 is heated *in vacuo* in a furnace that is completely free from air leaks, the only composition change that can occur is along the reaction path passing through that composition with loss of CO. If, on the other hand, leakage of oxygen into the apparatus occurs and the sample picks up oxygen, the composition will be displaced in the direction of the oxygen corner, moving from its original reaction path across paths that cross the C–UO_2 join at higher O/(C + O) ratios.

Most of the effects of oxygen on the stability relationships of the uranium oxycarbides can be related to the occurrence of one or both of these processes. Thus if a composition that is hyperstoicheiometric relative to the U(CO) phase is heated in a good vacuum with continuous pumping, single-phase $U(C \cdot O)$ is most likely to be formed if the composition lies on a reaction path passing close to point *e* where the CO pressure at the $U(C \cdot O)$ boundary is highest. Compositions on reaction paths lying above *e*, but still intersecting the boundary of the $U(C \cdot O)$ field, will become increasingly difficult to convert to single-phase $U(C \cdot O)$ as the distance of the point intersection from *e* increases, because of the rapid fall in the CO pressure along the boundary. They are more likely to yield $U(C \cdot O)$ containing $U(C \cdot O)_2$ or U_2C_3. According to the diagrams the former will first appear at considerably higher CO pressures than those required for the formation of U_2C_3. The relationships indicated, therefore, provide a possible explanation for the fact that, in hyperstoicheiometric $U(C \cdot O)$, the second phase is generally $U(C \cdot O)_2$, not U_2C_3.

Conversely if oxygen-free UC is heated in a continuously evacuated system into which a slight leak of oxygen is occurring, oxygen pickup without evolution of CO will occur initially and the composition will be displaced in the direction of the oxygen corner of Fig. 21, giving $U(C \cdot O)$ and U_2C_3 and later $U(C \cdot O)$ and $U(C \cdot O)_2$. If the oxygen potential is high enough, oxygen pickup will continue until the CO pressure in equilibrium with the oxycarbides reaches a point at which CO is being pumped off and the composition will then change along a reaction path towards the boundary of the $U(C \cdot O)$ phase. If the oxidation process is not too rapid relative to the evolution of CO, there would probably be a tendency for the boundary to be reached slightly to the left of *e*, which may explain why uranium monocarbide, made in a good vacuum frequently contains 0.03–0.05 g atom of oxygen per g mole.

If oxidation continues after the boundary is reached, the net result of the two processes would be to cause progressive replacement of C by O in the U(C·O). If the pressure of CO in the system had reached a constant value, the change in composition would then tend to change along a line of constant CO pressure (isobar), which, within the single phase field of U(C·O), will run approximately parallel to the length of the field. [The reason for this is that, in Fig. 21, the isobars in the $U(C\cdot O)_2$–U(C·O) field must, when extended, pass through the U(C·O) field and emerge again in the U(C·O)–UO_2 field, where they run to the boundary of the UO_2 phase. Hence, if replacement of C by O in the U(C·O) continues, UO_2 will ultimately separate as a phase when the boundary *ef* is crossed at an O/C ratio determined by the vacuum.

On the other hand when UC is heated in an argon atmosphere in a closed system, so that any CO formed is not continuously pumped off, the composition, with continued oxidation, will continue to change in the direction of the oxygen corner and UO_2 will appear when the boundary *ec* is crossed. Under these conditions, therefore, the absence of UO_2 is probably an indication that the oxygen content of the U(C·O) does not exceed 0.05 g atom/g mole, approximately.

Another significant feature of the relationships indicated is the relatively high CO pressure (about 10^{-3} Torr at 1600°C) at which liquid uranium coexists with U(C·O) and UO_2. Since this pressure increases with temperature, metallic uranium should separate when UO_2 contaminated with carbon is heated *in vacuo* at high temperatures. The occurrence of inclusions of uranium metal in arc-melted UO_2 can possibly be attributed to this cause, in some cases at least.

They also explain why UO_2 granules coated with impermeable pyrolytic carbon can be used as fuel elements. As shown in Fig. 22, the equilibrium CO pressure, above which carbon and UO_2 will not react at 1600°C, is of the order of 30 Torr, which is insufficient to disrupt the coating, and it will be lower at lower temperatures. Hence, once the equilibrium pressure has been reached within the granule, the carbon and UO_2 will be capable of existing in contact with each other, without further reaction.

Similar diagrams can be constructed to describe the situation found by Henney at 1500°C when UO_2 and U_2C_3 were found to be compatible, although the thermodynamic data are less complete. The tie triangles $U(C\cdot O)_2$–UO_2–*e* and U_2C_3–*b*–*d* in Fig. 21 will then be replaced by a tie triangle $U(C\cdot O)_2$–U_2C_3–UO_2 and a tie triangle joining U_2C_3, UO_2 and a point on the U(C·O) boundary analogous to *e*. In Fig. 22, the horizontal line corresponding to the former equilibrium will occur at a higher CO pressure than that corresponding to the latter equilibrium, i.e., U_2C_3 will now be formed at a higher CO pressure than will the U (C·O) phase.

Recently diagrams similar to Figs. 21 and 22 have been constructed for the system Pu–C–O by Potter (1967). CO pressures were calculated from known values of the free energies of formation of the plutonium oxides and carbides of CO, and it was assumed that solid solution between PuC and PuO the hypothetical end member of the monoxycarbide series, was ideal. In this system the latter series extends to approximately $PuC_{0.33}O_{0.67}$, i.e., in the ternary diagram, it crosses a line joining the composition of CO to the Pu corner of the diagram. (Compare Fig. 21.) Consequently, from the requirement that, in a ternary system at constant temperature, the CO pressure must always fall along a reaction path in the direction of CO evolution when it crosses an area where only two condensed phases coexist, the CO pressure along the boundary of the Pu(C·O) phase in equilibrium with Pu must pass through a maximum at 50% replacement of C by O. On the other boundary of this phase the CO pressure is a maximum at approximately 35% replacement of C by O at a point analogous to *e* in Fig. 21. which in this case coexists with Pu_2C_3 and Pu_2O_3.

References

Agamawi, Y. M., and White, J. (1952). *Trans. Brit. Ceram. Soc.* **51**, 293.

Agamawi, Y. M., and White, J. (1953). *Trans. Brit. Ceram. Soc.* **52**, 271.

Agamawi, Y. M., and White, J. (1954). *Trans. Brit. Ceram. Soc.* **53**, 1.

Allison, E. B., Brock, P., and White, J. (1959). *Trans. Brit. Ceram. Soc.* **58**, 496.

Alper, A. M., McNally, R. N., Ribbe, P. H., and Doman, R. C. (1962). *J. Am. Ceram. Soc.* **45**, 263.

Alper, A. M., McNally, R. N., Doman, R. C., and Keihn, F. G. (1964). *J. Am. Ceram. Soc.* **47**, 30.

Anselin, F., Dean, G., Lorenzelli R., and Pascard, R. (1964). *Symp. Carbides Nucl. Energy* **1**, 113. Macmillan, New York.

Atlas, I. M., and Sumida, W. K. (1958). *J. Am. Ceram. Soc.* **41**, 150.

Berry, T. F., Allen, W. G., and Snow, R. B. (1950). *J. Am. Ceram. Soc.* **33**, 121.

Brett, N. H., Harper, E. A., Hedger, H. J., and Pottinger, J. S. (1964). *Symp. Carbides Nucl. Energy* **1**, 162. Macmillan, New York.

Buist, D. S., Jackson, B., Stephenson, I. M., and White, J. (1965). *Trans. Brit. Ceram. Soc.* **64**, 173.

Curtis, C. D. (1963). *Geochim Cosmochim. Acta* **28**, 389.

De Vries, R. C., Roy, R., and Osborn, E. F. (1954). *Trans. Brit. Ceram. Soc.* **53**, 525.

De Vries, R. C., Roy, R., and Osborn, E. F. (1955). *J. Am. Ceram. Soc.* **38**, 158.

Doman, R. C., Barr, J. B., McNally, R. N., and Alper, A. M. (1963). *J. Am. Ceram. Soc.* **46**, 313.

DUTTA, S. K., and WHITE, J. (1967). *Proc. Brit. Ceram. Soc.* **7**, 177.
EL-SHAHAT, R. M., and WHITE, J. (1964). *Trans. Brit. Ceram. Soc.* **63**, 313.
EL-SHAHAT, R. M., and WHITE, J. (1966a). *Trans. Brit. Ceram. Soc.* **65**, 309.
EL-SHAHAT, R. M., and WHITE, J. (1966b). *Trans. Brit. Ceram. Soc.* **65**, 407.
FORD, W. F., HAYHURST, A., and WHITE, J. (1961). *Trans. Brit. Ceram. Soc.* **60**, 581.
FULLMAN, R. L. (1953). *Trans. AIME.* **197**, 447.
GLASSER, F. P., WARSHAW, J., and ROY, R. (1960). *Phys. Chem. Glasses* **1**, 39.
GOODISON, J., and WHITE, J. (1961). "Agglomeration," p. 251. Wiley (Interscience), New York.
GREENWOOD, G. W. (1956). *Acta Metallurgica* **4**, 243.
GREIG, J. W. (1927). *Am. J. Science* **13**, 1, 133.
GUTT, W. (1961). *Nature* **90**, (4773), 339.
GUTT, W. (1965). *Nature* **207**, (4993), 2184.
HARBACH, J., and FORD, W. F. (1964). *Trans. Brit. Ceram. Soc.* **63**, 145.
HARKER, D., and PARKER, E. R. (1945). *Trans. ASM* **34**, 156.
HAYHURST, A. (1961). PhD thesis, Sheffield University.
HAYHURST, A., and LAMING, J. (1963). *Refractories J.* **39**, 80.
HENNEY, J. (1966). *AERE Rept.* 4661.
HERRING, C. (1951). "The Physics of Powder Metallurgy" (W. E. Kington, ed), Chapter 8, McGraw-Hill, New York.
JACKSON, B., and FORD, W. F. (1966). *Trans. Brit. Ceram. Soc.* **65**, 19.
JACKSON, B., FORD, W. F., and WHITE, J. (1963). *Trans. Brit. Ceram. Soc.* **62**, 577.
JOHNSON, W. (1967). Private communication.
KOOY, C. (1964). *Pure Appl. Chem.* **9**, 441.
KUCZYNSKI, G. C. (1965). *Proc. Intern. Symp. Reactivity Solids, 5th* p. 352. Elsevier, Amsterdam.
LIFSCHITZ, J. M., and SLEZOV, V. V. (1961). *J. Phys. Chem. Solids* **19**, 35.
MAGNIER, P., TROUVÉ, J., and ACCARY, A. (1964). *Symp. Carbides Nucl. Energy*, **1**, p. 95, Macmillan, New York.
MELFORD, D. A. (1967). Private communication.
MUAN, A., and GEE, C. L. (1956). *J. Amer. Ceram. Soc.*, **39**, 207.
MYKURA, H. (1966). "Solid Surfaces and Interfaces," Routledge and Kegan Paul, London.
PARIKH, N. M., and HUMENIK, M. (1957). *J. Am. Ceram. Soc.* **40**, 315.
PHILLIPS, B., and MUAN, A. (1960). *J. Phys. Chem.* **64**, 1451.
PHILLIPS, B., SOMIYA, S., and MUAN, A. (1961). *J. Am. Ceram. Soc.* **44**, 167.
PIAZZA, J. R., and SINNOTT, M. J. (1962). *Chem. Eng. Data* **7**, 451.
POTTER, P. E. (1967). *AERE Rept.* 5544.
RAIT, J. R. (1950). "Basic Refractories, Their Chemistry and Performance." Iliffe, London.
RAMBERG, H. (1954). *Am. Min.* **39**, 256.
REIJNEN, P. (1965). *Proc. Intern. Symp. Reactivity Solids, 5th* p. 562, Elsevier, Amsterdam.
RICHARDSON, F. D., and JEFFES, J. H. E. (1948). *J. Iron Steel Inst.* **160**, 261.
RICHARDSON, F. D., and JEFFES, J. H. E. (1949). *J. Iron Steel Inst.* **161**, 229.
RICKER, R. W., and OSBORN, E. F. (1954). *J. Am. Ceram. Soc.* **37**, 136.
RIEGGER, O. K., and VAN VLACK, L. H. (1960). *Trans. AIME* **218**, 1933.
RIEGGER, O. K., MADDEN, G. I., and VAN VLACK, L. H. (1963). *Trans. AIME* **227**, 971.
SCHLAUDT, C. M., and ROY, D. M. (1965). *J. Am. Ceram. Soc.* **48**, 248.
SCHLAUDT, C. M., and ROY, D. M. (1966). *J. Am. Ceram. Soc.* **49**, 430.
SEYBOLT, A. U. (1960). *J. Electrochem. Soc.* **107**, 147.
SMITH, C. S. (1948). *Trans. AIME* **175**, 15.
SOLACOLU, S. (1960). *Ber. Dent. Keram. Ges.* **37**, 266.

STEPHENSON, I. M., and WHITE, J. (1967). *Trans. Brit. Ceram. Soc.* **66**, 443.
STOOPS, R. F., and HAMME, J. V. (1964). *J. Am. Ceram. Soc.* **47**, 59, 60.
STUBICAN, V. S., and DE MENEZES, I. (1966). *J. Am. Ceram. Soc.* **49**, 535, 55.
USP (1944). US Patent No. 2,351, 204.
VAN VLACK, L. H. (1960). *J. Am. Ceram. Soc.* **43**, 140, 39.
WAGNER, C. (1961). *Z. Electrochem.* **65**, 581, 41.
WHITE, J. (1962). *J. Iron. Steel Inst.* **200**, 611.
WHITE, J. (1968). "Ceramic Microstructures" (R. M. Fulrath and J. A. Pask, ed.), p. 728. Wiley, New York.
WILLSHEE, J. C., and WHITE, J. (1967). *Trans. Brit. Ceram. Soc.* **66**, 541.
WILLSHEE, J. C., and WHITE, J. (1968). *Trans. Brit. Ceram. Soc.* **67**, 271.
WOODHOUSE, D., and WHITE, J. (1955). *Trans. Brit. Ceram. Soc.* **54**, 333.

III

The Use of Phase Diagrams in the Development and Use of Refractories

HOBART M. KRANER

CONSULTANT
BETHLEHEM, PENNYSYLVANIA

I. INTRODUCTION

Refractories are nonmetallic heat-resisting materials that are generally used as components of structures that operate at high temperatures. These structures usually provide chambers in which thermochemical processes are carried out. The principal purpose of the refractory is to conserve heat so that the process can be carried out economically; at the same time, however, the refractory is expected to maintain its identity and stability as a structural member. If the process requires special atmospheres or develops erosive slags, a refractory must be selected that will resist these as well.

Refractories may be made into the form of bricks or special shapes; they may be used as grains charged loosely onto the furnace bottom and banks.

They may be made up as a castable refractory concrete, a ramming plastic, or as a slurry that can be sprayed onto a wall that is to be preserved. Bricks and other preformed shapes are generally laid in a refractory mortar.

The most common refractories are comprised of oxides of aluminum, silicon, magnesium, calcium, and chromium. These oxides can be used individually, but they are usually employed in the form of naturally occurring minerals in which they are associated with other oxides. The latter in many cases are objectionable from the standpoint of refractory performance.

Alumina and silica are common refractories. They are used separately or in combination with each other. Useful members of the alumina–silica series are found in nature as bauxite,* diaspore ($Al_2O_3 \cdot H_2O$), sillimanite, andalusite, and kyanite ($Al_2O_3 \cdot SiO_2$); pyrophyllite ($Al_2O_3 \cdot 4\,SiO_2 \cdot H_2O$); kaolins, and fireclays. Kaolins and fireclays are hydrated aluminum silicates that have the approximate formula $Al_2O_3 \cdot 2\,SiO_2 \cdot 2\,H_2O$.

Silica occurs in abundance in the earth's crust and often is found in remarkable purity. Silica bricks made of high-purity raw material exhibit excellent hot strengths up to temperatures very close to the melting point of pure silica. Of the compounds present in silica-brick raw materials, alumina and alkalies are the most detrimental to the hot-load-bearing ability of the product.

Lime and magnesia are found in nature as carbonate rocks as calcite, magnesite, and dolomite; also in seawater as chlorides. In the process of extracting magnesia from seawater, a substantial amount of the magnesia of the final product is usually derived from the calcined dolomite used in the process. Lime from calcined dolomite is required by the process to precipitate $Mg(OH)_2$. Magnesia obtained by this process is usually a high-purity product that is becoming increasingly desirable in the trend toward the use of purer raw materials for refractories.

Basic refractories are generally made of magnesia and chrome ore and mixtures of the two. Burned dolomite is also an important basic refractory. The refractory properties of chrome ore, and of dolomite, are generally improved by the inclusion of some magnesia in their respective products.

Technology has provided knowledge that has enabled us to use these minerals to best advantage to produce valuable refractory products for industry. Much of this has been derived directly from phase diagrams. The first phase diagrams available to ceramists were developed by scientists who were interested in a better understanding of the earth's geological processes. Most of these diagrams concerned silicates and therefore covered areas of composition that were generally richer in silica than the orthosilicates in

* Most bauxites used in refractories are probably complex oxyhydroxides of aluminum with compositions between diaspore, $Al_2O_3 \cdot H_2O$; and gibbsite $Al_2O_3 \cdot 3H_2O$.

their respective systems. They were nevertheless of considerable value to ceramists working in fields other than refractories. Phase-diagram compilations by Hall and Insley (1947) and those that followed by McMurdie and Hall (1949) and Levin *et al.* (1956, 1964) were extremely valuable to ceramists and others working in the field of nonmetallics and high temperatures.

The work of the Earth Sciences Department at Pennsylvania State University, particularly the research sponsored by the American Iron & Steel Institute, has resulted in the publication of many papers and phase diagrams that have expanded our knowledge of phase equilibria of special interest to technologists engaged in refractories manufacture. It is also valuable to technologists in the steel and other metal producing industries where most refractories are used. This sponsor group indicated its desire for specific phase diagrams which would concentrate on basic oxide equilibria so that the new data could be used in the design and use of refractories, slags, and inclusion technology of the steel industry. The foresight and understanding of Osborn and Muan in the planning, the research, and their direction of the programs for steelmaking refractories are effectively illustrated in their book, "Phase Equilibria among Oxides in Steelmaking" (1965). This book is written for the advanced student of ceramics. It deals with the thermodynamics of many oxide systems and is highly recommended as a valuable reference book in this field.

For many years refractory compositions were limited to individual minerals as mined. Combinations of these followed, and now beneficiation is frequently resorted to in order to obtain high purity. With the availability of new phase diagrams, we have come to realize that nature did not proportion the oxide contents of the minerals used for refractories in the most suitable ratios for optimum refractory performance. The use of pure alumina to fortify aluminous refractories and other ceramic compositions is no doubt a direct result of having the alumina–silica phase diagram. The intelligent use of chrome ores and many other natural raw materials has benefited by available application phase diagrams. Research is now underway, therefore, to synthesize more suitable compositions for particular applications from the oxides themselves. This will make it possible to broaden the horizons of refractory potentials far beyond those which are possible when only natural raw materials were used for this purpose. This type of research is aided materially, therefore, by applicable phase diagrams, which have been only recently made available to the technologist working in this field.

In their service as structural members in high-temperature units, the performance of a refractory is related to its resistance to heat as it is necessary that a refractory in service must maintain volume stability. Even partial melting at high temperature results in subsidence under load. A structure built of a refractory that becomes pyroplastic in service fails to fulfill its purpose as a stable structural member. This volume stability at high temperature depends in large measure on purity of composition.

As will be demonstrated in detail later, refractories that are part of polycomponent systems usually have melting ranges rather than precise melting points. Exceptions to this might be those compositions which are either eutectic compositions or congruent melting pure compounds. When such systems are heated to certain temperatures, liquid develops in their structures by partial and progressive melting that constitute gradual solution of the undissolved components. The amount of liquid in the refractory increases as the temperature increases. This lubricates the remaining undissolved refractory components thereby decreasing its volume stability.

A limited amount of dissolving action takes place in a refractory during the firing process. This usually causes the preformed refractory to shrink and lose some of its porosity due to coalescence by surface-tension forces. As the firing process is terminated at some point short of actual deformation, a large amount of liquid is seldom developed in the refractory during this process. In those cases in which the firing process is carried on to a greater degree, the product may be glassy and will be sensitive to heat shock when heated again as a part of an operating furnace.

The effect of the furnace environment during use is an important factor in the life of a refractory. In addition to the effect of high temperature during use, there are the effects of slags, metal, and the composition of the atmosphere to consider. Slags containing high percentages or iron oxide and lime can be very damaging to basis refractories and even more so to refractories of the alumino silica series.

Volume stability of a refractory is particularly important to the user of a refractory when it is expected to function continuously as a structural member at high temperatures for months or years. If the refractory is in the form of bricks, softening due to liquid development in a small portion of its composition permits gradual distortion or collapse. In most cases only the hot face of a furnace wall is in danger, as the temperature gradient through the wall leaves the cooler, rigid sections of the wall to carry the load. Nevertheless, the hot faces of individual bricks will usually shrink because of the action of surface-tension forces within them. This causes unit separation, poor loading patterns, and a weakening of the wall.

A ladle that holds liquid molten steel during the teeming of the heat is generally lined with a low grade of clay bricks that become soft at the temperature of liquid steel. During use these bricks bloat on their hot faces. This seals the joints, thereby preventing intrusion of metal into them. The holding time in the ladle is short, and the temperature drop through the brick wall is such that not all the brick reach critical bloating temperature. Such a brick is quite suitable in this service but would not be suitable in those locations in which a substantial cross section of the brick would reach softening or bloating temperature.

Monolithic units that are built of rammed or cast refractories are subject to the same destructive forces as bricks. The action of heat on the hot face may cause shrinkage from liquid formation, and coalescence from surface-tension forces if sufficiently high temperatures prevail.

Phase diagrams are very helpful in estimating the potential liquid development in a refractory. To this end they make it possible in many instances to calculate quite accurately the amount of liquid that will be contained in a refractory at a given temperature.

The principles of the phase rule have been set forth at considerable length in earlier chapter of this book. These principles will not be repeated here as only the *use* of phase diagrams as a tool in the study of refractories problems will be dealt with in this chapter. In this chapter examples of the practical use of phase diagrams in the design and use of refractories will be cited. In each case it will be necessary to use and explain the application of certain phase diagrams that are available in the compilation of authors Hall and Insley (1933), McMurdie and Hall (1949), Levin *et al.* (1956, 1964), Muan and Osborn (1965), and others.

The metallurgist views phase diagrams in a different light than does the ceramist. His product usually starts out as molten metal in which state it may be considered to be perfectly homogeneous. Upon cooling it crystallizes and is usually a wholly crystalline solid solution or crystalline and eutectoid. The ceramist, however, begins with granular materials often of several different grain sizes and compositions. These are pressed or otherwise assembled before firing. The firing process is arrested before much of the mixture becomes molten as would be the case in the manufacture of glass, enamels, or metals. The phases of the product cannot, therefore, be regarded as having been brought to equilibrium. Limited areas will be found to have been melted, and new crystal assemblages may appear locally in the structure. Often local areas of original unaltered raw material may be evident. Higher firing will bring the product nearer to complete equilibrium through the solution of original crystalline phases. If the refractory, or primary phase is present in large amount, the product will retain its form, otherwise it will slump, owing to the development of a large proportion of liquid. This of course is not a feasible manufacturing possibility.

The user often subjects the refractory to temperatures higher than those attained during the firing process. His refractory problem is very often therefore simply one of liquid development, and it is the author's opinion that this is the most important of the several causes of failure of a refractory during use. As the solid, crystalline portion of the refractory is more resistant to the solvent action of fluxes than the liquid portion, it is generally true that the refractory that will serve best is one in which the liquid development is minimal. To accomplish this the user intelligently must select a composition that will meet this requirement.

Crystal development has a secondary influence on volume stability during use of a polycomponent refractory such as those of the aluminum silica series in which alkalies play a specially significant role.

In refractories that are essentially one-component systems, crystal development and crystal stability can be important factors in their performance during use. Examples of this class are zirconia, silica, 2 $CaO{\cdot}SiO_2$, and certain others that have generally not been considered as refractories. In the latter group are manganese oxide, iron oxide, and titania.

Zirconia is a very refractory oxide with some very desirable characteristics and many potential uses. At room temperature it occurs as a monoclinic mineral. When heated to approximately 1040°C, it inverts to its tetragonal form with a 9% volume change. On cooling again the reversal volume change causes it to powder, making it unsatisfactory as a refractory structural unit. ZrO_2 can be stabilized with lime by which a stable refractory $CaO–ZrO_2$ cubic solid solution is formed. It can also be stabilized by MgO and cations of certain rare earths (Duwez *et al.*, 1952; Lynch *et al.*, 1961).

When hematite (Fe_2O_3) is heated in air above 1390°C it is converted to magnetite (Fe_3O_4). Mn_3O_4 (Muan and Somiya, 1961) is tetragonal up to 1160°C, but when heated above this temperature it adopts a spinel structure. As these oxides have high densities and are quite refractory, they could conceivably be used as heat-exchange media in furnace regenerators if it were not for their instability due to their polymorphic inversion volume changes. The conversion of hematite to magnetite can be extended upward to a limited extent by combining it with about 10% titania (MacChesney and Muan, 1960; Muan and Osborn, 1965). Alumina exists in several polymorphic modifications, but they appear to offer no problem in its use as a refractory (Dana, 1944; Kohn *et al.*, 1957; Muan and Osborn, 1965).

II. THE SYSTEM Al_2O_3–SiO_2

Although this chapter is devoted to a great extent to the liquidus of phase diagrams and to the effects of liquid development in refractories, these examples demonstrate the importance of subliquidus equilibria as well.

The alumina–silica phase diagram, Fig. 1* (Aramaki and Roy, 1959)

* This is the newly revised diagram of Aramaki and Roy.

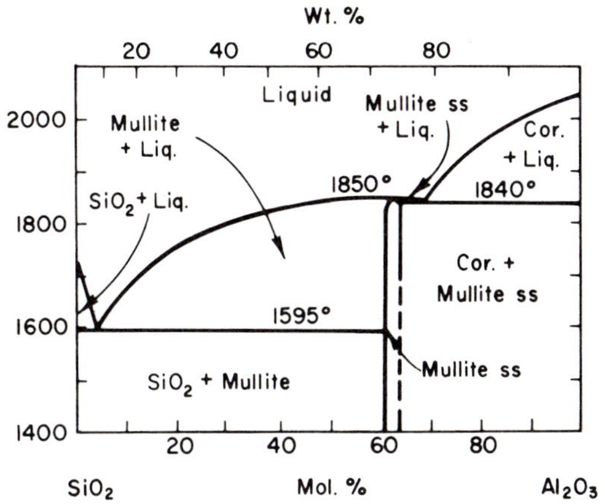

Fig. 1. The alumina-silica system. Revised, reprinted from Levin *et al.* (1956).

has wide application in refractories technology as it explains the thermochemical characteristics of many refractories whose compositions are substantially all alumina and/or silica. It bears a similar relationship and usefulness to the ceramist as does the iron–carbon diagram to the metallurgist. Clays are seldom pure and generally contain excess silica over that of the kaolinite molecule, but many of them serve useful purposes as refractories in a wide variety of industrial furnaces. These and other compositions that are substantially all alumina and silica can be looked upon as members of the system, and much of their pyrochemical behavior can be attributed to relationships expressed in the phase diagram.

This system is characterized by its two end members alumina and silica and an intermediate compound mullite having the formula $3\ Al_2O_3 \cdot 2\ SiO_2$. There are two eutectics in the system: one between cristobalite and mullite, melting at 1595°C, and the other between mullite and corundum, which melts at 1840°C. Mullite takes a limited amount of alumina into solid solution. In practice when iron oxide and titania are present in the raw material, mullite may take a limited amount of these also into solid solution. Compositions between mullite and cristobalite will develop liquid at 1595°C. As this eutectic contains only 5% alumina, it is evident that alumina is a powerful flux for silica.

A large number of aluminosilica minerals including clays whose compositions contain less than 72% alumina when calcined fall between mullite and silica on this diagram. Their fields of service in industry vary owing to their impurities, but each, nevertheless, has many useful applications. In this category are several types of fireclays and kaolins containing up to 45% alumina when fired, and minerals such as pyrophyllite, kyanite, andalusite, and sillimanite. Each of the last three contains approximately 63% alumina.

The phase diagram shows that the melting points of compositions between 5 and 72% alumina increase with their alumina contents; however, all compositions in this range will be liquid in some degree at temperatures above 1595°C. Compositions containing more than 72% alumina do not develop liquids until 1840°C is reached. Any of these compositions that lie between mullite and corundum may therefore be used with assurance up to 1840°C. Refractories whose compositions lie close to pure alumina may be used at temperatures well above 1840°C.

These facts were the basis of a solution of an industrial refractories problem which now may seem to be elementary but the solution can be considered to be somewhat of a classroom example as follows.

A high-temperature steel reheating furnace 125 ft long was required to maintain a temperature between 1650° and 1700°C. This furnace was essentially a tunnel 125 ft long, through which steel strip was heated to welding temperature at a rate 500 ft/min. In each of the side walls, there were 100 burner openings through which high-temperature burners operated to supply the heat. For each burner there was a refractory burner block which extended from the hot face half-way through the wall. The face of the burner block

Fig. 2. 68% alumina burner blocks after two weeks service at 1700°C.

was exposed to the high temperature of the furnace (1700°C), but the remainder was somewhat cooler owing to the temperature gradient through the wall. The burner blocks were required to support some of the weight of the roof as well as the wall above them. These blocks were originally made of a 68% alumina material.

As will be seen in the alumina–silica diagram, the 68% Al_2O_3 blocks developed liquid at 1590°C and continued to increase liquid content at temperatures above this. Under these conditions they were pyroplastic and the result of their use is shown in Fig. 2. The life of these blocks was less than two weeks. As the walls were made of the same material, the replacement was extensive and time consuming. Hence two furnaces were required for continuous production schedules.

Fig. 3. 90% alumina burner blocks after four weeks service at 1700°C.

The walls and burner blocks were then replaced with a 90% alumina brick whose composition was well above the mullite composition on the diagram and in which no liquid developed below 1840°C. Figure 3 shows the condition of these high-alumina blocks after four weeks of service. They were not affected, and the furnace required only spot repairs. Furnace availability was greatly increased owing to the greater length of campaign—decreases in repair delays were substantial because of the decrease in frequency and extent of repairs. Although the unit cost of these refractories seemed very high at the time, their cost per ton of product heated was substantially lower than that for the previous installation.

Clays for refractories related to this system are generally used as mined. They therefore bring with them constituents other than alumina and silica, which are indigenous to their natural environment. Titania, iron oxide, and the alkalies are the principal impurities. American fireclays are coal formation clays and are classified as flint, semiflint, and plastic fireclays. This connotation is derived from the fact that they occur with coals and acquire chemical constituents from the same source as do coals, i.e., the geological source of each being co-related. Of particular significance here is the presence of alkalies, especially potash, which is derived as it is in the biogenesis of coal, from plant life.

TABLE I

COMPOSITIONS OF TYPICAL FLINT AND PLASTIC FIRECLAYS[a]

	Flint fireclay	Plastic fireclay
Percent raw clay basis	0	0
Loss at 105°C	1.35	2.28
Ignition loss	12.90	7.69
SiO_2	45.20	60.07
Al_2O_3	36.77	22.57
Fe_2O_3	1.30	2.38
CaO	0.50	0.68
MgO	0.08	0.64
TiO_2	1.80	1.60
P_2O_5	0.02	0.07
Na_2O	0.00	0.11
K_2O	0.52	1.71
MnO	Trace	Trace
S	0.04	0.95
C	0.35	0.22
PCE	33	27
Product	Firebrick	Firebrick

[a] Stout *et al.* (1923).

Flint fireclays and kaolins contain only small amounts of potash whereas plastic fireclays contain considerably more as will be seen in the Table I.

The high potash content of plastic fireclay is largely responsible in this case for the low refractoriness. The principal effect of the alkalies is to lower the fusion point, as indicated by their PCE (pyrometric-cone-equivalent) test values or other refractoriness criteria. Although iron oxide, titania, and high silica contents of some of these clays are factors in their firing behavior and

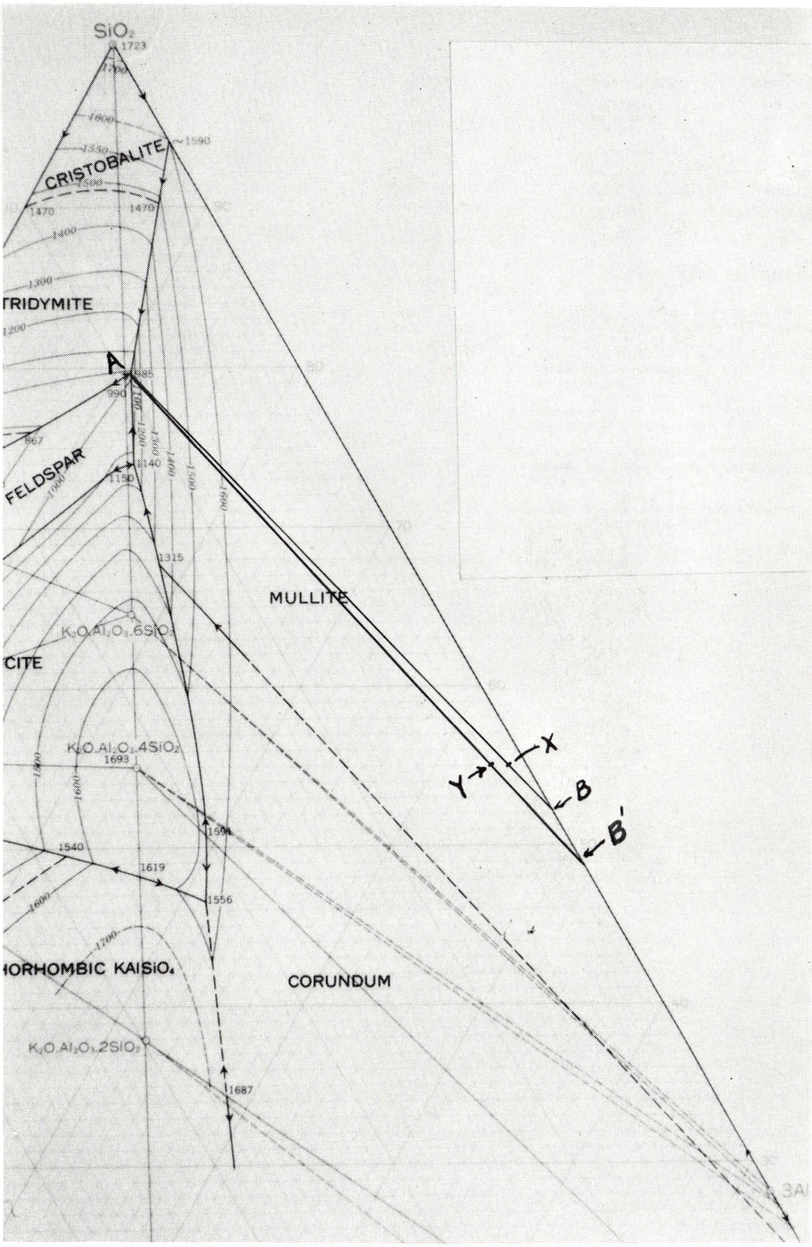

SiO₂
1723
1590
CRISTOBALITE
1470
TRIDYMITE
A
985
990
867
FELDSPAR
1140
1150
1315
MULLITE
K₂O.Al₂O₃.6SiO₂
CITE
K₂O.Al₂O₃.4SiO₂
1693
Y
X
B
B'
1540
1619
1556
1594
ORHOMBIC KAlSiO₄
CORUNDUM
K₂O.Al₂O₃.2SiO₂
1687
3Al

Fig. 4.

of the final product, it is considered that the alkalies even in quite pure raw materials such as kaolins are a major problem in attaining high-temperature stability and good hot load bearing ability in alumina–silica refractories.

Most refractory clays are considered to have the approximate composition of the mineral kaolinite ($Al_2O_3 \cdot 2\,SiO_2 \cdot 2\,H_2O$). When this mineral is fired, it first loses its water. Further firing results in the formation of mullite and free silica as follows:

$$3(Al_2O_3 + 2\,SiO_2) = 3\,Al_2O_3 \cdot 2\,SiO_2 + 4\,SiO_2$$

Figure 4 is a part of the phase diagram of the system K_2O–Al_2O_3–SiO_2 (Schairer and Bowen, 1955). Compositions X and Y spotted thereon represent compositions of two otherwise pure clays in their fired condition. X contains 1% K_2O whereas Y contains 2% K_2O. They lie in the mullite field. The first liquid to form in these compositions when they are fired is the 985°C eutectic at point A. At that temperature composition X will be 11.5% liquid; composition Y will be 22.5% liquid. A brick of composition Y would deform to a greater extent during use at 985°C or above than would composition X. The latter would therefore be regarded to be more refractory and generally more desirable. The calculation is made as follows:

$$XB/AB \times 100 = \%\text{ liquid in composition X at 985°C}$$

and

$$YB'/AB' \times 100 = \%\text{ liquid in composition Y at 985°C}$$

Mullite is the refractory or primary phase in these clays. When they are fired, large crystals of mullite are not developed until enough liquid has been formed to allow crystal assemblages to form. When firing temperatures progressively exceed 985°C, the composition of the liquid departs from that of the eutectic, following the boundary line between tridymite (or cristobalite) and mullite until all of the excess silica has been consumed and only mullite remains. Then as heating progresses further, mullite will go into solution until the composition of the brick is reached and melting is complete. Normal firing temperatures for clays seldom exceed 1550°C. However, this could be critical in the case of the two compositions X and Y. The one containing 2% K_2O would require less fire and firing at temperatures below 1470°F would suffice. In the case of the 1% K_2O composition, firing above 1500°C might be required. In the former, silica would be present as tridymite whereas in the latter it would be present as cristobalite and would not be as resistant to heat shock as the tridymite-containing brick.

The foregoing comments on the thermochemical characteristics of clay refractories is related only to the alkalies contained in them. Equally severe is the effect of alkalies on clay refractories when they are used in glass-melting furnaces. In these the alkali vapor phase and dust react to form nephelite ($Na_2O{\cdot}Al_2O_3{\cdot}2\,SiO_2$) or perhaps to drain from the refractory a liquid phase richer in silica than this with the result that nephelite and coarsely crystalline mullite appears. Additional fluxing action by soda leaves only corundum (Thompson and Kraner, 1933).

Nephelite is frequently found as stones in glass products. It is also found on the surface of the blocks of a glass tank in which soda-lime glass was produced. Corundum is also often found as stones in glass as well as in various areas of the glass furnace. All of these observations attest the trend of reaction between the alkalies of the process and the clay or other alumina-silica refractory used in the furnace.

The iron oxide in clay refractories should not be totally ignored. In clay refractories it is present in relatively small percentages around 1.5 to 2%. In plastic fireclays that are somewhat less pure, it may run as high as 3 or 4%. Under oxidizing conditions this may not be seriously detrimental to the refractory. The diagram for $FeO{\cdot}Fe_2O_3$–Al_2O_3–SiO_2, Fig. 5 (Muan

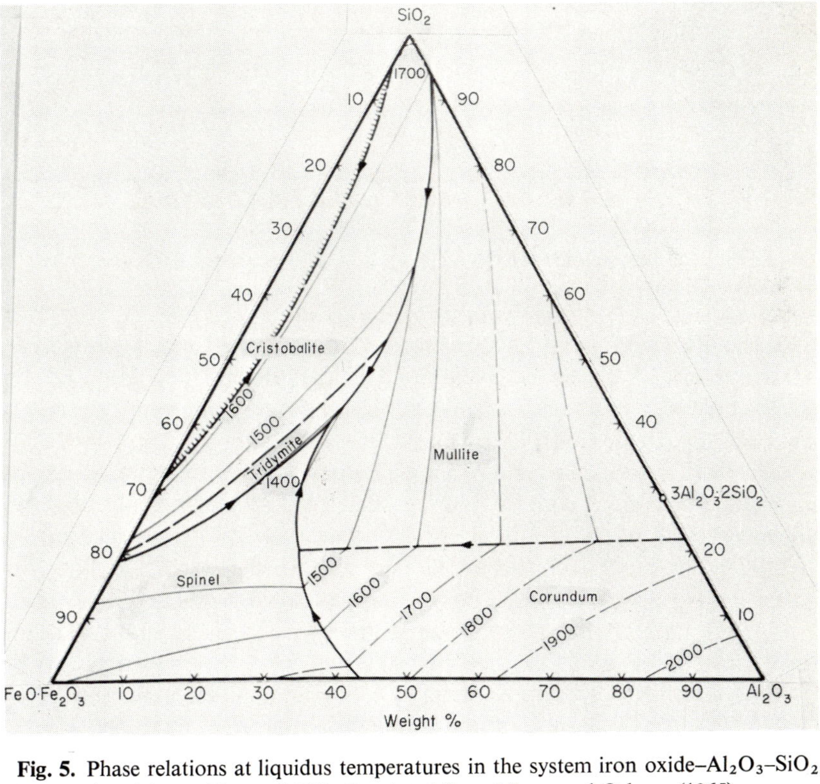

Fig. 5. Phase relations at liquidus temperatures in the system iron oxide–Al_2O_3–SiO_2 in air. Based mainly on Muan, (1957). Reprinted from Muan and Osborn (1965).

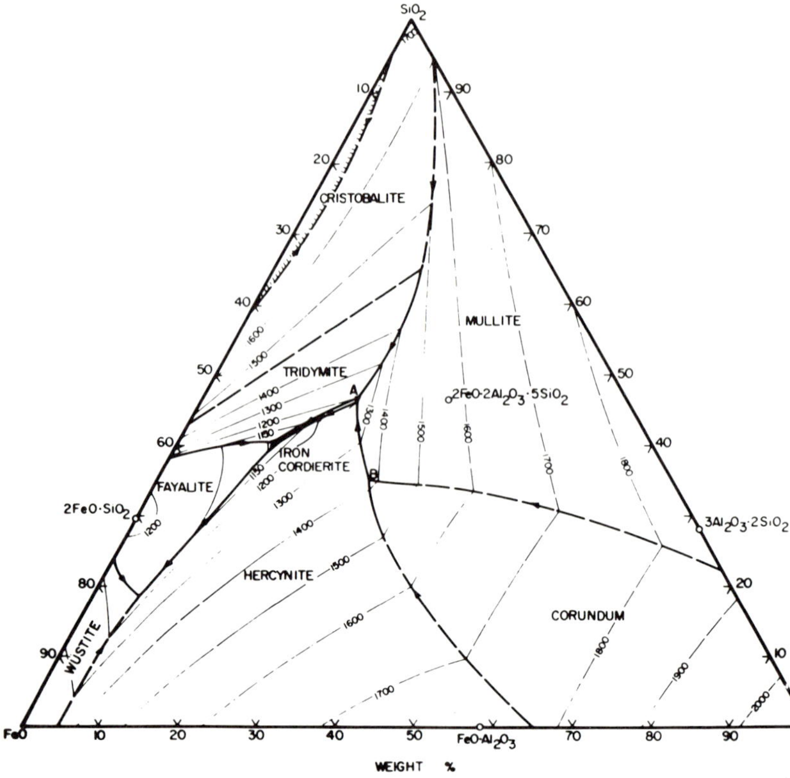

Fig. 6. The system $FeO–Al_2O_3–SiO_2$. (Bowen and Schairer, 1932). Reprinted from Muan and Osborn (1965).

1951), shows that there is no liquid formed in this system much below 1400°C. However, the diagram for $FeO–Al_2O_3–SiO_2$, Fig. 6 (Schairer and Yagi, 1952), shows that under reducing conditions iron-oxide-bearing brick will develop liquid below 1150°C.

Clay brick and other alumina–silica compositions are often employed in furnaces where iron oxide appears as scale or dust. Such conditions prevail in open-hearth checkers or reheating furnaces in the steel industry. Under oxidizing conditions clay firebricks perform satisfactorily. If reducing conditions prevail, clay firebricks will be eroded rapidly by ferrous oxide. This is especially true in the case of siliceous fireclay bricks which contain as much as 75 or 80% silica.

The liquidus isotherms of these two diagrams indicate quite clearly the advantage of high-alumina brick when iron oxide is contained or when iron oxide exposure is involved. Liquidus temperatures are generally higher in all compositions containing substantial amounts of alumina.

Metallurgical and glass-making processes require alumina–silica refractories that are more refractory and volume-stable than clays without fortification by adding alumina. Sillimanite and kyanite refractories and others containing as much as 95% alumina are available for this purpose.

Even high-alumina compositions are seldom entirely free of alkalies and are adversely affected by small amounts of them. They are also seldom entirely free of silica. Alkali and silica, even in amounts in the order of tenths of a percent, provide the components of small amounts of liquid at high temperature that will affect the hot-load and creep-test values. Although corundum and mullite are very refractory, they may be compatible with liquids that have low melting points.

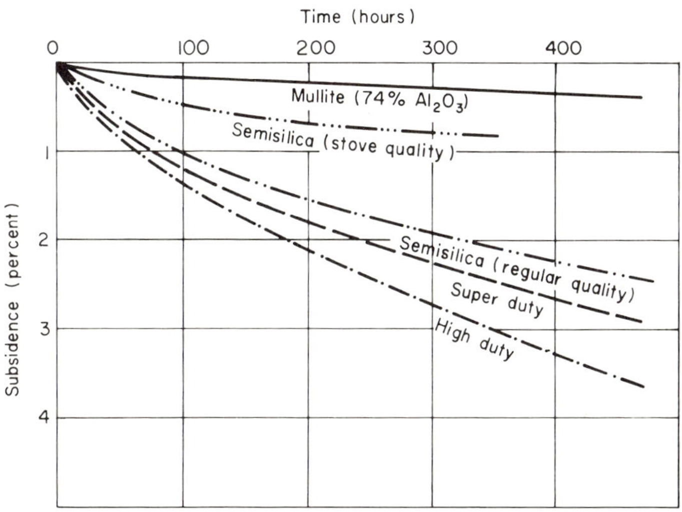

Fig. 7. Deformation of stove refractories in creep test under 25 psi load at 2350°F (1288°C).

The demands made on high-alumina refractories are sometimes expressed in terms of creep (Kraner, 1962). Stringent specifications for hot-load-bearing abilities may often be met only by very careful consideration of the chemistry of the raw materials of which they are comprised with particular attention being given to their alkali contents. See Fig. 7. In this figure the mullite bricks and the semisilica bricks contain approximately only 0.4% alkalies. Of the two semisilica bricks the stove quality product contains a slightly higher silica content and lower impurities generally. Superduty clay bricks may contain up to 1% alkalies. The high-duty clay bricks contain 1 to 2.5% alkali.

Creep is often experienced even in the use of very high melting point materials such as has been described for corundum and mullite. Highly refractory materials such as periclase (in some systems), magnesia spinel, and baddeleyite may also be found in equilibrium with liquids that have low melting points. This represents a serious problem to the producer of the refractory when limited creep and maximum load-bearing ability are requirements of the user. In Fig. 7 only the effect of liquid development in a refractory is assumed to be the cause of the hot-load bearing characteristics of the products represented. It is recognized that creep within crystals of highly volume-stable materials also may be a factor in such cases.

III. KINETIC LIMITATIONS

It was mentioned in earlier paragraphs that the use of phase diagrams in ceramics must be viewed somewhat differently than would be the case if we were dealing with completely homogeneous melts as in the case of metals. Mixing and firing granular solids often creates conditions wherein only local equilibrium is obtained because of kinetic limitations in the reaction that is predicted by the phase diagram. The following paragraphs discuss a few examples of kinetic limitation and show that the local area equilibrium in oxide mixtures may still be explained by the diagrams involved.

In order to improve the quality of clay refractories or other alumina–silica compositions, alumina is often added to increase refractoriness and mullite formation. When stoichiometric proportions of alumina and silica are designed to produce mullite, it would be expected that the alumina would be consumed in the process. The product would be fired above 1595°C, the eutectic melting temperature at 5% alumina, but firing would usually stop below complete melting.

X-ray or petrographic studies of the product would probably reveal considerable uncombined alumina as corundum indicating that not all of this addition had reacted with the siliceous liquid below the mullite composition. This appears to be an anomaly, but it is readily explained by the diagram itself. Some mullite forms in the solid state even below the melting of the 1595°C eutectic composition. As soon as liquid develops by the melting of this composition, reaction takes place between it and the corundum grains to form mullite. Corundum should continue to react with this liquid as the temperature is increased to form more mullite. However, complete formation of mullite is prevented by the development of dense coatings of mullite on the corundum grains. This seriously impedes further formation of mullite as the mullite already formed is in equilibrium with the corundum and is also in equilibrium with the siliceous liquid that surrounds it.

Finer grinding of the corundum would seem to be of considerable benefit in the solution of this problem but raising the temperature to redissolve the mullite or complete melting would be more practical.

Another similar example of such an "anomaly" is known in the open-hearth process of steelmaking. The slags produced in the progress of the heat are composed of lime, iron oxide, and silica. Early in the heat the lime content is low. The liquid slag is a ferrous silicate with a lime–silica ratio of about 1:1. The ultimate slag will have a lime–silica ratio of 3:1 to 4:1. The additional lime comes from the solution of lumps of lime from the limestone charged with the scrap. When these lumps are released by the melting scrap, they enter the slag and begin to react with it to form dicalcium silicate. This coating of $2CaO \cdot SiO_2$ on the larger lumps slows the diffusion of the slag into lime and soon becomes a buffer layer that retards further reaction and solution of lime in the ferrous silicate slag.

The process here is similar to that described above for the buffer layer of mullite on corundum. The dicalcium silicate is in equilibrium with lime on the one hand and in equilibrium with the liquid slag that surrounds it. In this slag problem the buffer layer is broken up by local application of CaF_2. Higher furnace temperature and the use of iron oxide locally are also solutions to the problem, but they are less desirable than the CaF_2 method of removing the dicalcium silicate.

IV. SILICA BRICK

Silicon is the most abundant element next to oxygen, and silica is therefore found widely in nature, usually as quartz. Its characteristics are varied and interesting to the technologist. Quartz is converted to several polymorphic forms depending upon temperature. The physical properties of these forms as well as those of the fused material are distinctive and merit the efforts of those who have amassed volumes of scientific information on this very useful and interesting mineral.

Silica bricks are one of the unique products that have been used very successfully in industry for many years. In the last decade, however, steelmaking processes have radically changed. This has affected refractory requirements, with the result that the market for silica brick has been reduced to less than half the volume that was formerly used. There still exists, however, a sizable market for silica brick in the glass-melting furnaces of the nation and in other high temperature industrial units. In the year 1967, 71,214,000 silica bricks were sold in the United States for refractory purposes. The value of these bricks was $14,702,000 (U.S. Bureau of Census, 1967).

For many years silica bricks were used very successfully in open-hearth roofs and other important applications in the steel industry. Their success was due to the fact that they could be used under load at high temperatures close to the melting point of pure silica. The low expansion characteristics of the high-temperature polymorphs of silica was another practical advantage in the open-hearth roof where wide swings of temperature take place during the operation cycle. Not until the industry required temperatures above the melting temperature of silica were they replaced.

The raw material for silica bricks is derived from well-bonded sandstones, quartzite, and quartz pebbles such as the Sharon Conglomerate. These raw materials must have high purity, particularly in respect to their content of alumina and alkalies. Titania and iron oxide are objectionable but not as much so as alumina and the alkalies. In the manufacture of silica brick the rock is crushed and sized. This is then bonded with 2 to 4% CaO as slaked lime. The bonded bricks are fired at temperatures around 1500°C.

The process of firing permanently converts almost all of the quartz to other polymorphic forms of silica. Quartz has a density of 2.65 g/cc, whereas the density of a silica brick is approximately 2.33 to 2.34 g/cc. The brick expand during the firing process. Most of the fired brick is cristobalite and tridymite, but a small amount of glass and some residual quartz are always present. When a substantial portion of the final product is tridymite, it will have good strength and somewhat better heat-shock resistance. Special attention is therefore directed toward developing as much tridymite as is practicable in preference to cristobalite.

The diagrams Fig. 8 by Kingery (1960) and Fig. 9 by LeChatelier (1913, 1914) illustrate the equilibrium conditions in the silica system as applied to silica bricks during firing. See also Sosman (1927a), Kingery (1960), and Chesters (1957). They do not in themselves indicate the rate of inversion from one polymorphic form to another: this being important, as will be seen in the following paragraphs.

When the brick is heated to 573°C, alpha quartz (the common form) converts rapidly to beta quartz. If beta quartz is cooled, it reverts to alpha quartz readily, but when heating is continued to 870°C, another inversion point (beta quartz to beta$_2$ tridymite) is reached. This inversion is sluggish and does not occur readily; instead the next high-temperature form develops. This is beta cristobalite. Between 873°C and 1470°C, cristobalite may be considered to be metastable for, if the brick were held at temperatures between 870° and 1470°C for a longer period of time, the previously formed cristobalite would eventually invert to tridymite. This would entail considerable furnace time and expense and it is therefore not often done. When the brick are cooled, quartz is not reproduced and the ratio of tridymite and cristobalite existing at high temperature remains upon cooling. The high-temperature forms of tridymite and cristobalite do, however, invert to their respective lower-temperature forms—265°C for the low cristabolite form and 163° and 117°C for the low tridymite inversions. The low inversion for cristobalite is

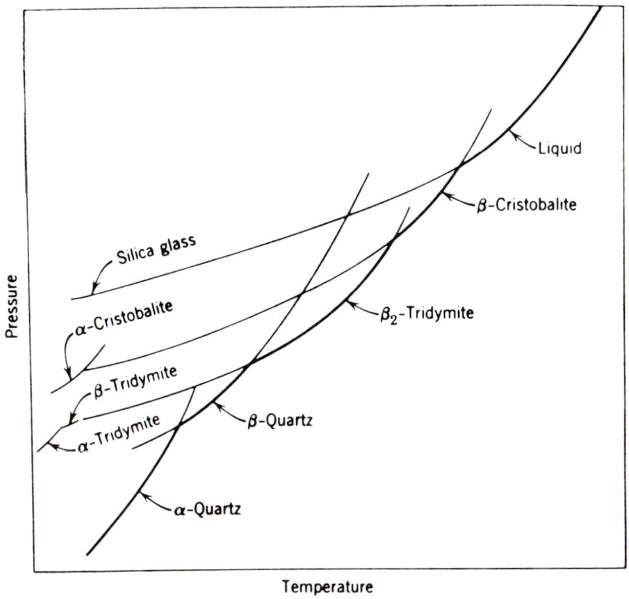

Fig. 8. Equilibrium relations of the minerals in the silica system. Reprinted from Kingery (1960).

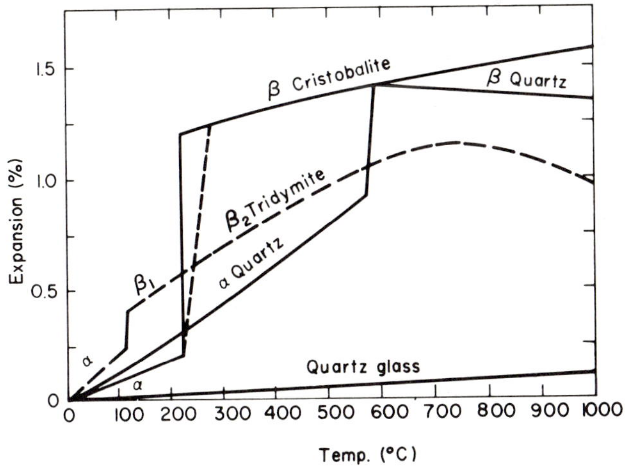

Fig. 9. Thermal expansion of silica minerals (after LeChatelier). Reprinted from Chesters (1957).

accompanied by a 2.8% volume change. Its reconversion must be dealt with carefully when heating furnaces that contain structual elements of silica bricks.

Heating shedules have been developed which are successful. These employ slow heating rates in the temperature range of the high expansion inversions (Kraner and Jewart, 1952). As the high-temperature form of cristobalite has a very low coefficient of expansion, it is largely responsible for the almost zero expansion of silica brick above 500°C, as shown in Fig. 10. This is the typical expansion curve of a commercial silica brick.

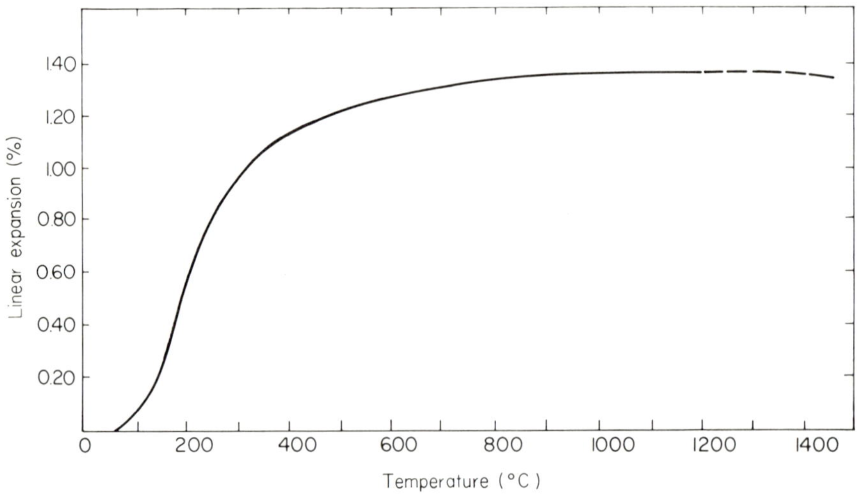

Fig. 10. Typical thermal expansion curve for silica brick.

The large expansion of the brick below 400°C makes it unsuitable for use in units such as blast furnace stoves and other regenerators in the range between room temperature and 500°C.

As cristobalite does not melt until 1723°C and, as good silica raw materials are rather plentiful, silica bricks can be used at temperatures very close to the melting point of pure silica. In a used silica brick taken from an open-hearth roof, one will find the hot face to have been completely converted to cristobalite.

In a vigorously operated open hearth furnace, the efficient operator will always maintain hot face temperatures at or very near the melting point of the brick. A brick containing 1% of Al_2O_3 plus alkalies will not withstand as high a temperature as one containing only $\frac{1}{2}$% or less. It is known that as much as 25% more steel has been made in a given length of time in a furnace under a 0.5% alumina silica roof as compared with a furnace that was operated with a 1% alumina roof. If a furnace is operated with a split roof with one-half being a 0.5% alumina brick and the other half being 1% alumina, the low-alumina half will last approximately twice as long as the higher-alumina half. This has been demonstrated in practice. The use of a complete roof of low-alumina brick allows the operator to run his furnace at higher temperatures thereby producing steel at a faster rate. In split-roof trials, the producing rate is limited by the lower refractoriness of the higher-alumina brick.

The ability of the brick to withstand temperatures so close to its melting point has been attributed in a general way to the purity of the rock. More specifically, as Greig (1927a, b) pointed out in 1927, it was due in large measure to the characteristic of silica to form immiscible liquids with lime and with iron oxide.

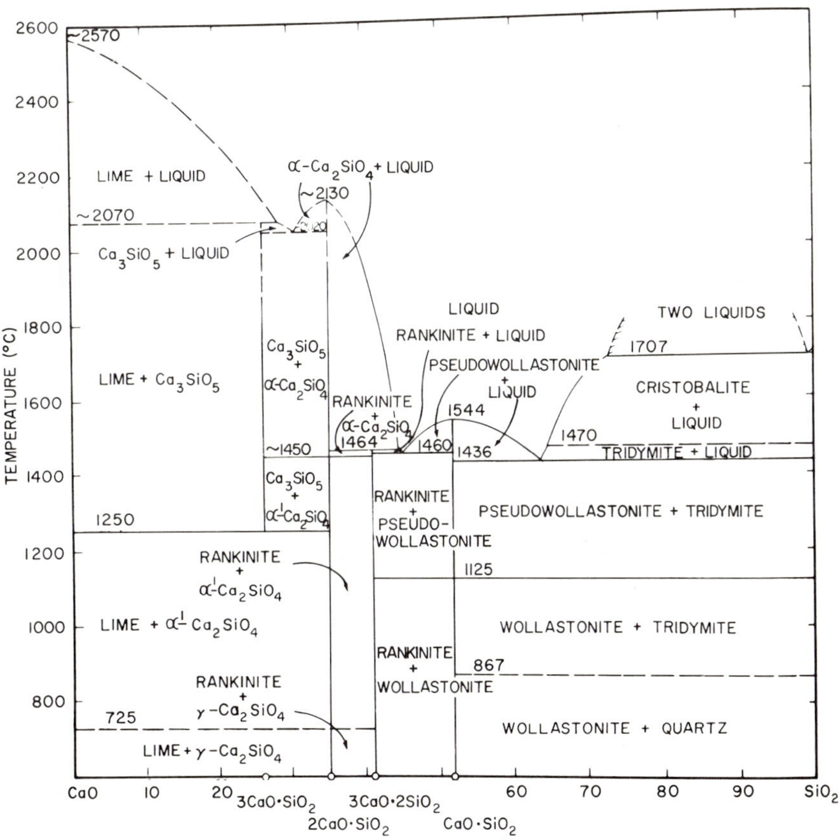

Fig. 11. Phase diagram for the system CaO–SiO_2. Based mainly on the work of Day *et al.* (1906). Reprinted from Muan and Osborn (1965).

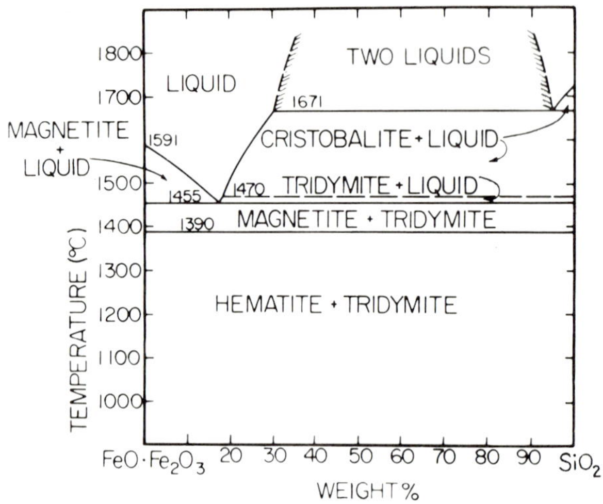

Fig. 12. The phase relations for iron oxide–silica in air plotted on the Fe_3O_4–SiO_2 join (Muan, 1958). Reprinted from Muan and Osborn (1965).

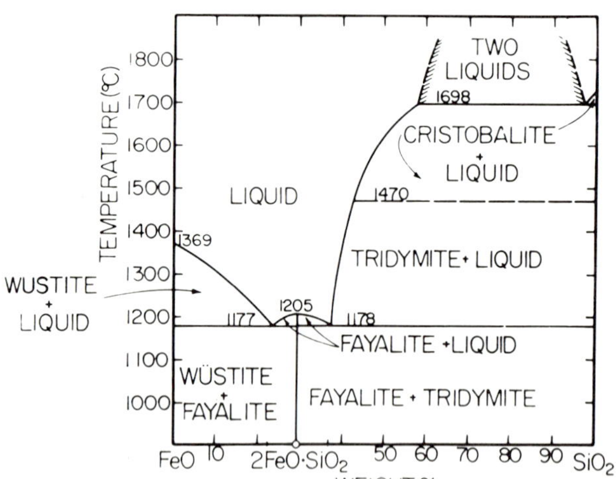

Fig. 13. The phase relations for iron oxide–SiO_2 in contact with metallic iron plotted on the FeO–SiO join (Bowen and Schairer, 1932). Reprinted from Muan and Osborn (1965).

The lime–silica diagram, Fig. 11 (Day *et al.*, 1906), shows that silica forms two immiscible liquids with lime and that cristobalite is stable in the presence of these at 1707°C, this being very near to the melting point of cristobalite. The lime content of the composition reaches approximately 28 %. The diagrams for iron oxides and silica, Figs. 12 and 13 indicate that the same situation prevails wherein immiscible liquids are favorable to the refractory behavior of cristobalite. The ternary systems of silica and common divalent oxides such as FeO, MnO, and MgO also show this immiscibility area as an extensive plateau. See Fig. 17. Silica does not form immiscible liquids with alumina. This was shown in the alumina-silica diagram, Fig. 1, and in the $CaO–Al_2O_3–SiO_2$ phase diagram, Fig. 14 (Rankin and Wright, 1915). The latter shows the relationships that apply to silica brick. Even small amounts of alumina in this system eliminate the immiscibility field.

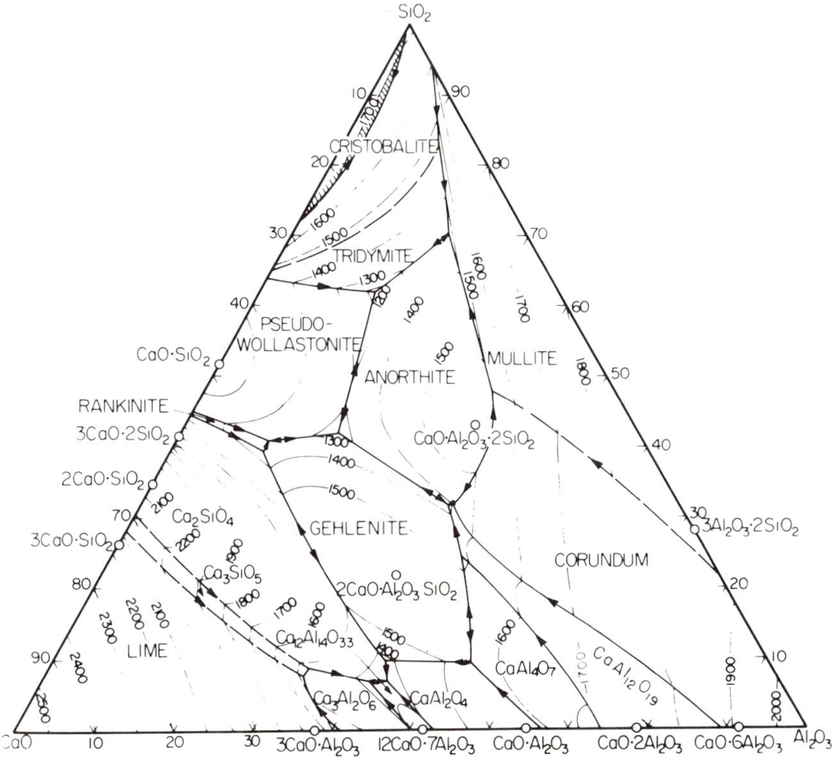

Fig. 14. Phase diagram for the system $CaO–Al_2O_3–SiO_2$. Mainly from Rankin and Wright (1915) and Grieg (1927, pp. 1, 133). Reprinted from Muan and Osborn (1965).

Silica does not form immiscible liquids in the presence of alkalies, hence there is no immiscibility with silica in the systems Na_2O–Al_2O_3–SiO_2 and K_2O–Al_2O_3–SiO_2.

The extent of liquid phase development in silica brick at high temperatures is determined by the use of the lime–alumina–silica phase-equilibrium diagram, Fig. 14 (Rankin and Wright, 1915; Kraner, 1944). The brick is

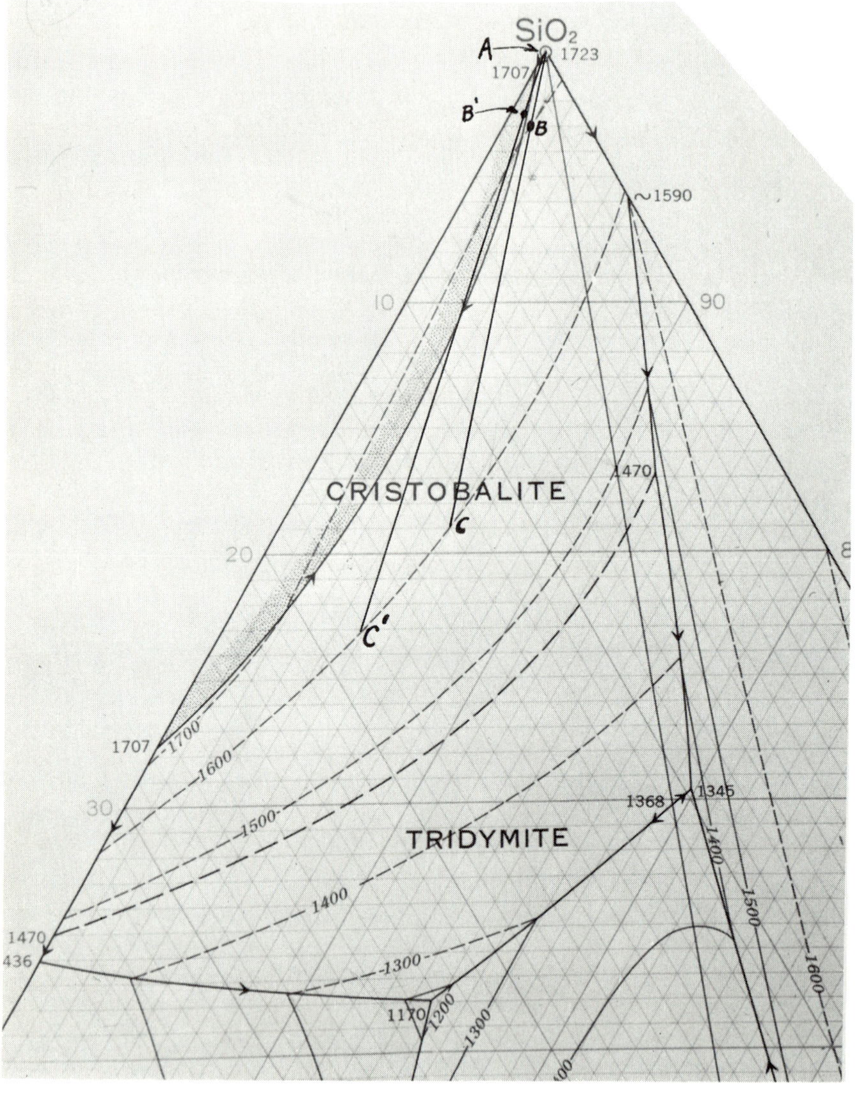

Fig. 15. Partial diagram of the system lime–alumina–silica.

largely silica. With the very small amount of alumina that is present plus about 2% lime, the first liquid to form when the brick is fired will be the eutectic composition between tridymite, anorthite, and pseudowollastonite at 1170°C. With increased temperature the composition of the liquid changes following the boundary line between pesudowollastonite and tridymite dissolving lime and silica simultaneously. When the lime–alumina ratio of this liquid reaches that of the brick composition, further heating carries only silica into solution until the maximum temperature of the firing process is reached. Meanwhile the mineral constitution of the brick has been established for the temperature attained.

The phase diagram of the system lime–alumina and silica is very useful in studying the effect of alumina on the liquid development in a silica brick during firing and use, and for the calculation of the amount of liquid and solid present at temperatures of use. For example, consider compositions B and B′ on Fig. 15. These contain 1% and 0.5% alumina, respectively, and 2% lime. To determine the amount of liquid in these compositions as an example, a line is drawn from the apex A (silica) through B or B′ and then on to the 1600°C isotherm, (point C or C′). The proportions of the liquid and solid are then determined from distance measurements AB, AC, and BC wherein

$$AB/AC \times 100 = \%\ \text{liquid in composition B at } 1600°\text{C}$$

$$BC/AC \times 100 = \%\ \text{solid in composition B at } 1600°\text{C}$$

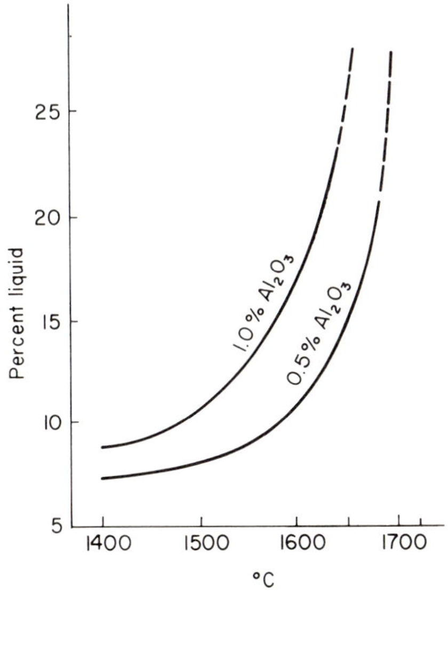

Fig. 16. Liquid content of silica bricks at elevated temperatures.

From such measurements and calculations one may obtain data as shown in Fig. 16 for bricks containing other percentages of alumina and at other temperatures.

The sharp increase in the liquid development at temperatures above 1600°C as illustrated in this figure very impressively indicates the effect of alumina on the refractoriness of silica brick at temperatures of use.

Liquids containing high percentages of silica are quite viscous, and this may be a factor in retarding crystallization. It is also sufficient to rule out the PCE test as a practical one for testing the refractoriness of silica brick (Kraner, 1944). The viscosity is not sufficient, however, to improve the hot load bearing ability of silica brick at high temperatures. In work performed on silica bricks of various alumina contents, it was shown that when the liquid content of a silica brick reached 18%, failure would occur in the hot-load test under a load of 25 psi (Kraner, 1944).

V. THE SYSEM $FeO–Fe_2O_3–SiO_2$*

At high temperatures such as those that prevail in the open-hearth furnace, iron oxide may exist as FeO or Fe_3O_4. When in contact with the metal bath, it is present largely in the form of FeO. In the roof where the oxygen pressure is higher, it is present in the form of magnetite (Fe_3O_4). In this operation iron oxide is found in all parts of the furnace arriving there by splashing from the molten-metal bath, by dust from the ores charged, and as vapor from melting scrap. When the roof is made of silica bricks, the bricks are usually saturated with iron oxide to some depth from the hot face. Although the bricks are quite porous and act like a sponge in absorbing liquid oxides, the depth of penetration is limited by the temperatures prevailing at distances from the hot face. The hot face may be dripping with molten oxide after a few heats have been made. In the first few millimetres of the hot face, the percentage of iron oxide will be very high and the average in the first inch may exceed 70%. Further back in the brick, the amount tapers off gradually to the normal amount of iron oxide that is usually present in the original brick, this being from $\frac{1}{2}$% to 1%.

Under normal operating conditions this iron oxide is present as magnetite and has little effect upon the refractoriness of a silica brick up to rather high percentages. However, if the atmosphere undergoes a change toward lower oxygen pressure, the face of the brick is rapidly eroded or sloughed off at furnace operating temperatures. This is illustrated in the phase diagram $FeO–Fe_2O_3–SiO_2$ (Muan, 1955, 1965).

* Muan (1955, 1965).

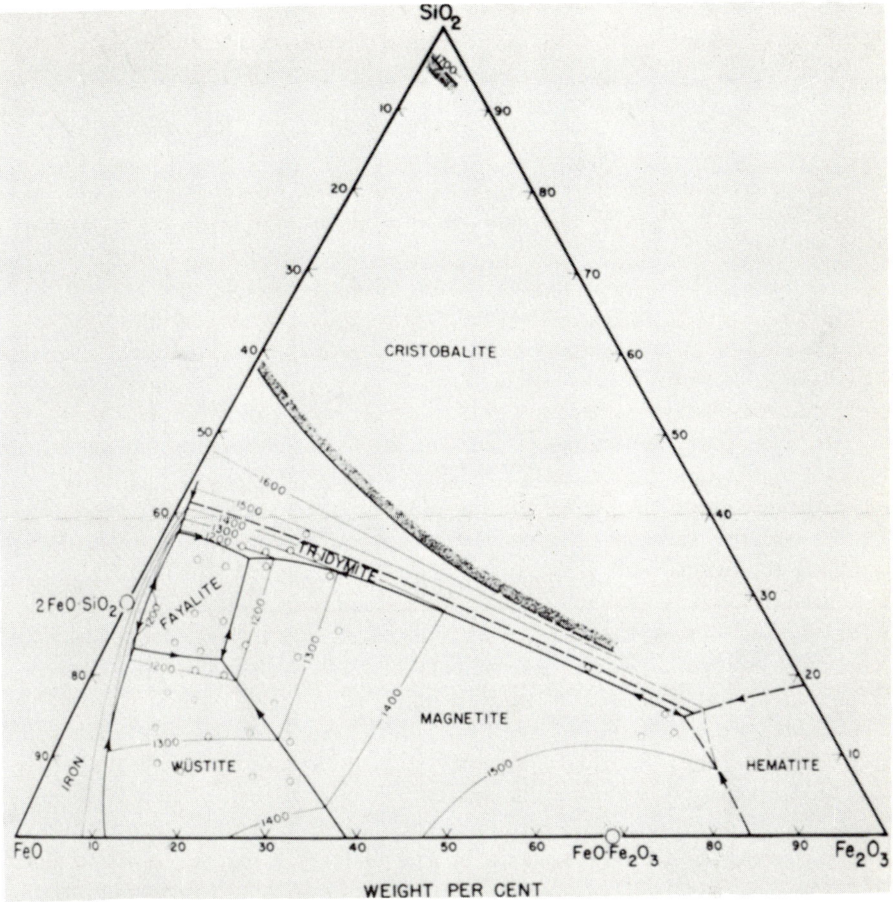

Fig. 17. Phase diagram for the system $FeO–Fe_2O_3–SiO_2$. After Muan (1955b). Reprinted from Muan and Osborn (1965).

The phase relations for iron oxide–silica in air as plotted on the Fe_3O_4–SiO_2 join were shown in Fig. 12. A composition on this join at 70% iron oxide and 30% SiO_2 is very refractory, its liquidus being above the 1600°C isotherm.

If the atmosphere changes with respect to oxygen pressure and there is less oxygen available, the iron oxide gradually becomes FeO with the result that the melting point of the hot face is lowered and therefore cannot exist at 1600°C normally and necessarily maintained in the open hearth furnace. The brick face, therefore, rapidly sloughs away to a point in the brick in which the iron oxide content is only 50% or less. Figure 13 is the phase diagram for iron oxide–silica in contact with metallic iron plotted on the $FeO–SiO_2$ join (Bowen and Schairer, 1932). This shows that compositions which contain approximately 70% iron oxide as FeO and 30% silica will melt at approximately 1200°C.

One open hearth furnace operator observed that in a few minutes during a period of malfunction of the combustion system more than an inch of silica roof was lost (Smith, 1935). To a lesser extent, when conditions may be only slightly "reducing," this may account for short roof life not otherwise explainable.

The use of the phase diagram for FeO–Fe_2O_3–SiO_2 extends beyond the fields that are ordinarily referred to as ceramics. Recent use of iron ore pellets has proved them to be very valuable as a charging material in modern blast-furnace operation. Their manufacture involves much oxide chemistry, and for that reason ceramic technologists often contribute in this field.

Success in the manufacture of iron ore pellets depends in no small measure on the understanding and use of the FeO–Fe_2O_3–SiO_2 phase diagram. It is the desire of both the maker and the user to have a pellet that is nearly all hematite. It must be strong in order to withstand the rough handling in shipping to the user as well as the abrasion and crushing action it receives in its descent in the blast furnace. To meet these requirements it must have a uniform continuous crystal structure throughout its entire cross section, and it may not be brittle owing to the presence of glassy constituents, particularly when these glassy constituents are the bond.

Pellet raw materials are often ores that must be ground very fine for beneficiation. The principal constituent that is objectionable and that it is aimed to remove by beneficiation is silica. The beneficiated product, however, often contains as much as 8% silica but recently this has been reduced to as little as 3 or 4%. In most cases the ferrous material is magnetite, which can be readily converted to hematite in oxidizing conditions of the induration process. This is usually carried out at temperatures not over 1370°C. Occasionally a producer will attempt to incorporate some solid fuel as part of the pellet composition when it is being formed. This creates reducing conditions within the pellet. The solid fuel must then be oxidized at low temperatures in order to avoid the formation of ferrous oxide. If this is ignored a low-temperature liquid phase will develop in which part of the iron oxide fluxes the free silica. The phase diagram, Fig. 13, shows that this can take place at temperatures as low as 1177°C under such reducing conditions. The result is a core of an impervious glassy phase in which the iron oxide is in the ferrous condition. The silica content of this may be only 25 to 30% and iron oxide 70 to 75%. Combined in this way it is not easily reduced. This is an undesirable situation from the standpoint of blast furnace efficiency as well as from the standpoint of strength of the pellet.

On the other hand a pellet that contains no solid fuel when it is indurated in oxidizing conditions will have a final mineral constitution of hematite and quartz. Under oxidizing conditions there is theoretically no liquid in this system below approximately 1450°C. These pellets derive high strength from an intergrowth of hematite crystals and are resistant to impact and abrasion.

In the manufacture of reduced pellets it is the aim of the maker to produce pellets in which a large proportion of the oxide is converted to metallic iron. In order to avoid the problem of producing a glassy (liquid) phase involving FeO that has fluxed silica and from which it is difficult to reduce most of the iron oxide to metal, the reduction must be carried out at temperatures below those at which ferrous silicates form. Although this may require a somewhat longer time than is desired, it is possible, by adhering to this procedure, to obtain a higher percentage of reduction than would otherwise be attained when conditions favor the formation of the ferrous silicate liquid. Certainly it would mean that the use of solid fuel in the pellet is not recommended.

In the production of oxidized pellets in a rotary kiln, the heat required is sometimes supplied with fuel oil. In this process "rings" that are particularly troublesome to the operators may form in the kiln. These rings are annular buildup of pasty, semifused product. They can occur in this operation owing to excessive kiln temperatures or to local reducing conditions in the luminous oil flame. Both ore dust and coarse ore particles may be at least partially reduced to low-melting-point ferrous silicates thus providing potentially suitable conditions for the buildup.

VI. BASIC REFRACTORIES

Basic refractories are made almost exclusively from three raw materials: magnesia, chrome ore, and dolomite. Lime is a basic material; but because of its tendency to hydrate in moist atmospheres and because it reacts with iron oxide at relatively low temperatures, its use as a refractory is very limited. About 85% of our production of basic refractories is consumed in the steel industry in its open hearths, basic oxygen vessels, and electric furnaces. The other 15% is used largely by the copper industry. A small amount is used in the glass industry in regenerators and other spot-furnace locations.

Blast furnaces reduce iron oxide of ores charged to produce metallic iron. This and steel scrap are charged into steelmaking furnaces. The blast furnace hearth is lined with carbon refractories but the remainder of the furnace is lined with clay or other alumina–silica refractories. Basic refractories are unsuitable for use in the blast furnace as they usually contain oxides that are reducible in its environment. Although magnesia and dolomite are very refractory, they are seldom used in the blast furnace because they hydrate when exposed to water vapor or water that might result from a leak in the cooling system. Carbon hearths are subject to oxidation by water or water vapor, but adequate measures are taken in the design and operation to prevent this.

Iron from the blast furnace contains approximately 4% carbon, 1% silicon, and small amounts of manganese, sulfur, and phosphorous. Each of these is brought to lower concentration levels in the steelmaking process to meet required properties and specifications as steel. The open hearth, the basic oxygen process, and electric furnaces are employed to carry out the process of refining and these are essentially oxidizing processes. The open hearth uses air and oxygen to burn its fuel, and oxygen is frequently used in the open hearth directly applied to the bath through lances inserted through the roof. The basic oxygen process is carried out in special refractory-lined vessels. It also employs oxygen applied to the bath through water-cooled lances that are lowered into the vessel to a point near the metal level. This burns the carbon, silicon and manganese of the hot metal from the blast furnace. In both of these processes basic slags are developed that are liquid at operating temperatures. These slags are composed essentially of lime, iron oxide, manganese oxide, magnesium oxide, and silica. The silica, the MnO, and some of the iron oxide of the slags are oxidized from their respective metals in the process. The lime and some of the iron oxide are part of the furnace charge; magnesia is picked up from the hearth refractories.

Phase diagrams for lime–iron oxides and silica (Bowen *et al.*, 1933; Phillips and Muan, 1959), Figs. 18 and 19 (Muan and Osborne, 1965), show that large composition areas in these systems are liquid at steelmaking temperatures and are the areas of composition for metallurgical slags.

The slags which develop early in open hearth and BOF heats contain 35% or more iron and manganese oxides. They have a low lime–silica ratio of 1:1 or less. During the progress of the heat, the lime that had been charged enters the slag thereby diluting the other constituents. The composition of the slag changes along the dotted line shown schematically in Fig. 18. The final slags may have a lime-silica ratio as high as 3.5 or 4, but usually it is around 3.

As iron oxide is considered to be a universal high-temperature solvent for most mineral matter, the refractory problem in these refining units is a difficult one due to the solvent potentials of the iron-rich basic slags that are produced in them. Increases in temperature that are the result of endeavors to speed operations increase the rates of refractory wear by accelerating this solvent action.

Magnesium and chromium oxides are resistant to the solvent action of iron oxide at high temperatures and hence are important as refractories in the steel industry. Each of these is capable of absorbing 60% or more of

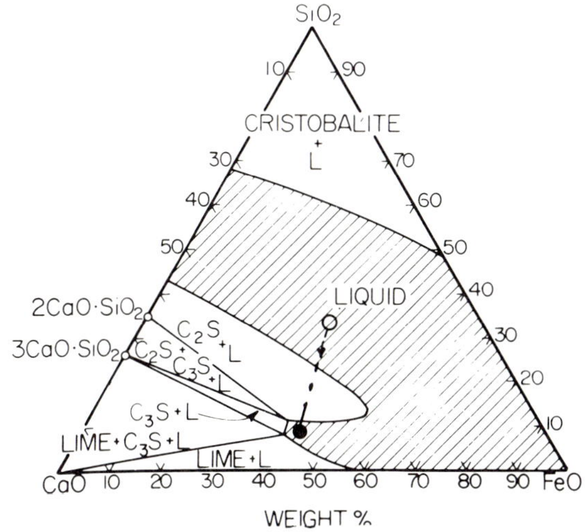

Fig. 18. 1600°C isothermal section in the system CaO–FeO–SiO_2 in contact with metallic iron. After Bowen and Schairer (1932). The dotted line represents the course of approximate composition changes in basic open hearth and BOF slags from early slag to finish slag. Section reprinted from Muan and Osborn (1965). Dotted line and early slag composition superimposed by present author.

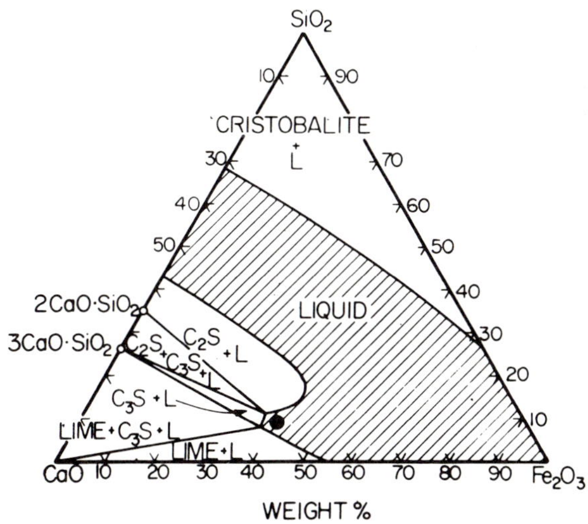

Fig. 19. 1600°C isotherm section in the system CaO–iron oxide–silica in air. After Phillips and Muan (1959). Section reprinted from Muan and Osborn (1965).

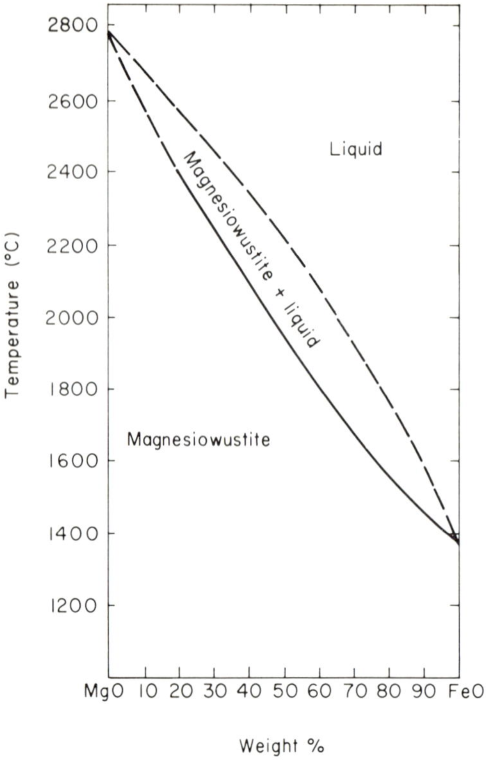

Fig. 20. The system MgO–FeO. Mainly from Bowen and Schairer (1935). Reprinted from Johnson and Muan (1965).

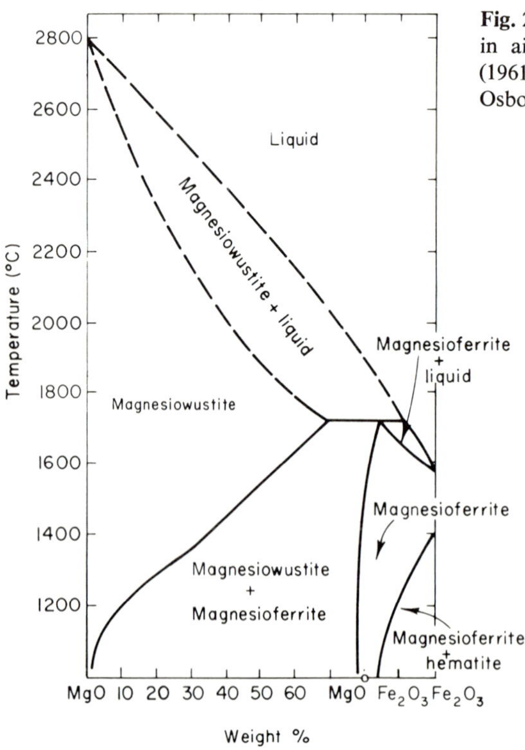

Fig. 21. The system MgO–iron oxide in air. Mainly from Phillips *et al.* (1961). Reprinted from Muan and Osborn (1965).

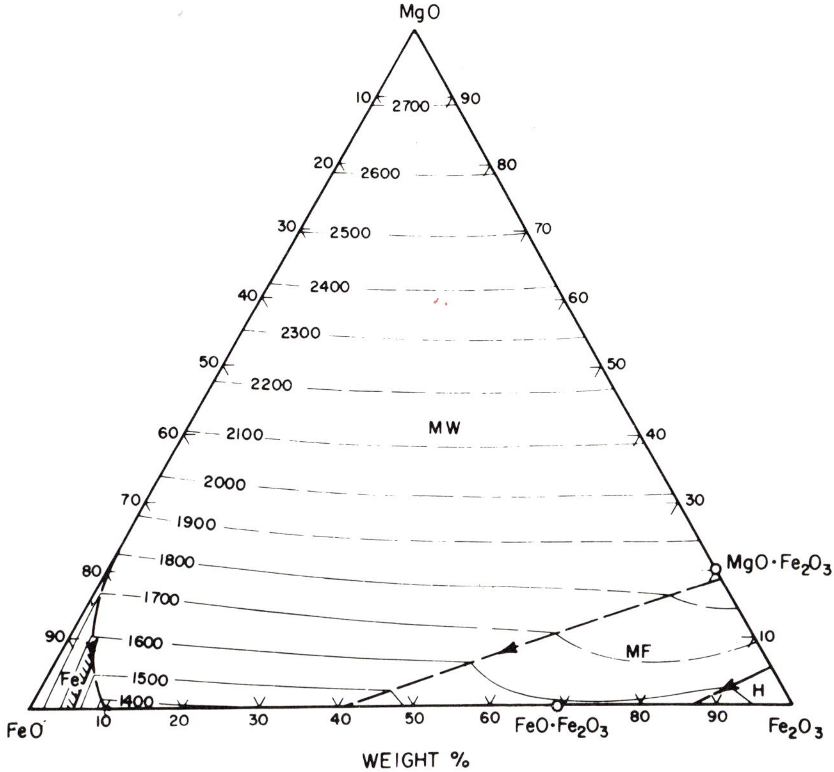

Fig. 22. Approximate liquidus surface for the system MgO–FeO–Fe_2O_3. After Phillips and Muan (1962). Reprinted from Muan and Osborn (1965).

iron oxide in solid solution and at the same time remain refractory. The phase diagrams for MgO–FeO (Bowen and Schairer, 1935), Fig. 20, and MgO–Fe_2O_3 (Phillips *et al.*, 1961), Fig. 21; and that for MgO–FeO–Fe_2O_3 (Phillips and Muan, 1962), Fig. 22, attest to this as the liquidus isotherms at 60% iron oxide are above 2000°C. These diagrams also show that liquidus temperatures in this system do not fall below 1600°C until compositions containing more than 80% iron oxide are reached.

Chromium oxide ably resists the fluxing action of iron oxide at high temperature. The inferred phase diagram for FeO–Fe_2O_3–Cr_2O_3 by Muan and Somiya (1959, 1960), Fig. 23, shows that Cr_2O_3 raises the liquidus continuously from that of Fe_3O_4 to well over 2200°C, the melting point of Cr_2O_3. Because of its high cost, Cr_2O_3 is seldom used alone as a refractory. In those cases where it has been used by the author for spot-location

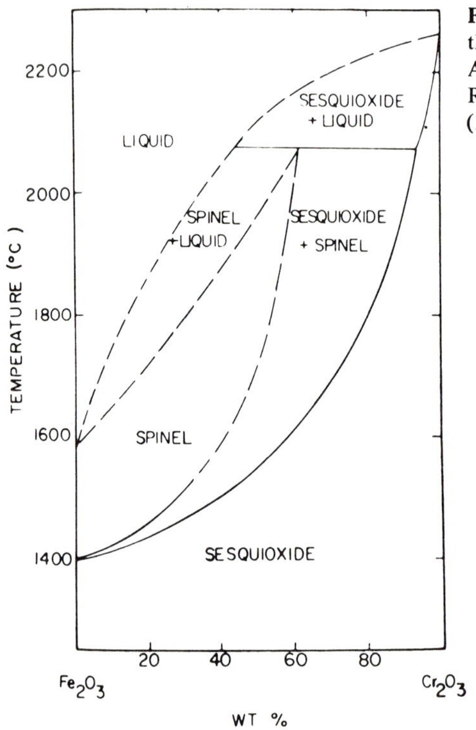

Fig. 23. Approximate phase relations in the system iron oxide–Cr_2O_3 in air. After Muan and Somiya, (1959, 1960). Reprinted from Muan and Somiya (1960).

applications, a fused cast product of FeO–Cr_2O_3 performed very well in continuous contact with molten iron oxide at around 1700°C (Field, 1942).

The cheapest form in which Cr_2O_3 can be obtained for refractory purposes is as chrome ore. The principal phase in chrome ore is a spinel. Spinels have the typical formula $RO{\cdot}R_2O_3$ and have a characteristic of being able to take into its constitution or structure the oxides MgO and FeO as the bivalent (2+) oxides, and Al_2O_3, Fe_2O_3, and Cr_2O_3 as the trivalent (3+) oxides to form the typical compound. All spinels have the same crystal structure, are isomorphous with each other and even though they may have entirely different chemical compositions they form extensive solid solutions with one another. Their refractoriness and performances may differ widely. Table II represents two compositions of commonly used chrome ores.

It is important to note that the principal difference between the Phillippine and the Transvaal ores is in their Cr_2O_3 and alumina contents. The Philippine ore also has a higher magnesia content than the African ore and hence contains a larger proportion of the $MgO{\cdot}Al_2O_3$ molecule. The African ore contains more iron oxide and chromium oxide and hence contains more of the $FeO{\cdot}Cr_2O_3$ molecule. As both have about equal melting points, they

TABLE II

TYPICAL REFRACTORY GRADE CHROME ORES[a]

	Philippine Masinlac	African Transvaal[b]
Cr_2O_3	32.1	50.7
Al_2O_3	29.2	13.0
FeO	12.6	16.4
MgO	18.8	13.2
CaO	0.6	0.8
SiO_2	5.5	4.3
Ignition loss	1.0	0.9

[a] "Modern Refractory Practice." Harbison–Walker Refractories Co.
[b] "Industrial Minerals and Rocks." 1960.

can be substituted for one another in refractory brick without loss of refractoriness.

For many years only the Philippine ore was regarded to be refractory grade. Recently, however, the higher-chrome-oxide ores have found use in refractories.

Chrome ores usually contain a minor amount of associated mineral such as serpentine ($3\,MgO{\cdot}2\,SiO_2{\cdot}2\,H_2O$) that is distributed in such a way that it cannot economically be removed. This silicate is the source of low-temperature liquid development of the refractory in which the ore is used and must be taken into account in the production of good basic refractories.

Magnetite $FeFe_2O_4$ (Fe_3O_4) is a spinel that has the lowest melting point of the spinels in which we have interest in refractories technology. Hercynite $FeAl_2O_4$ ($FeO{\cdot}Al_2O_3$) has the next higher liquidus temperature. The spinels of greater interest to us are those containing Cr_2O_3 and MgO such as picrochromite [$MgCr_2O_4$ ($MgO{\cdot}Cr_2O_3$)] and $MgAl_2O_4$ ($MgO{\cdot}Al_2O_3$).

Most of the spinels of interest in refractories can be formed in the solid state from their constituent oxides but at rates that would be too slow to be economically practical in conventional refractory firing practices. Fusion casting may be necessary to develop the full potential of new spinel products which are to be synthesized from oxides.

Several fused cast basic products are now being produced but it appears that they are made from commercial chrome ore and magnesia. Their producers have not ventured into potential oxide synthesis as a result of presently available applicable phase diagrams.

Conventional basic bricks are made of chrome ore or magnesia and mixtures of the two. Those that contain more than 50% magnesia are called magnesite–chrome bricks, whereas those that contain less than 50% are called chrome–magnesite. They may be chemically bonded, in which case they are not fired, or they may be fired in kilns before delivery.

About 10 years ago basic bricks used in open hearth roof in this country were predominantly of the unfired chemically bonded type. Magnesites* of low lime and silica contents for use in these products were not then available. Typical analyses of magnesites used in basic refractories over the years are shown in Table III.†

TABLE III

	1945	1959	1962	1962
MgO	86.1	85.9	87.3	91.4
CaO	5.3	3.9	4.5	2.9
SiO_2	6.0	5.1	4.0	3.7
Al_2O_3	0.6	0.9	1.2	0.5
Fe_2O_3	2.3	4.3	3.4	1.5
Ignition loss	0.2			

Table IV gives analyses of popular basic bricks of 1962 and of today with particular reference to the total lime and silica content. Brick C is a modern product that is enjoying good service records.

The principal cause of failure in the basic roof brick of 1940 to 1964 vintage was a massive peeling or slabbing off in the substantial areas in the manner shown in Fig. 24 (Kraner, 1961).

Much research on this problem has been conducted in numerous laboratories since 1957. The cure that appears to have been effected apparently was

TABLE IV

	A	B	C
	1962	1962	1968
SiO_2	5.3	4.1	1.5
Al_2O_3	19.7	8.9	9.1
Fe_2O_3	0.1	0.2	4.9
FeO	8.7	4.1	—
CaO	0.8	1.5	0.8
MgO	44.1	71.2	73.3
Cr_2O_3	21.3	10.0	10.4

* "Magnesite" as used here refers to hard burned or "dead"-burned Magnesia. This is common usage in the refractories field.

† Author's records.

Fig. 24. Typical failure in basic open-hearth roof before improved brick chemistry and direct bonding (Kraner, 1961).

not the golden nugget discovered by any one individual, as there seemed to have been a general feeling by a number of researchers that the blame could be laid to the inadequacy of the silicate bond that had a relatively low melting temperature and was not able to hold the hot face of the brick in place. Although it was in equilibrium with periclase and chrome spinel at fairly high temperatures, it deteriorated rapidly with the intrusion of lime and iron oxide during use.

To reduce the silicate phases required the use of higher-purity magnesites and chrome ores. Seawater magnesites solved the magnesia question, and better selection and preparation were required for the chrome ore problem. Available phase diagrams to assist in liquid-phase studies can be credited with bringing improvements to the present stage.

The magnesia used in these bricks may be derived from carbonate rocks, Brucite $Mg(OH)_2$, and from seawater. The purity of the chrome ores and magnesites varies sufficiently to warrant making intelligent choices of raw materials and to adjust the proportions of the minor constituent compositions to obtain best properties in the final product.

The presence of MgO, Al_2O_3, FeO, Fe_2O_3, CaO, and silica in a basic refractory composition might seem to make the problem of applying phase-diagram data to these products an impossibility. Considering that chrome spinel and magnesia are not particularly reactive at low temperature and

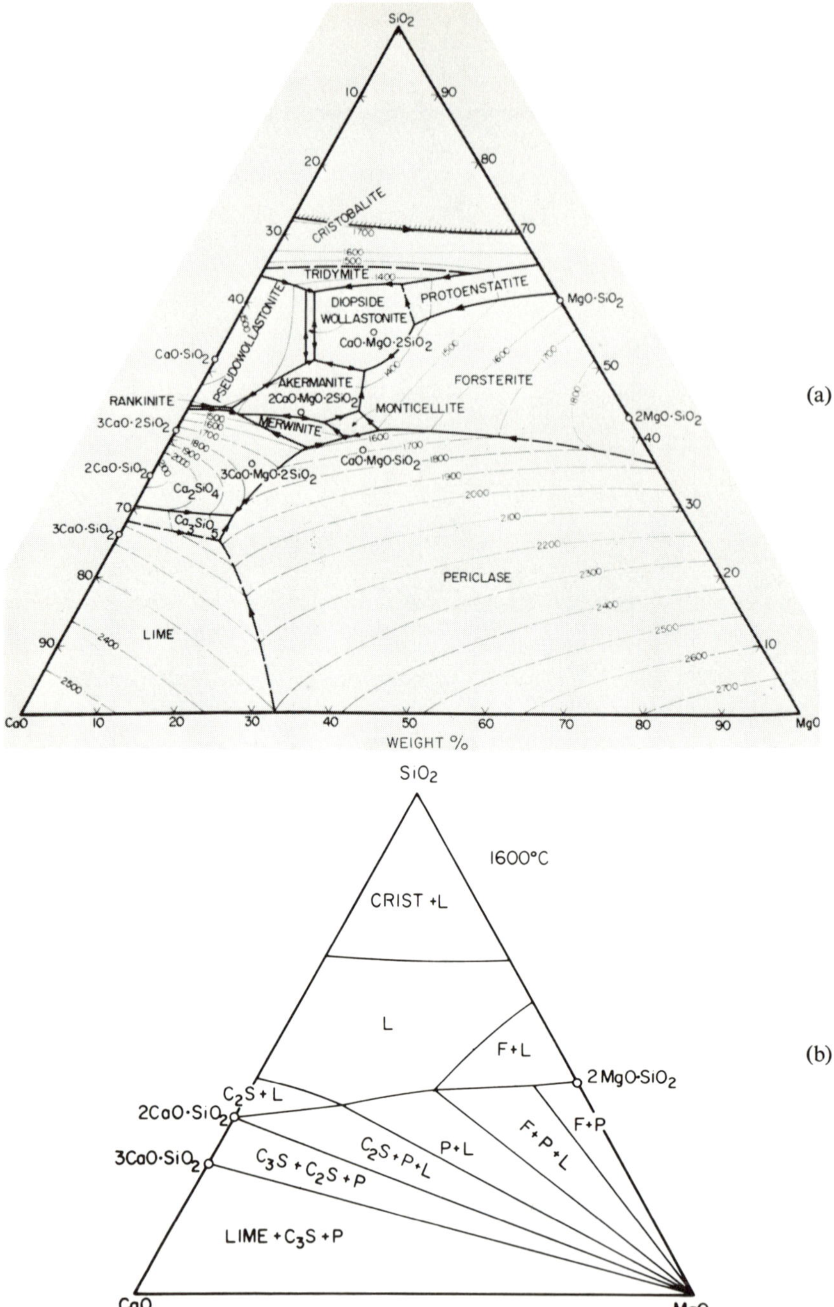

Fig. 25. (a) The system CaO–MgO–SiO_2 (Osborn and Muan, 1960). Reprinted from Muan and Osborn (1965).

(b) 1600°C isothermal section of the system CaO–MgO–SiO_2 (Osborn and Muan, 1960). By Muan and Osborn (1965). F is forsterite, 2 MgO·SiO_2; M is monticellite, CaO·MgO ·SiO_2; C_2S is 2 CaO·SiO_2; C_3S is 3 CaO·SiO_2; P is periclase, MgO; and L is liquid.

that each of them may be in equilibrium with liquids of these polycomponent systems at low temperatures, certain liberties can be taken to simplify the problems by studying the minor constituents in the systems alone as follows.

Magnesites formerly used in refractories contained up to 5% CaO and 5% SiO_2. The amounts of these were responsible for the amount of liquid that developed in the grain when it was fired in its production and during its use. The ratio of lime-to-silica determines the character and composition of the liquid that develops and its relation to periclase (MgO).

Figure 25 shows the phase equilibrium diagram for the system CaO–MgO–SiO_2 (Osborn and Muan, 1960). Our principal interest here is in the characteristics and potential performance of compositions very close to the MgO (periclase) corner. Those of interest would contain 90% or more MgO. The periclase field of the diagram shows falling temperatures toward the center of the diagram and indicates that periclase can exist in equilibrium with fairly siliceous low temperature liquids at temperatures which are attained in steel furnaces. This diagram also shows that improvements can be derived in the high MgO compositions if their CaO:SiO_2 ratios are controlled.

Figure 25a (Muan and Osborn, 1965), the 1600°C isothermal section for this system clearly indicates that if a periclase refractory is to be chosen which contains no liquid at temperatures up to 1600°C, compositions would be selected from areas F + P and $C_3S + C_2P + P$ or lime + C_3S + P. Forsterite is preferred over lime or calcium silicates as an interstitial mineral as it is more resistant to the fluxing action of iron oxide than the calcium silicates involved. The following paragraphs will show how this conclusion is obtained.

Figure 26 shows that in the system CaO–FeO–SiO_2 (Bowen *et al.*, 1933) in contact with metallic iron, a mixture of 50% dicalcium silicate and 50% iron oxide is completely liquid at 1600°C. Figure 27 shows that in the system CaO–iron oxide–SiO_2 in air (Phillips and Muan, 1959) a mixture of 50% dicalcium silicate and 50% iron oxide is completely liquid at 1400°C.

In contrast, Fig. 28 for the system MgO–FeO–SiO_2 (Bowen and Schairer, 1935) in contact with metallic iron, shows that a mixture of 50% forsterite and 50% iron oxide is not completely liquid until 1700°C is reached; and in the system MgO–iron oxide–SiO_2 (Muan and Osborn, 1956) in air, Fig. 29, a mixture of 50% forsterite and 50% iron oxide is not melted until heated to 1675°C.

From the standpoint of volume stability at operating temperatures and resistance to iron oxide, forsterite should be desired as the accessory mineral in basic refractories in which periclase is a major constituent.

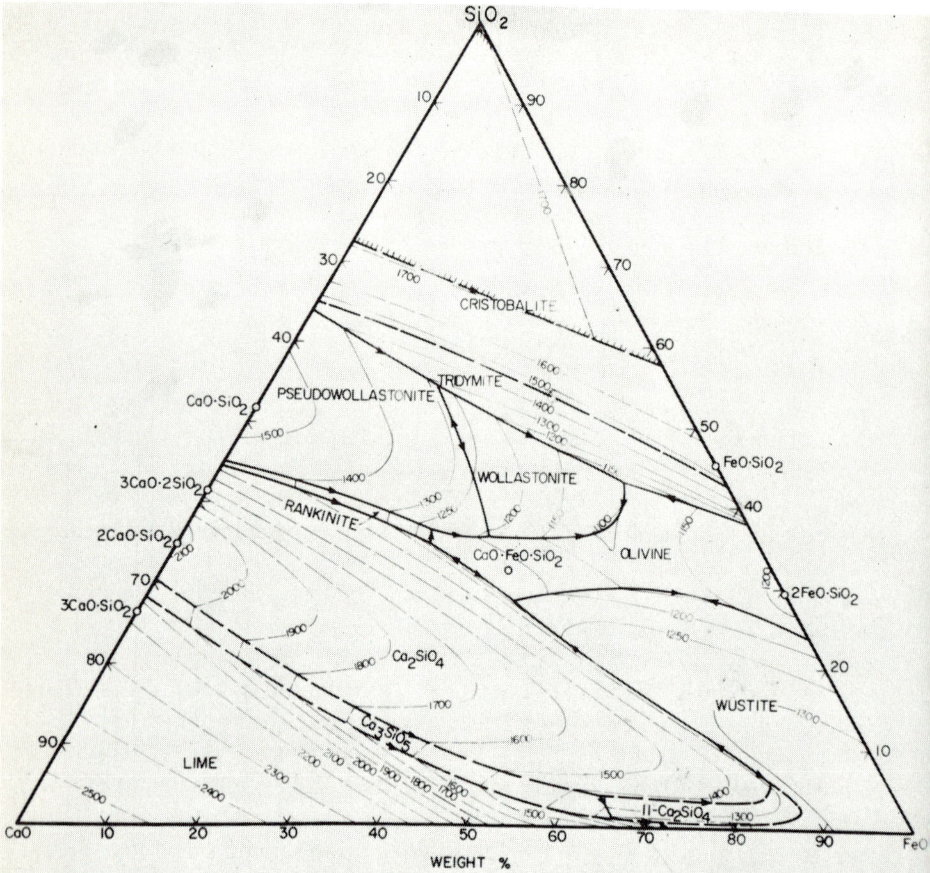

Fig. 26. The system $CaO–FeO–SiO_2$ in contact with metallic iron (Bowen *et al.*, 1933). Reprinted from Muan and Osborn (1965).

A. Direct Bonding

The questions of volume stability and hot strength of basic bricks at high temperature have been the subject of considerable research here and abroad. Answers to these questions it was hoped, would solve the problem of "shelling" or "peeling" in open hearth roofs during use. The works of Lambing (1959), Richardson *et al.* (1960), White (1960), and Hayhurst (1963) all were directed to the problem of determining the effects upon basic refractory bricks of high-temperature firing treatment. In this country considerable work was done on the effect of hard firing upon hot strength.

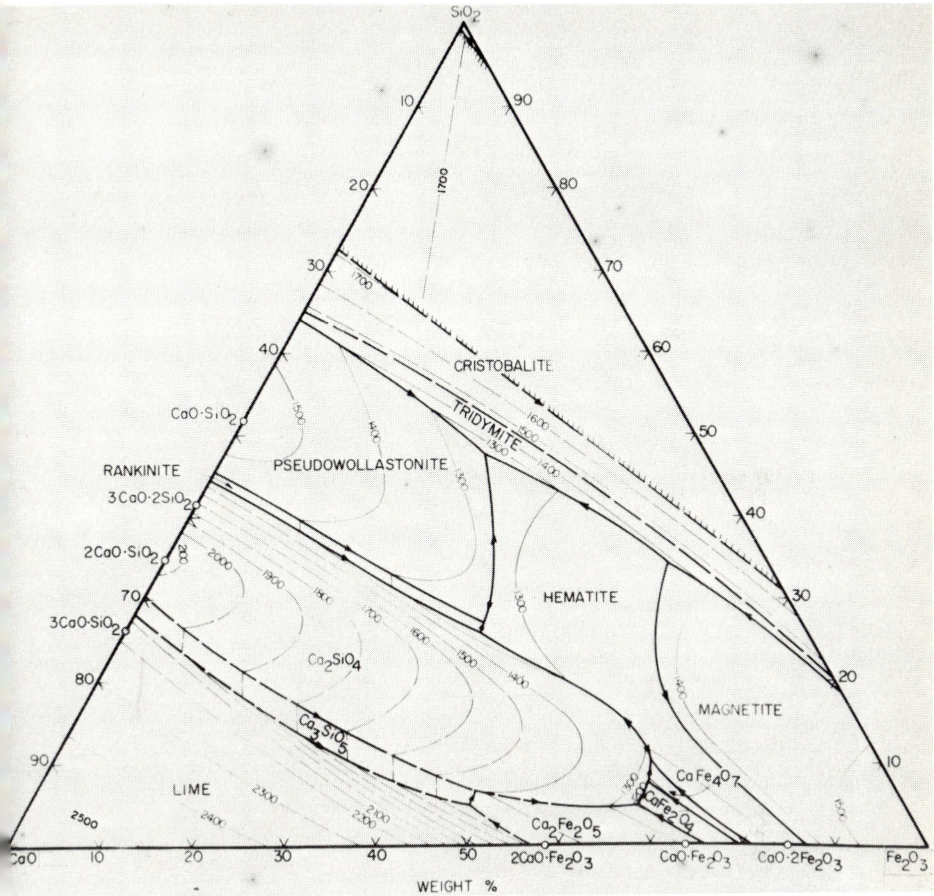

Fig. 27. The system CaO–iron oxide–silica in air (Phillips and Muan, 1959). Reprinted from Muan and Osborn (1965).

A combination of chemistry and firing treatment has produced the present "direct bonded" basic bricks (Hubble and Powers, 1963; Davies and Walther, 1964) that have high hot strength and have materially decreased the peeling problem in open hearth roof refractories. The results are specifically due (a) to a reduction of impurities, particularly lime and silica; (b) control of the lime and silica ratio in the brick; and (c) increased firing temperature.

Purer raw materials have reduced the amount of interstitial liquid phases; control of the lime–silica ratios has provided more refractory bonds,

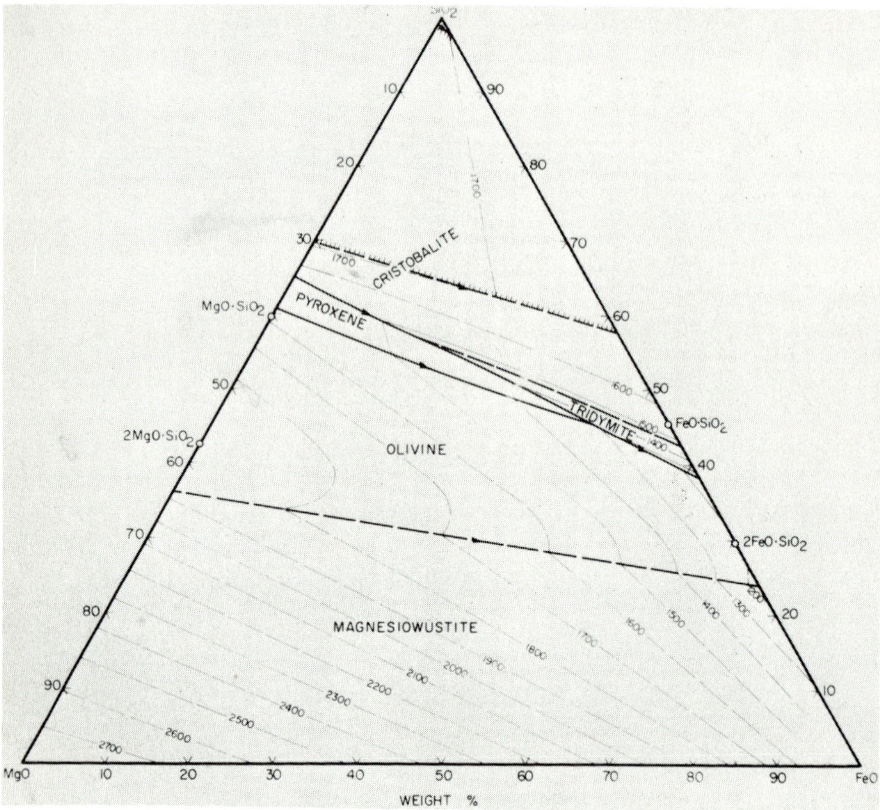

Fig. 28. The system magnesia–iron oxide–silica in contact with metallic iron (Bowen and Schairer, 1935b). Reprinted from Muan and Osborn (1956).

and hard firing has developed refractory solid solution bonds between contacting grains of periclase and spinels.

The work of Alper *et al.* in the development of phase diagrams for the systems MgO–$MgAl_2O_4$ (1962), Fig. 30, and MgO–$MgCr_2O_4$ (1964), Fig. 31, show that the formation of solid solution of spinels and periclase is appreciable when samples are fired in the temperature range of normal refractory firing processes, namely, 1400°C to 1600°C.

The contact of the grains of periclase and the spinel permits ion migration and the formation of strong solid solution bonds.

These developments constitute valuable contributions to refractories technology and must in large measure be attributed to the availability of applicable phase-equilibrium diagrams.

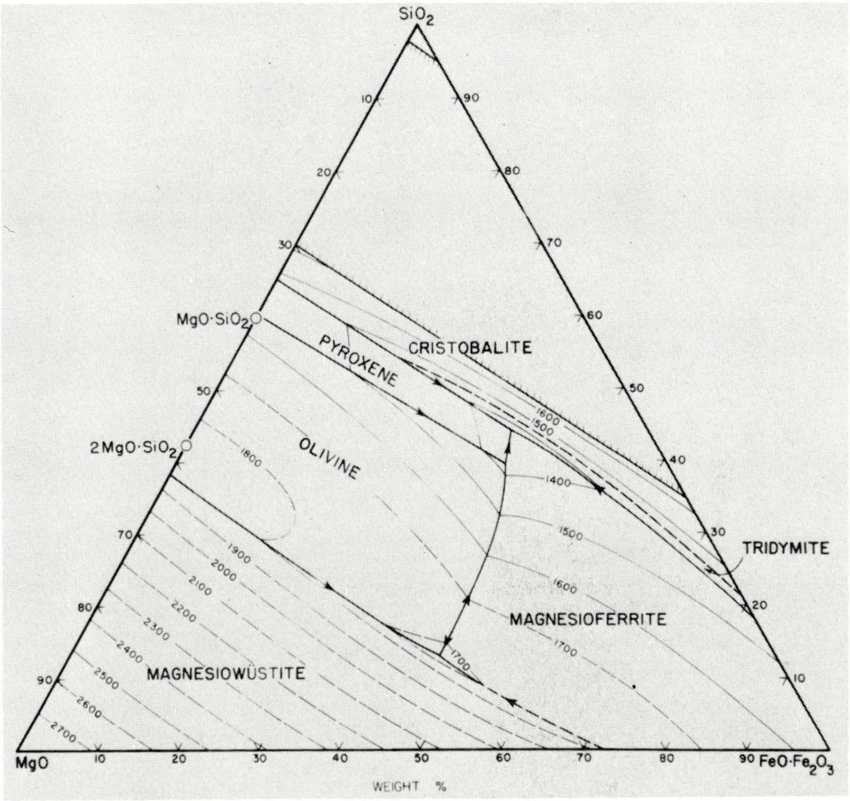

Fig. 29. The system MgO–iron oxide–silica in air (Muan and Osborn, 1956). Reprinted from Muan and Osborn (1965).

B. Dolomite Refractories

Dolomite refractories are used on open-hearth banks either in the form of raw, uncalcined grains or a "burnt" dolomite. The latter is prepared by firing the raw granules in a rotary kiln with enough iron oxide to produce a product that contains prescribed amounts of Fe_2O_3 up to 8%. Dicalcium ferrite is formed around and through the granules providing a certain amount of hydration resistance and a degree of "setting" quality. The dicalcium ferrite melts at 1449°C. Eight percent of iron oxide seems to be the maximum that may be used; more will flux so much lime that manufacturing problems in the kiln will develop. More than 8% iron oxide in this application would also shorten its service in the open hearth. This and the raw

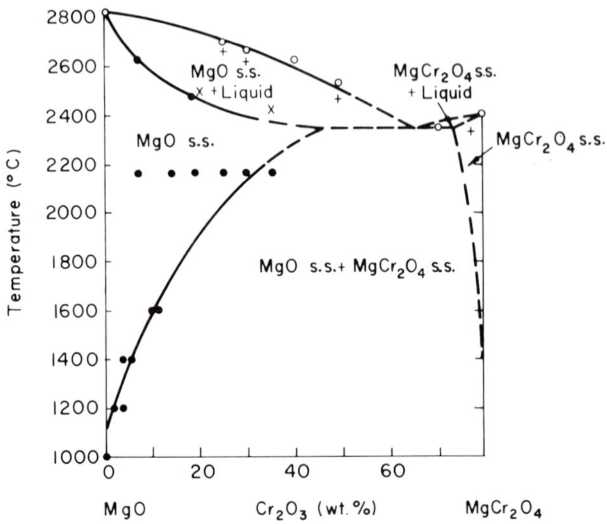

Fig. 30. Phase equilibrium diagram for the system MgO–$MgCr_2O_4$. Reprinted from Alper *et al.* (1964).

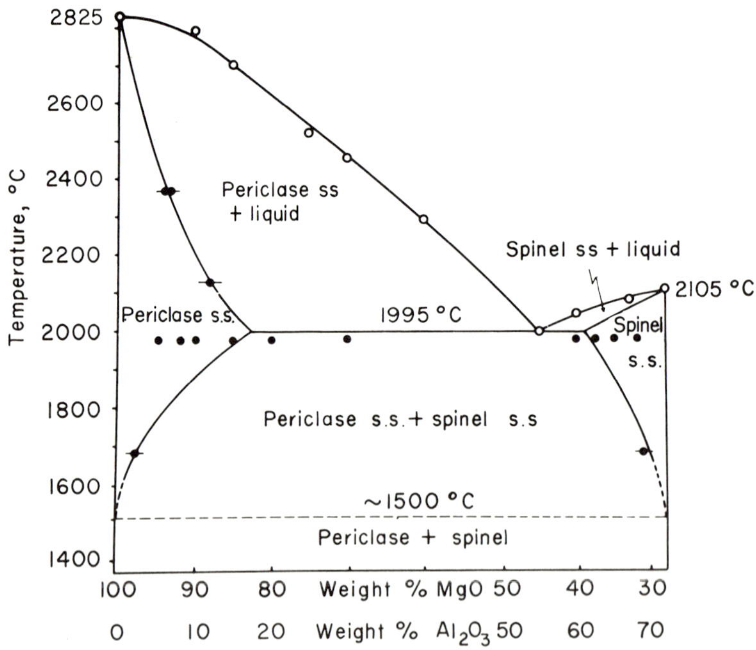

Fig. 31. Phase equilibrium diagram for the system MgO–$MgAl_2O_4$. Reprinted from Alper *et al.* (1962).

dolomite as applied to the open hearth are expendable and eventually find their way into the slag.

C. The Basic Oxygen Practice

In the basic oxygen process, dolomite refractories are employed very successfully. In this process calcined dolomite grain is used alone or with magnesia grain supplementing it in bricks bonded with tar and pitch.

This process produces steel at a spectacular rate of over 300 tons/hr. It is carried out in the refractory-lined-vessels resembling a top-blown Bessemer converter. The vessel is tilted for charging and for discharging the final product. The charge is approximately 30% scrap and 70% liquid metal from a blast furnace or other source. Limestone containing a small percentage of dolomite is also charged as a requisite to form slag.

Oxygen is blown through water cooled lances to burn the silicon and carbon and some iron to carry on the process.

The refractory lining is "balanced," which means that for each part of the vessel a specific type of basic refractory is employed in order to obtain a uniform rate of wear in all parts of the vessel. The most severe conditions are found in the barrel from the trunion level downward to the hot metal level or slag line. The area onto which the charge is deposited receives hard service from the impact of scrap and hot metal. In some vessels the lip or cone opening receives severe wear.

As the process requires basic slags to remove sulfur and phosphorous, the slags and their chemistry and functions are almost identical to those of the open hearth slags. The refractories that line the furnace must withstand the chemical attack of basic slags. Unlike the open hearth, however, the BOF process does not use chrome-bearing refractories as the metal is at one time or another in contact with the sidewall refractories and would be contaminated with chromium. In addition dolomite is a much less expensive refractory.

The BOF employs calcined or sintered dolomite and magnesite or combinations of the two. These combinations may be fired or not, but for most of the lining, tar or pitch bonds play an important part in the life of the vessel lining. The following may be used in a single vessel.

1. Hard-fired ceramic processed magnesite bricks for the safety lining against the shell.
2. Tar- or pitch-bonded pressed blocks of precalcined dolomite or magnesite grain, or mixtures of the two.
3. A product produced as in No. 2 may be tempered by heating to a low temperature to remove the pitch volatiles in order to decrease softening of

the block that might occur because of the low-temperature bond. This hardens the block to better withstand the blow-in heatup.

4. A high-strength fired block produced by conventional ceramic methods as in No. 1 may be impregnated with tar and used in areas such as on the charging pad where high strength is required.

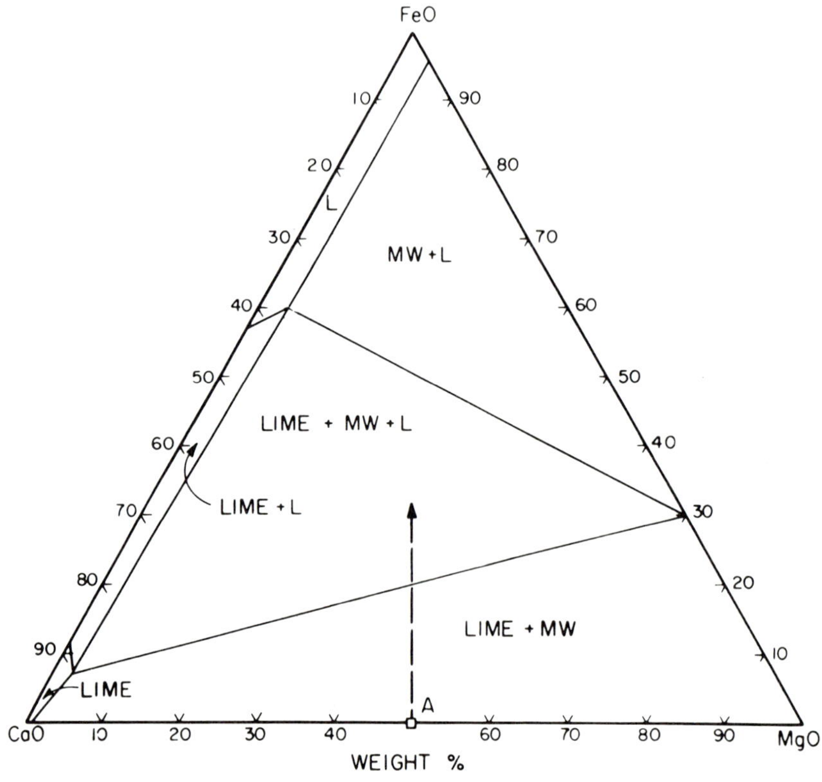

Fig. 32. 1500°C isothermal section in the system CaO–MgO–iron oxide at an oxygen pressure of 10^{-9} atm. (Johnson and Muan, 1965). Reprinted from Muan and Osborn (1965).

Figures 32 and 33 (Johnson and Muan, 1965) explain the important role in the refractory of the residual carbon from tar or pitch. It will be seen that under these conditions the area lime + MW (lime + magnesia wustite) is very broad whereas most of the other areas of the diagrams contain some liquid. The large area lime + MW (lime plus magnesia wustite) are the result of the reducing conditions provided by the carbon residue from the tar and pitch. As iron oxide-rich slags penetrate the pores of the refractory, the iron oxide

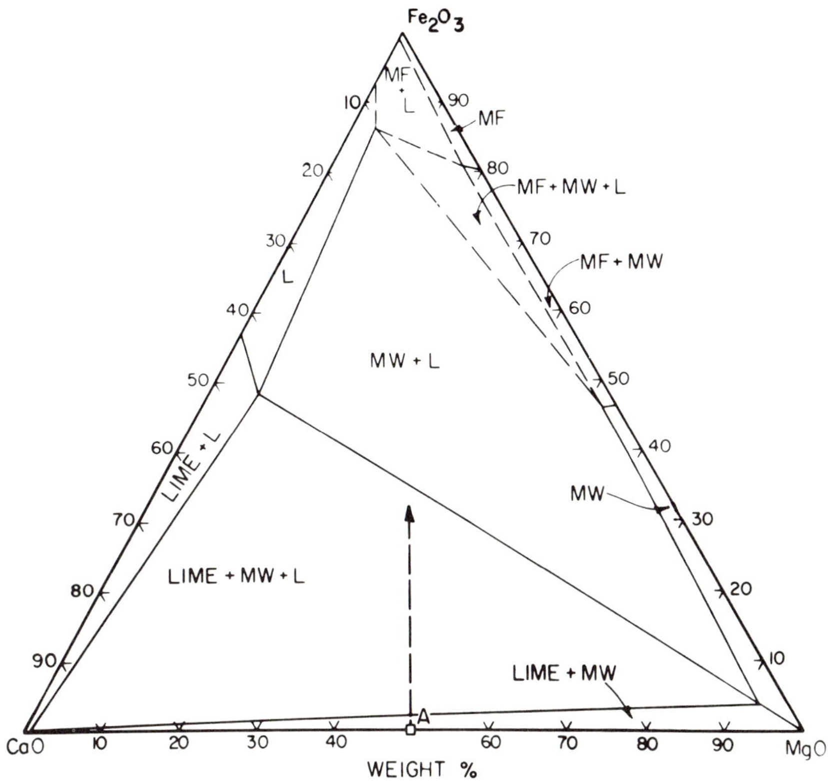

Fig. 33. 1500°C isothermal section in the system CaO–MgO–iron oxide in air (Johnson and Muan, (1965). Reprinted from Muan and Osborn (1965).

is reduced to FeO and metal by the carbon coatings on the refractory grains.

Without these carbon coatings, conditions within the refractory as indicated in Fig. 31 would prevail. In that case, the lime + magnesia wustite area is very restricted and all other major areas of the system contain liquid. These conditions favor rapid decline of the refractoriness of dolomitic refractories.

In addition to its role as a reducing agent as described above, the carbon residue from the pitch and tar acts as a mechanical barrier lowering the permability to slag intrusion. It also is not wet by the slag and thus resists penetration of the slag.

These factors work toward the success of dolomite refractories in this process. Magnesite is known for its resistance to the action of iron oxide at high temperature, but its use is governed by overall cost. The excellent performance of impregnated dolomite refractories makes them attractive competitors to higher-priced magnesia.

The first phase equilibria diagrams of mineral systems were produced by scientists who were interested in geological genesis of rocks and mineral assemblages. These men and others who frequently use phase diagrams are quite conscious of the effects of time, temperature, viscosity and homogeneity upon the attainment of equilibrium. The use of phase diagrams just as in the use of any tool, can be valuable if its limitations are understood.

When one uses a phase diagram, he must understand that it represents equilibrium conditions, whereas in many industrial or even in research situations equilibrium conditions do not exist. For these cases the user will find the diagrams to be helpful in indicating tendencies.

References

Alper, A. M., McNally, R. N., Ribbe, P. H., and Doman, R. C. (1962). *J. Am. Ceram. Soc.* **45**, 263.

Alper, A. M., McNally, R. N., Doman, R. C., and Keighn (1964). *J. Am. Ceram. Soc.* **47**, 30.

Aramaki, S., and Roy, R. (1959). *J. Am. Ceram. Soc.* **42**, 644.

Bowen, N. L., and Schairer, J. F. (1932). *Am. J. Sci.* **24**, 177.

Bowen, N. L., and Schairer, J. F. (1935). *Am. J. Sci.* **29**, 151.

Bowen, N. L., Schairer, J. F., and Posnjak, E. (1933). *Am. J. Sci.* **26**, 193. Also Muan and Osborn (1965), p. 113.

Chesters, J. H. (1957). "Steel Plant Refractories." Percy Lund, Humphries & Co., Ltd., London.

Dana, J. D. (1944). "The System Mineralogy." Wiley, New York.

Davies, B., and Walther, F. H. (1964). *J. Am. Ceram. Soc.* **47**, 116.

Day, A. L., Shepherd, E. S., and Wright, F. E. (1906). *Am. J. Sci.* **22**, 265.

Duwez, P., Odell, F., and Brown, F. H., Jr. (1952). *J. Am. Ceram. Soc.* **35**, 107.

Field, T. E. (1942). U.S. Patent 2,271,363.

Greig, J. W. (1927a). *Am. J. Sci.* **13**, 1.

Greig, J. W. (1927b). *Am. J. Sci.* **14**, 473.

Hall, F. P., and Insley, H. (1933). *J. Am. Ceram. Soc.* **16**, 458.

Hall, F. P., and Insley, H. (1947). *J. Am. Ceram. Soc.* **16**, 458.

Hayhurst, A., and Laming, J. (1963). *The Refractories J.*

Hubble, D. H., and Powers, W. H. (1963). *Am. Ceram. Soc. Bull.* **42**, 409.

Johnson, R. E., and Muan, A. (1965). *J. Am. Ceram. Soc.* **48**, 359.

Kingery, W. D. (1960). "Introduction to Ceramics." Wiley, New York.

Kohn, J. A., Katz, G., and Broder, J. D. (1957). *Am. Mineralogist* **42**, 398.

Kraner, H. M. (1944). *Proc. AIME* **27**, 303.

Kraner, H. M. (1961). *Proc AIME* **44**, 309.

Kraner, H. M. (1962). *Blast Furnace and Steel Plant* **50**, 323.

Kraner, H. M. (1966). *Blast Furnace and Steel Plant* **54**, 147.

Kraner, H. M., and Jewart, C. (1952). AISI Regional Meetings.

LAMBING, J. (1959). *Refractories J.* **35**, 116.
LE CHATELIER, H. (1913). *Rev. Univ. Mines* **1**, 90.
LE CHATELIER, H. (1914), "La Silice et les Silicates."
LEVIN, E. M., McMURDIE, H. F., and HALL, F. P. (1956). "Phase Diagrams for Ceramists." American Ceramic Society, Columbus, Ohio.
LEVIN, E. M., ROBBINS, C. R., and McMURDIE, H. F. (1964). "Phase Diagrams for the Ceramist." American Ceramic Society, Columbus, Ohio.
LYNCH, C. T., VAHLDICK, F. W., and ROBINSON, L. B. (1961). *J. Am. Ceram. Soc.* **44**, 147.
MACCHESNEY, J. B. and MUAN, A. (1960). *J. Am. Ceram. Soc.* **43**, 586.
McMURDIE, H. F., and HALL, F. P. (1949). *J. Am. Ceram. Soc.* **12**.
MUAN, A. (1955a). *J. Metals* **7**.
MUAN, A. (1955b). *Trans AIME* **263**, 965.
MUAN, A. (1957). *J. Am. Ceram. Soc.* **40**, 121.
MUAN, A. (1958). *Am. J. Sci.* **256**, 171.
MUAN, A., and OSBORN, E. F. (1956). *J. Am. Ceram. Soc.* **39**, 121.
MAUN, A., and OSBORN, E. F. (1965) "Phase Equilibria among Oxides in Steelmaking." Addison-Wesley, Reading, Massachusetts.
MUAN, A, and SOMIYA, S. (1959) *J. Am. Ceram. Soc.* **42**, 603.
MUAN, A. and SOMIYA, S. (1960). *J. Am. Ceram. Soc.* **43**, 204.
MUAN, A., and SOMIYA, S. (1961). *Am. Mineralogist* **46**, 364.
OSBORN, E. F., and MUAN, A. (1960). *Am. Ceram. Soc.*
PHILLIPS, B., and MUAN, A. (1959). *J. Am. Ceram. Soc.* **42**, 413.
PHILLIPS, B., and MUAN, A. (1962). *J. Am. Ceram. Soc.* **45**, 588.
PHILLIPS, B., SOMIYA, S., and MUAN, A. (1961). *J. Am. Ceram. Soc.* **44**, 167.
RANKIN, G. A., and WRIGHT, F. E. (1915). *Am. J. Sci.* **39**, 1.
RICHARDSON, H. M., FITCHETT, K., and LESTER, M. (1960). *Trans Brit. Ceram. Soc.* **59**, 483.
SCHAIRER, J. F., and BOWEN, N. L. (1955). *Am. J. Sci.* 253, 681.
SCHAIRER, J. F., and YAGI, K. (1952). *Am. J. Sci.* **471**.
SMITH, E. C. (1935). *Trans. AIME* **13**.
SOSMAN, R. B. (1927a). *Am. Chem. Soc. Monogr.* **37**, 836.
SOSMAN, R. B. (1927b). *Blast Furnace and Steel Plant* **17**, 1076.
STOUT, W., STULL, R.T., McCAUGHEY, W. J., and DEMOREST, D. J. (1923). "Coal Formation Clays of Ohio." Geological Survey of Ohio.
THOMPSON, F. S., and KRANER, H. M. (1933). *Indust. Eng. Chem.* **25**, 856.
TOROPOV, N. A., and GALAKHOV, F. Y. (1951). *Dokl. Akad. Nauk. SSSR* **78**, 245.
TOROPOV, N. A., and GALAKHOV, F. Y. (1953). *Vop. Petrog. Mineral Akad. Nauk SSSR* **2**, 245.
US BUREAU OF CENSUS (1967). Refractories—Current Industrial Reports.
WHITE, J. (1960). *Refractories J.* **36**, 483.

IV

The Use of Phase Diagrams in Fusion-Cast Refractory Materials Research

A. M. ALPER, R. C. DOMAN, R. N. McNALLY, and H. C. YEH†*

RESEARCH AND DEVELOPMENT LABORATORIES
CORNING GLASS WORKS, CORNING, NEW YORK

I. INTRODUCTION

To understand a fusion-cast refractory product and to control its microstructure, it is necessary to have a good grasp of the phase diagram of the system being studied. By understanding the phase diagram, one is able to

* Present address: Chemical and Metallurgical Division, Sylvania Electric Products Inc., Towanda, Pennsylvania.

† Present address: Department of Metallurgical Engineering, Cleveland State University, Cleveland, Ohio.

control the phase assemblage and distribution. The phase assemblage and microstructure usually determine the physical and chemical properties of the materials.

The microstructure of a given fusion-cast composition will vary greatly depending on whether the composition contains solid solution, eutectic, immiscibility, peritectic and/or metastable phases. The stability of the solid solution will also affect the number, composition, and distribution of phases. The extent to which equilibrium is obtained will also have a pronounced effect on phase assemblage and distribution. In many cases, this can be predicted from the phase diagram.

In recent years, much work has been done on crystallization of eutectic materials. Some eutectics can be crystallized so that they have a microstructure that resembles a single crystal with reinforced whiskers. That type results in very unusual mechanical and electronic properties. Combining a knowledge of the phase diagram with a knowledge of the entropy of fusion of the eutectic phases helps in predicting microstructure.

II. SYSTEMS CONTAINING COMPLETE SOLID SOLUTIONS

The MgO–FeO and MgO–NiO systems (Figs. 1a and 1b) are examples of complete solid solutions occurring in all compositions at all temperature levels. Care must be taken, of course, not to change the valence of the transitional metal cation. In fused, essentially monophase refractory systems where a composition is melted in an arc furnace and then chill cast into a cold mold, the phases appear to have a greater tendency to form a columnar structure. This structure occurs when long crystals form perpendicular to the isotherms in the casting. The low-melting impurities in systems that are essentially monophased (greater than 98% monophased) segregate and spread out along the grain boundaries. Consequently, even though the system might be very refractory, it is bonded by a low-melting phase, often a silicate phase in fusion-cast basic refractories. This low-melting phase lowers the high-temperature strength and corrosion/erosion resistance of the MgO-based refractories.

III. SYSTEMS CONTAINING A SIMPLE EUTECTIC AND PARTIAL SOLID SOLUTION

A. MgO–Spinel

The MgO–$MgAl_2O_4$ (Alper *et al.*, 1962) and the MgO–$MgCr_2O_4$ (Alper *et al.*, 1964) systems (Figs. 2 and 3) are examples in which the phases melt congruently and where a simple eutectic is present and a limited solid solution

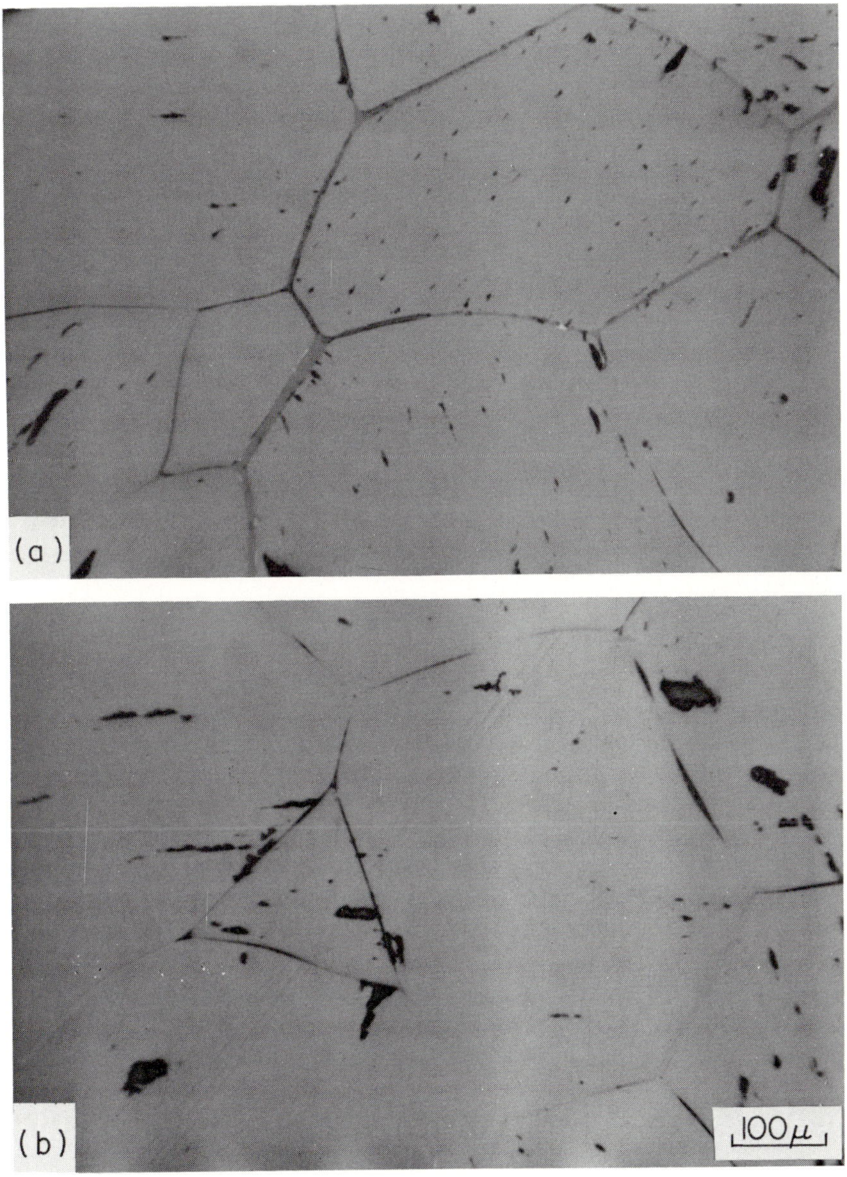

Fig. 1. Photomicrographs of (a) a polished section of fusion-cast (Mg, Fe) O and (b) a fusion-cast (Mg, Ni) O with a silicate film surrounding the periclase solid-solution crystals.

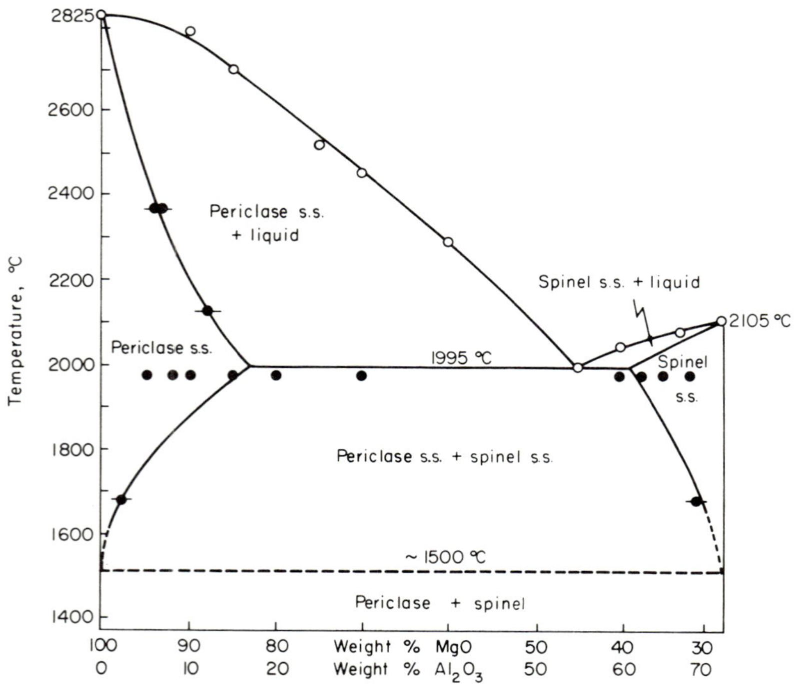

Fig. 2. Phase diagram of $MgO–MgAl_2O_4$ system (Alper *et al.*, 1962).

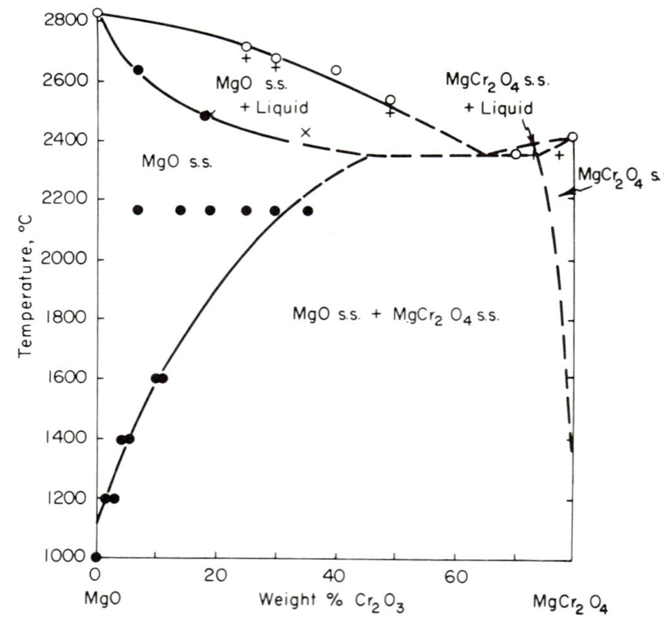

Fig. 3. Phase diagram of $MgO–MgCr_2O_4$ system (Alper *et al.*, 1962). ○, Melted; +, partially melted; ×, not melted; ●, solid-solution experiments.

occurs. Fusion-cast refractories in different parts of this type of system have essentially different microstructure and properties. Compositions that contain less than the maximum amount of solute ions have a microstructure composed mainly of large crystals of the host phase, which contains included crystals of the precipitated phase as well as some amount of the exsolved phase at grain boundaries. When a composition that contains more than the maximum amount of solute ions is fused and chill cast, the resulting microstructure contains an intergranular low-melting phase.

Figures 4 and 5 are photomicrographs of polished sections of fusion-cast

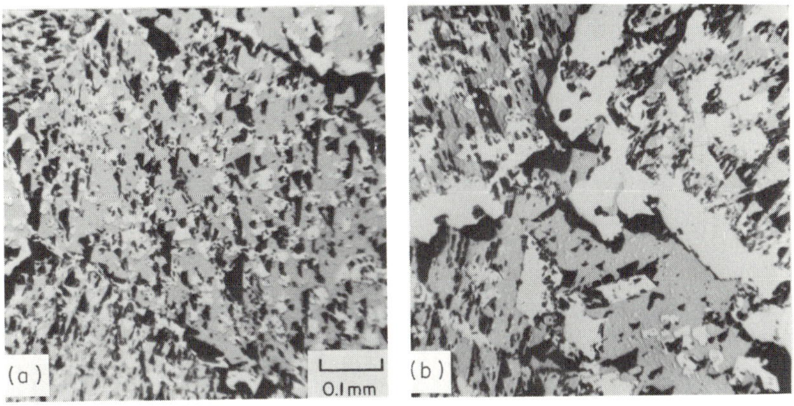

Fig. 4. (a) Photomicrograph of MgO–Cr_2O_3 (55/45) solidified ceramic. All of the spinel has been precipitated. (b) Photomicrograph of MgO–Cr_2O_3 (50/50) solidified ceramic that contains intergranular spinel.

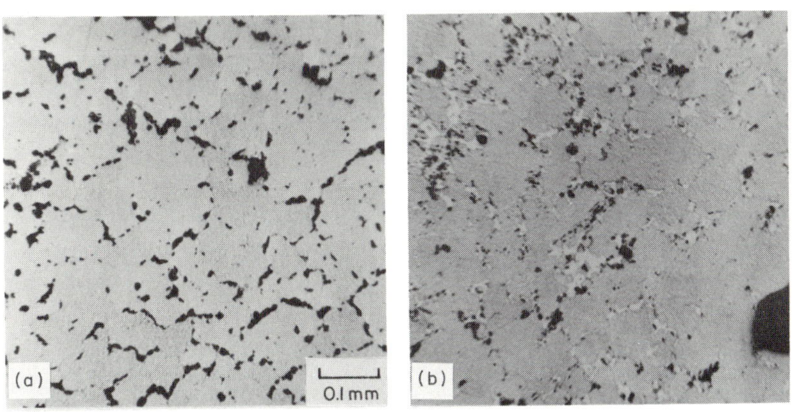

Fig. 5. (a) Photomicrograph of MgO–Cr_2O_3 (55/45) solidified ceramic. All of the spinel has been precipitated. Quenched sample slowly cooled. (b) Photomicrograph of MgO–Cr_2O_3 (50/50) solidified ceramic that contains intergranular spinel. Quenched sample slowly cooled.

$MgO–MgCr_2O_4$ compositions. The samples in Figs. 4a and 5a contain all the chromic oxide in solid solution in the periclase lattice at high temperatures. At temperatures above the eutectic temperature, only a monophase body is present. If the refractory starts to cool below the eutectic temperature, chromium ions start to exsolve from the periclase lattice, and precipitated chromium spinel (picrochromite) is formed. Figures 4a and 5a show that almost all of this precipitated phase occurs intragranularly. Figures 4b and 5b are examples of an MgO–chromic oxide ceramic that contains more than the maximum solubility of chromic oxide in the periclase lattice at high temperature. Consequently, intergranular picrochromite is present in this sample.

Aluminum ions are less soluble in the periclase lattice than are chromium ions. This is due to the larger differences in the ionic sizes between aluminum and magnesium when compared to the similar sizes of magnesium and chromium. A comparison of Figs. 2 and 3 shows that aluminum ions are less soluble than are chromium ions in the periclase lattice at high temperature. Figure 6 shows a composition, 65/35 MgO/Al_2O_3, which contains more aluminum ions than can go into solid solution at high temperatures. Upon crystallization, the periclase crystals are coated by a spinel phase. The presence of an intergranular second phase appears to decrease the size of the crystallizing primary phase. In most fusion-cast refractory oxide systems, a second intergranularly formed phase appears to decrease the amount of columnar crystals that form, and to decrease the average size of the primary crystals. In other words, the second phase appears to break up the microstructure and increase the formation of equiaxial primary crystals.

In the $MgO–MgAl_2O_4$ system, after the primary periclase phase crystallizes, the eutectic composition composed of periclase and spinel crystals forms intergranularly. However, sometimes, the primary periclase crystals appear to serve as crystallization sites for the periclase that is crystallizing out from the eutectic composition. Consequently, a refractory is formed that is composed of periclase crystals bonded by a spinel phase as seen in Fig. 6. In other systems the primary phase is surrounded by smaller crystals containing the eutectic phases.

In the $MgO–MgAl_2O_4$ system, the primary periclase phase in bodies with compositions near the eutectic point crystallizes dendritically. The spacing and size of the dendrites increase from skin to core of the casting as shown in Fig. 7. The periclase dendrites are surrounded by the eutectic structure of spinel and periclase, which can be seen more clearly in Fig. 8. Jackson *et al.* (1966) proposed that dendrites remelt at a certain stage of crystallization. Some evidence of remelting of the dendrite arms and the different degrees of remelting in the $MgO–MgAl_2O_4$ system are illustrated in Fig. 9. At X, a neck

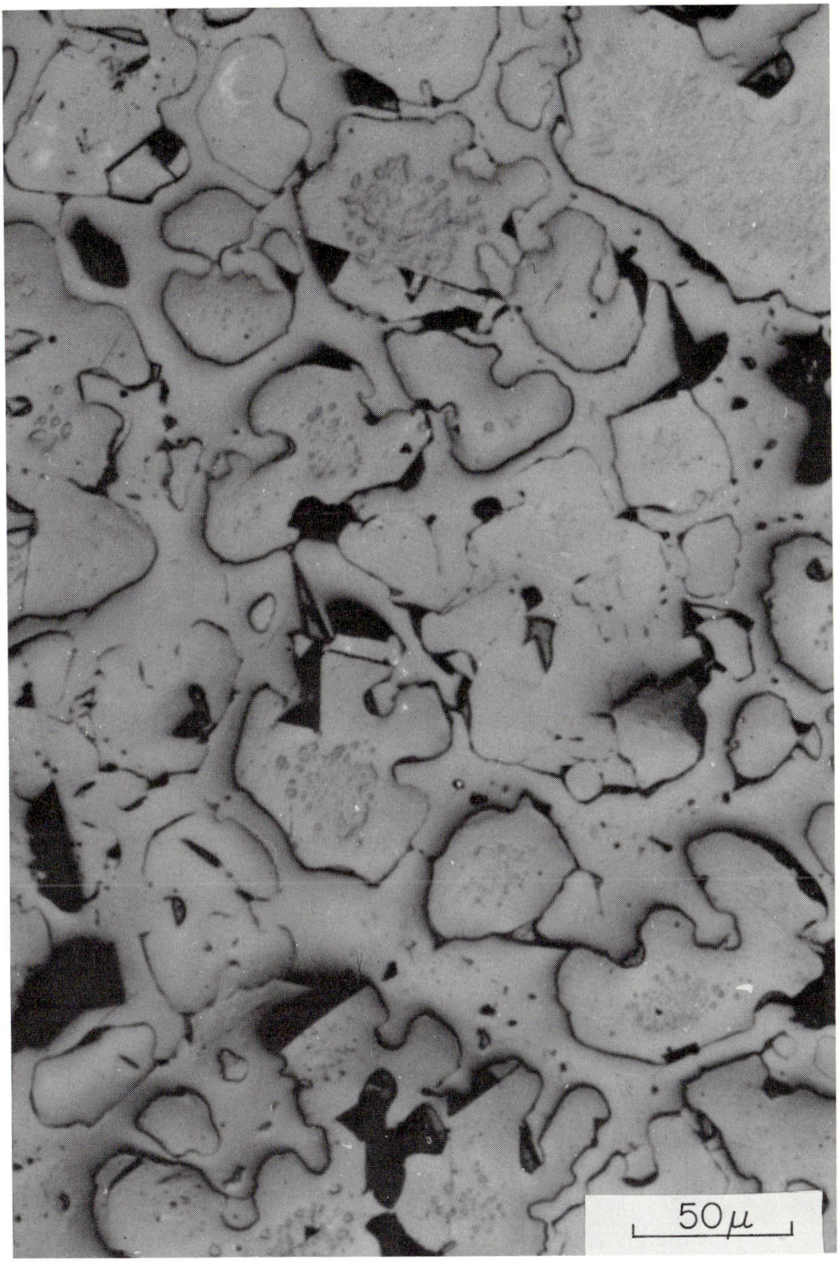

Fig. 6. Photomicrograph of 65/35 MgO/Al_2O_3 fusion-cast ceramic.

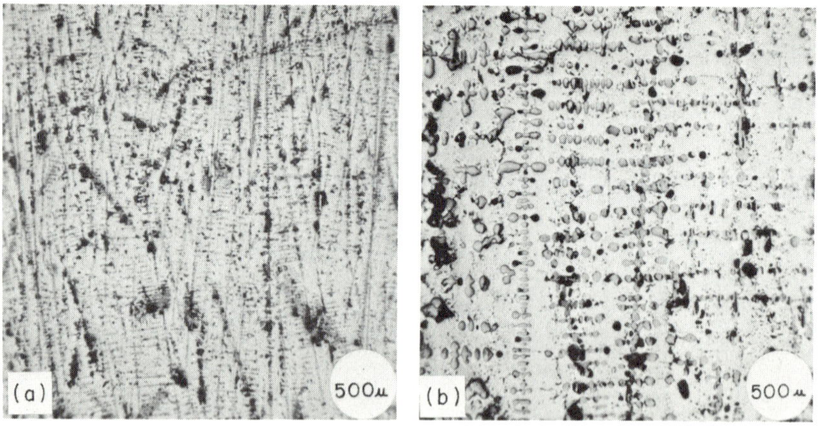

Fig. 7. Photomicrographs of polished sections showing the effect of cooling rate on the denrite structure of about 50% MgO·50% Al_2O_3 casting. (a) Skin (chilled region), (b) 1 in. from the skin.

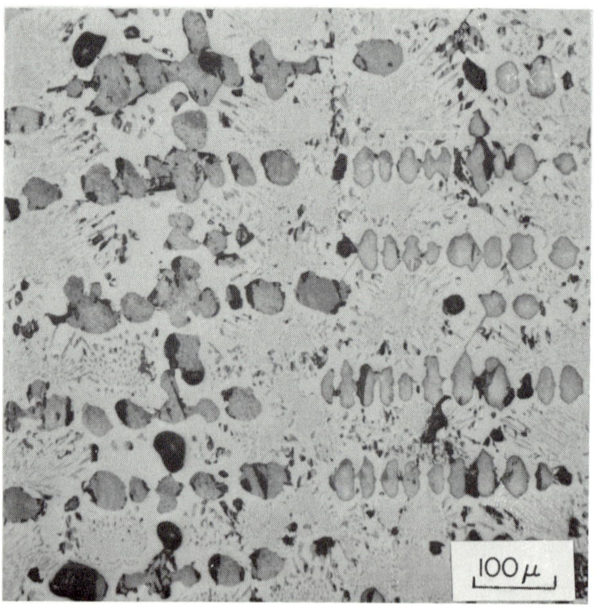

Fig. 8. Photomicrograph of a 50/50 MgO/Al_2O_3 fusion-cast composition composed of primary dendrites of periclase surrounded by eutectic periclase and spinel.

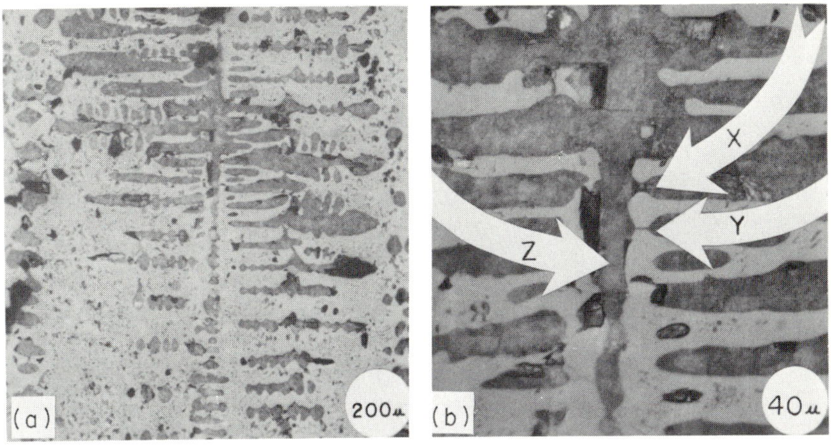

Fig. 9. Photomicrographs of polished sections of ~50 wt % MgO–50 wt % Al_2O_3 fusion-cast body. The dendrite (black phase) is periclase and the background is the eutectic structure. (b) is part of (a) at higher magnification to show evidence of dendrite remelts.

Fig. 10. Photomicrograph of 68% Al_2O_3–32% MgO fusion-cast refractory, which contains large primary spinel crystals surrounded by the eutectic composition.

if forming at the root of the arm, at Y the arm is about to detach from the main stem, and at Z the arm is completely detached from the main stem. Figure 10 is a photomicrograph taken near the core of a billet on the spinel-rich side of the eutectic. It shows the primary spinel crystals, characterized by the fourfold symmetry, surrounded by eutectic structure. Fusion-cast compositions that are essentially eutectic tend also to crystallize in a columnar fashion similar to compositions that are essentially monophased.

The heat treatment of solid-solution phases that decompose at lower temperatures is a very important area for future study of ceramic alloy systems. Magnesia–R_2O_3 fused ceramics, when rapidly solidified, can retain the R^{3+} ions in solid solution in the periclase lattice. Upon reheating at temperatures lower than the eutectic temperature, precipitation of the spinel phase occurs. This results in a periclase ceramic with small included spinel crystals. If a small amount of TiO_2 is added, the precipitated phase becomes lenticular rather than spherical. High temperatures and longer hold times tend to coarsen the precipitated phase.

B. MgO–CaO

Fusion-cast refractories composed of MgO–CaO appear to have a microstructure (less facets) that is more like that of low-entropy-of-fusion metallic systems than MgO–$MgAl_2O_4$ compositions. The phase diagram (Fig. 11) of the MgO–CaOsystem (Doman *et al.*, 1963) shows a simple eutectic and very limited solid solution of the end-member oxides at elevated temperatures. All the microstructures in Fig. 12 were collected from brick-sized specimens fused in an arc furnace and solidified in graphite molds. The microstructure of the compositions near the eurectic can be described as finely disseminated, rounded to rodlike, periclase crystals in a CaO matrix (Fig. 12a). As the MgO content increases (52.9% MgO–47.1% CaO), larger periclase crystals occur that are surrounded by the eutectic structure (Fig. 12b). With additional MgO (70.6% MgO–29.5% CaO), large periclase crystals with a quasicontinuous periclase–periclase bond are surrounded by a CaO matrix (Fig. 12c). Minor silicate impurities occur in this system as isolated pockets. This may be due to the presence of two refractory phases, which decrease the wetability of the silicate phase. White (see Chapter 2 in this volume) has presented data that substantiates this hypothesis. Fusion-cast CaO–MgO refractories are very resistant to corrosion by basic slags (lime/silica ratio greater than 2/1), and they have good high-temperature strength and thermal shock resistance.

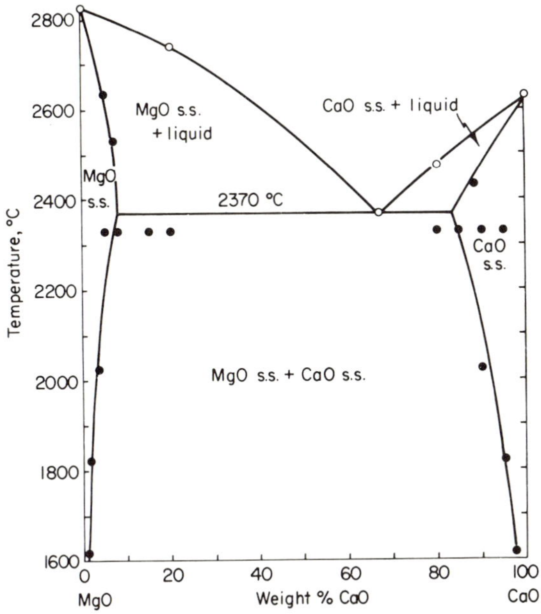

Fig. 11. Phase diagram of MgO–CaO system (Doman *et al.*, 1963).

C. Al_2O_3–ZrO_2 System

The alumina–zirconia system is important for two reasons: (1) fusion-cast alumina–zirconia ceramics form excellent abrasion-resistant grains used in snagging steel and other metal ingots, and (2) fusion-cast alumina–zirconia–silica refractories are the most widely used for lining glass melting furnaces. The alumina–zirconia-based fusion-cast refractories are very insoluble in molten glass, allowing production of high-quality glass.

Alumina–zirconia is a simple eutectic system that has very limited solid solution at high temperatures (Fig. 13). Pure alumina crystallizes in a columnar-oriented structure. Voids form between the columnar crystals; these are primarily due to shrinkage that occurs when alumina liquids solidify. Materials with this structure are very weak (modulus of rupture less than 1000 psi) and have very poor resistance to thermal shock. When ZrO_2 is added to alumina, the structure becomes more equiaxial. Compositions that fall between alpha-alumina and the eutectic contain primary crystals of alumina surrounded by the eutectic composition. The eutectic composition consists mainly of alumina crystals with included rodlike zirconia crystals.

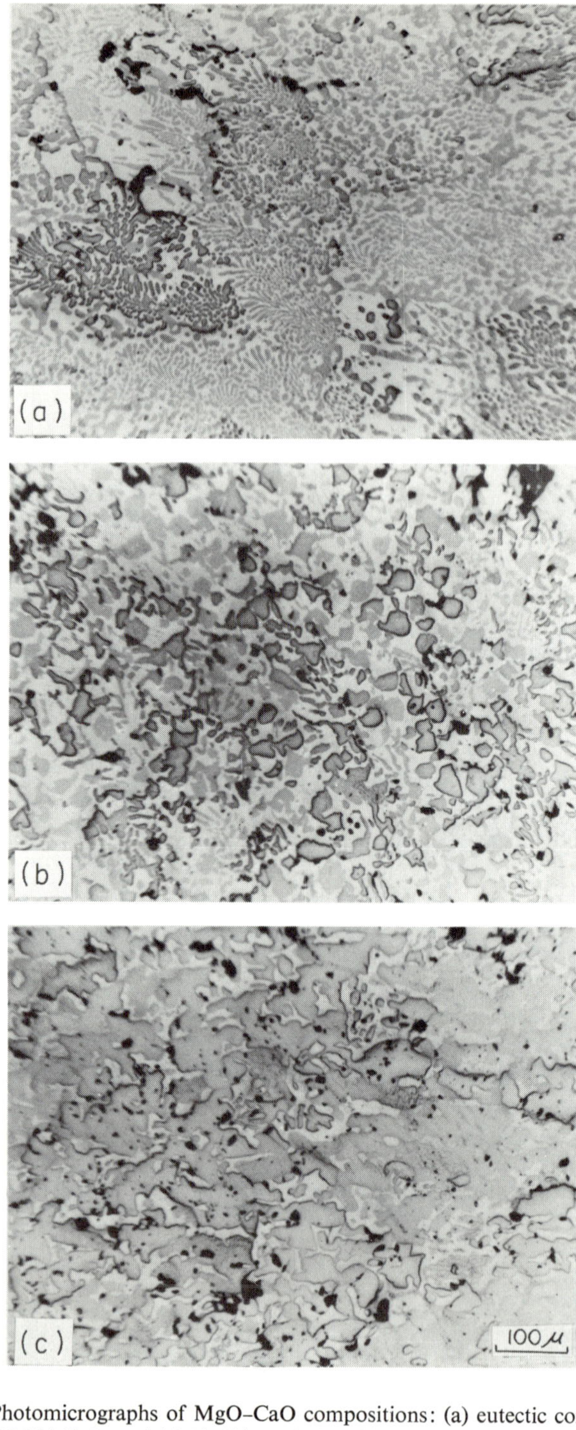

Fig. 12. Photomicrographs of MgO–CaO compositions: (a) eutectic composition, (b) 52.9% MgO–47.1% CaO, and (c) 70.6% MgO–29.4% CaO.

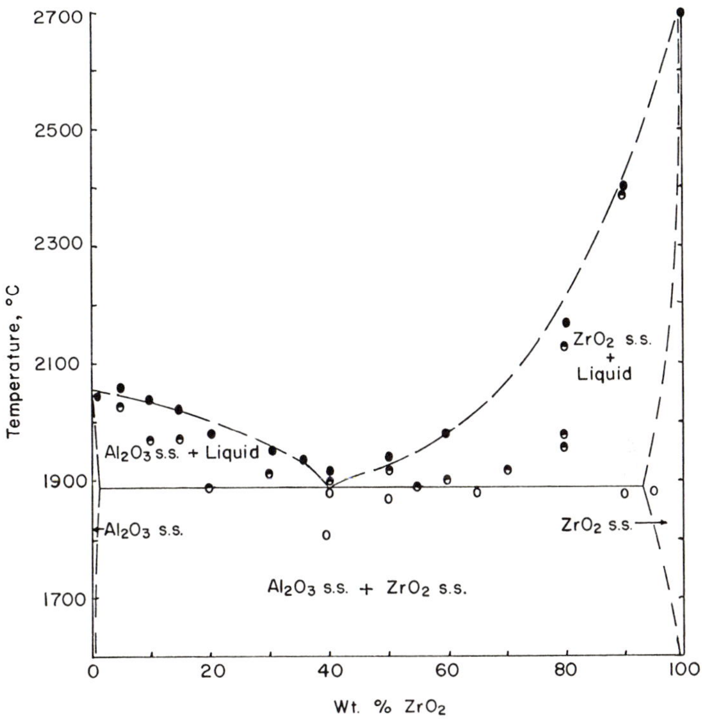

Fig. 13. Phase diagram of the ZrO_2–Al_2O_3 system. Argon atmosphere; ●, melted; ◓, partially melted; ○, not melted.

Figure 14 shows the eutectic structure of the Al_2O_3–ZrO_2 system where the white phase (rods) is ZrO_2 and the dark background is Al_2O_3. The eutectic grains grow along a threefold symmetry axis of the alpha-alumina, as revealed by the dark field photomicrograph in Fig. 15a. Both Figs. 14 and 15 were taken from a polished section that was parallel to the wall of a molybdenum crucible in which the melt was made. The crystallographic relation between the ZrO_2 and Al_2O_3 phases of the eutectic and the crystallization nature of this eutectic are being studied.

The finely divided eutectic structure gives these fusion-cast materials their high impact strength and toughness, which permits their use in snagging operations. It is desirable to solidify these alumina–zirconia compositions rapidly to toughen the refractory (Kistler and Rue, 1964; Marshall and Roschuk, 1965). Compositions that fall between the eutectic and pure zirconia have microstructure consisting of primary zirconia surrounded by the eutectic phases. The habit of the zirconia phase appears to be related to composition. Compositions high (greater than 60%) in zirconia tend to have

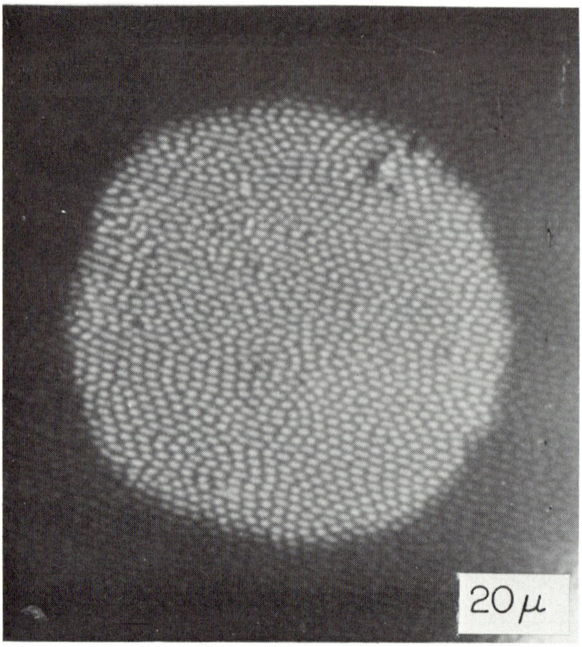

Fig. 14. Photomicrograph of polished section showing the eutectic structure of Al_2O_3–ZrO_2 system (40% ZrO_2–60% Al_2O_3).

primary crystals of zirconia with a bleblike or an equigranular shape. The zirconia crystals in compositions with less than 60% zirconia tend to crystallize in a dendritic and needlelike manner when they are arcmelted and immediately cast. However, if they are slightly cooled before casting, they tend toward a bleblike microstructure more characteristic of the higher-melting higher-zirconia compositions. Studies by Lewis (1968) indicate that the zirconia habit is related to the degree of superheat in the liquid. Consequently, compositions high in zirconia that are melted at high temperatures also cool more rapidly during casting, and the zirconia crystals solidify in the bleblike habit.

IV. COMPARISON OF EUTECTIC- AND PERITECTIC-CONTAINING CARBIDE–GRAPHITE SYSTEMS

A. Introduction

Systems that contain a eutectic have different microstructures than systems that contain a peritectic. The eutectic-containing systems permit the

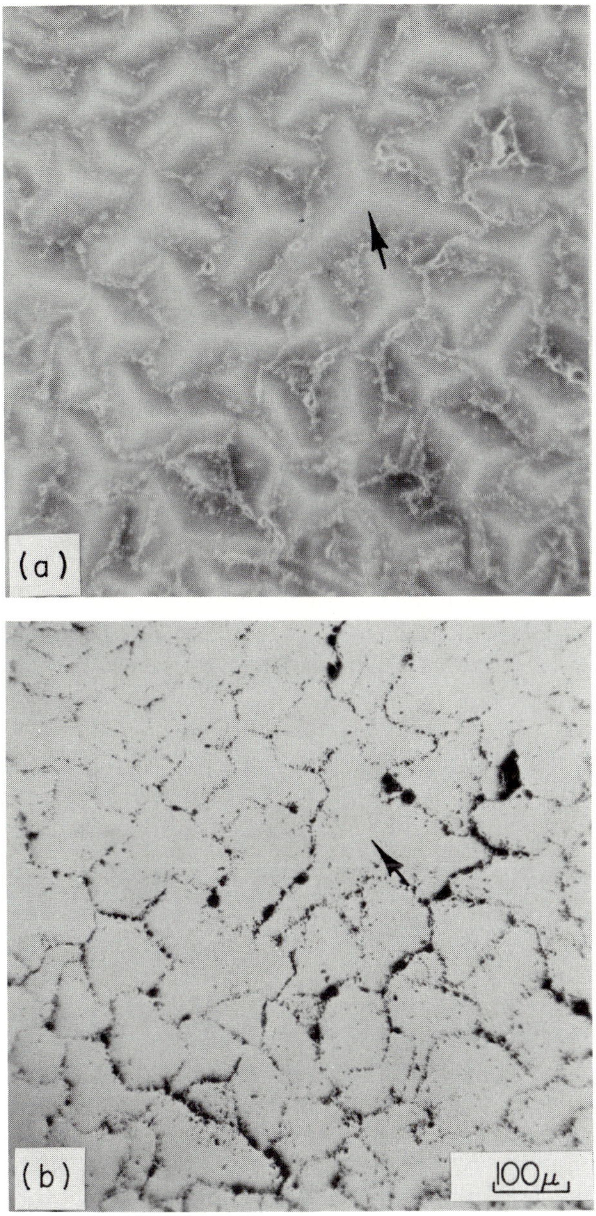

Fig. 15. Photomicrographs of the (a) dark field and (b) bright field of the Al_2O_3–ZrO_2 eutectic (40% ZrO_2–60% Al_2O_3) showing threefold symmetry of the alpha-alumina. The arrow is pointing toward the same grain under bright and dark field illuminations.

formation of finely divided eutectic phases; peritectic systems tend to form large crystals that are wholly or partly surrounded by a low-melting incongruent phase. In other words, the intergranular phase in a eutectic can be made much finer than is possible in a peritectic system that does not contain a eutectic. The fine structure is useful in improving the mechanical properties of the ceramic.

B. TiC–C System (Eutectic System)

Figure 16 is the phase diagram of the Ti–C system, showing that there is a simple eutectic between TiC and C. Figure 17 shows the microstructures of fusion-cast carbides in which the compositions change from TiC to the eutectic composition, to one rich in graphite. Compositions that fall between TiC and the eutectic contain equigranular TiC crystals surrounded by the eutectic phase. The eutectic composition is composed of titanium carbide and platelike graphite crystals. Compositions that contain slightly more graphite in the eutectic composition are composed of large noninterlocked primary plates of graphite surrounded by the eutectic phase. As the graphite is further increased, the primary crystals of graphite become interlocked, and these interlocked primary crystals of graphite are then surrounded by fine-grained eutectic phases. The properties of this system are quite varied and

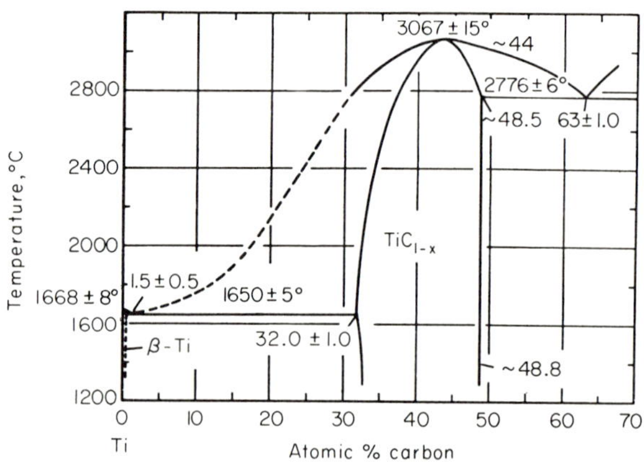

Fig. 16. Constitution diagram of the titanium–carbon system (79.95 wt% Ti, 20.05 wt% C) (Rudy *et al.*, 1965a).

Fig. 17. Microstructures of fusion-cast TiC-C materials and their relationship to thermal shock (1850°C to water). (a) 73% Ti–27%C, 5 cycles; (b) 67% Ti–33%C, 18 cycles; (c) 43% Ti–57%C, 50 cycles.

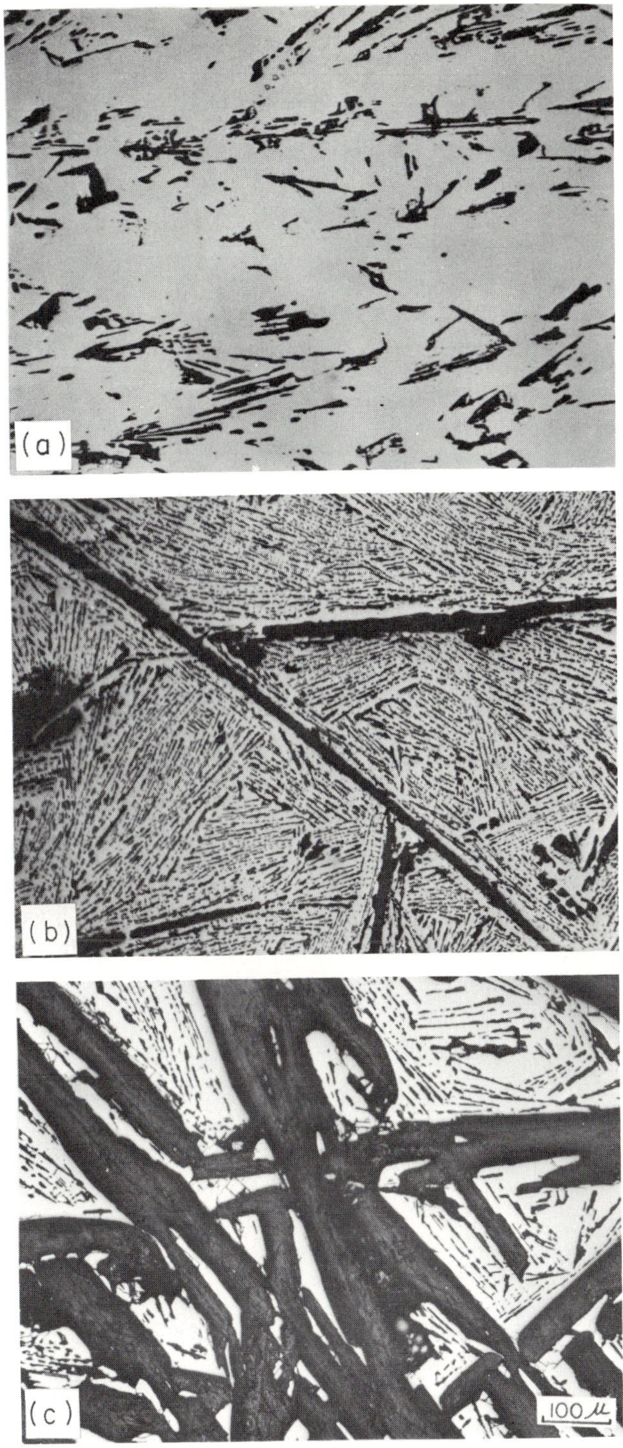

Fig 17.

dependent upon the composition and microstructure. Fused ceramics high in graphite have much better thermal-shock resistance but less mechanical strength than fusion-cast refractories high in titanium carbides. In applications where abrasion resistance or mechanical strength is desirable, the low-graphite materials are preferred.

C. ZrC–C System (Eutectic System)

Figure 18 is a constitution diagram of the zirconium carbon system (Sara *et al.*, 1963; Sara, 1965). It is very similar to the TiC–C system shown in Fig. 15a. A simple eutectic exists between the zirconium carbide and the graphite

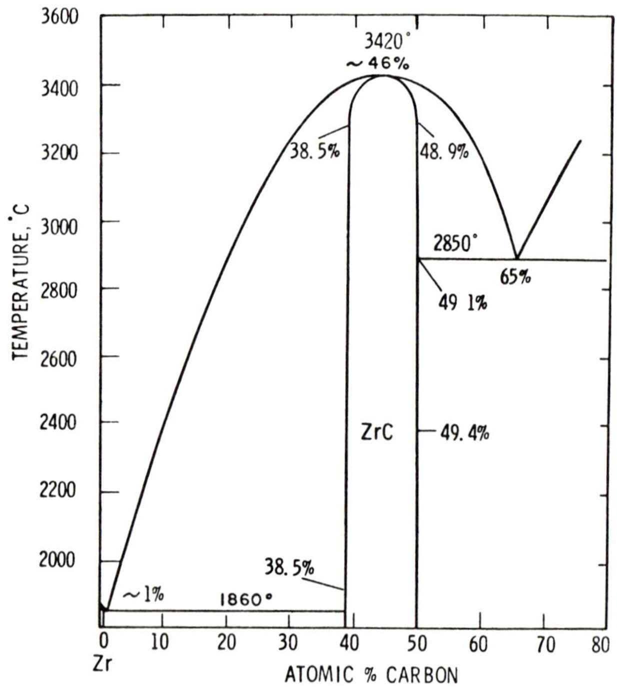

Fig. 18. Constitution diagram of the zirconium–carbon system (88.36 wt% Zr, 11.64 wt% C) (Sara *et al.*, 1963; Sara, 1965).

Fig. 19. Microstructures of fusion-cast ZrC–C ceramics. (a) 85% Zr–15%C, (b) eutectic structure, (c) 60% Zr–40%C.

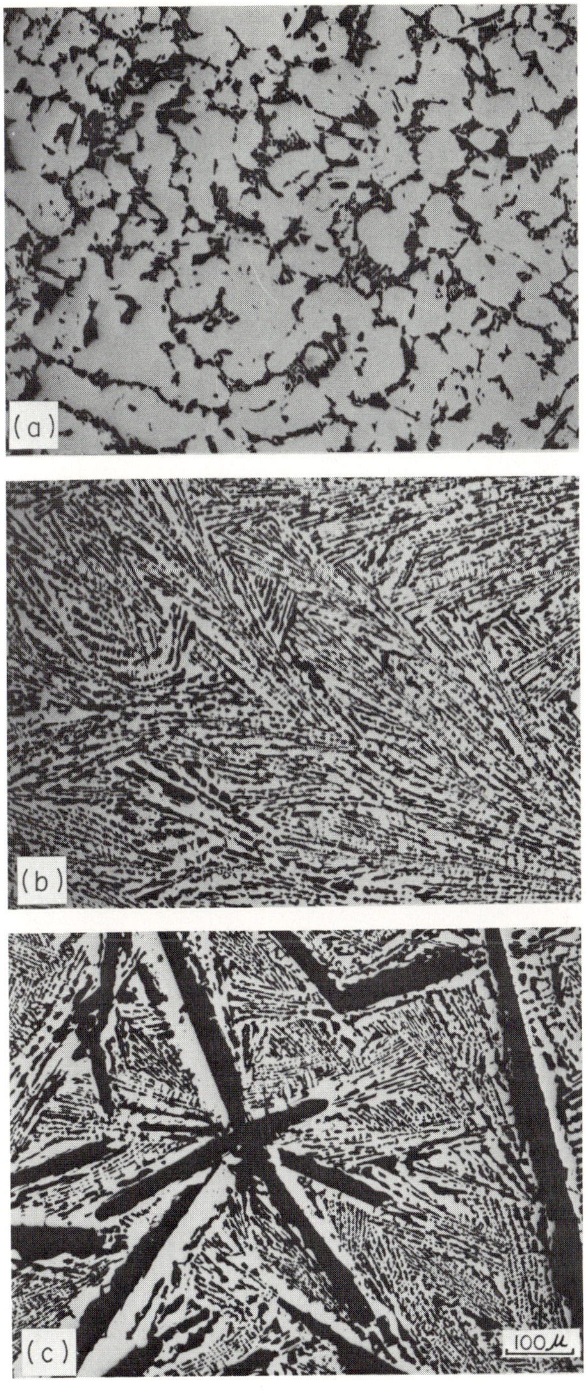

FIG. 19.

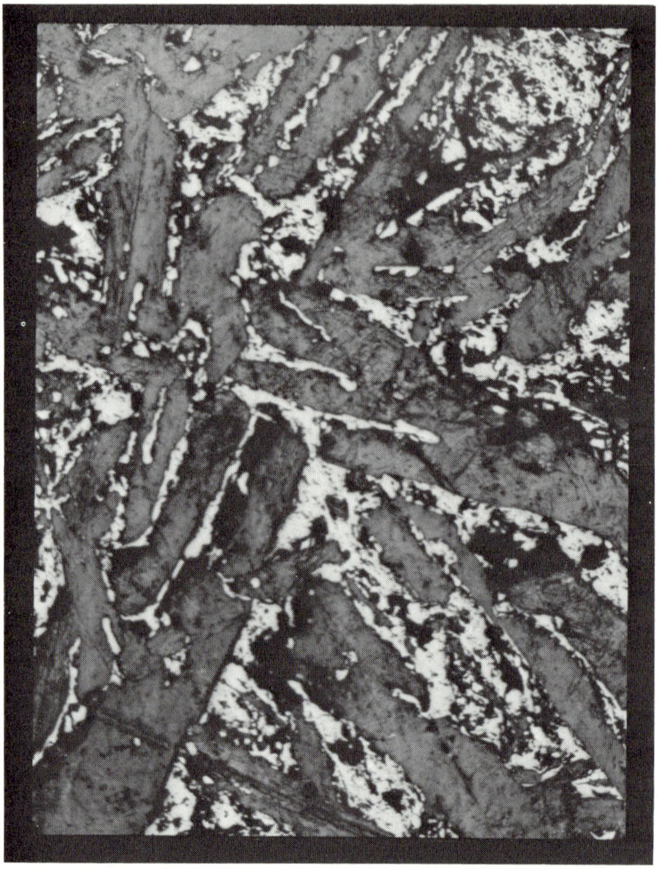

Fig. 20. Fusion-cast (C–ZrC–ZrB_2) ceramic containing a large quantity of interlocking graphite.

phases. Figure 19 shows how the microstructure changes in going from the primary field of the zirconium carbide to the eutectic and then to the primary field of the graphite. Compositions high in zirconium carbide contain interlocked zirconium carbide crystals with an intergranular eutectic structure, finer grained carbide and platelike graphite. Compositions high in graphite contain interlocked crystals of graphite surrounded by the eutectic structure. Figure 20 shows an extreme example, where a large quantity of carbon is added to the zirconium–carbon system. The graphite crystallized first and became almost completely interlocked, and the eutectic phase crystallized intergranularly. (One percent boron is also present in this sample.)

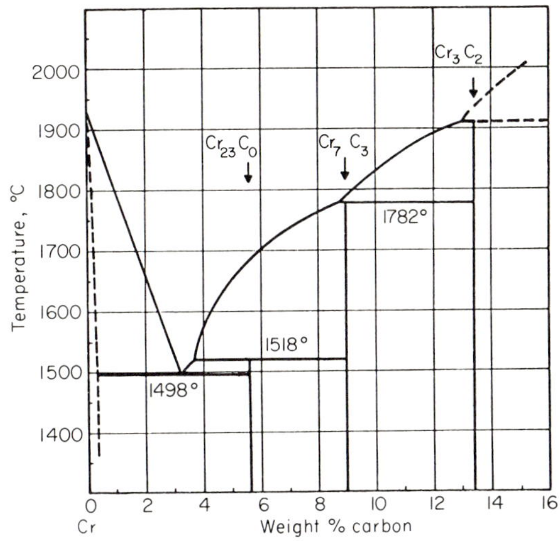

Fig. 21. Constitution diagram of the chromic-carbon system (Bloom and Grant, 1950).

D. Cr_3C_2–Carbon System (Peritectic System)

Figure 21 shows the constitution diagram of the chromium–carbon system (Bloom and Grant, 1950). Graphite crystallizes out first upon solidification in compositions that fall between Cr_3C_2 and pure graphite. Consequently, these compositions have a much different microstructure than do compositions that contain a eutectic between the carbide and graphite phases. In systems that contain an incongruent phaselike Cr_3C_2, it is much more difficult to reach equilibrium than it is in systems that contain congruently melting phases. To form Cr_3C_2, some of the primary crystals of graphite have to be reabsorbed and react with chromium-rich liquids to form Cr_3C_2. Since solidification takes place rapidly in chill castings, nonequilibrium phases such as Cr_7C_3 are often present. Figure 22 compares the microstructure of a chromium carbide–graphite sample to one of TiC–graphite. These photomicrographs show that in systems without a eutectic there are no fine crystals of graphite and carbide between the large graphite platelets. Systems that contain the eutectic structure tend to have better thermal shock resistance.

Figure 23 shows how the microstructure of a TiC–C fusion-cast carbide is modified as chromium is added. As chromium is increased from a to d, it enters the TiC lattice in solid solution. The fine platelets of graphite decrease until they are completely absent when 18% retained chromium is present (Fig. 23d). Fusion-cast refractories that contain chromium resist oxidation better than do pure TiC–C refractories.

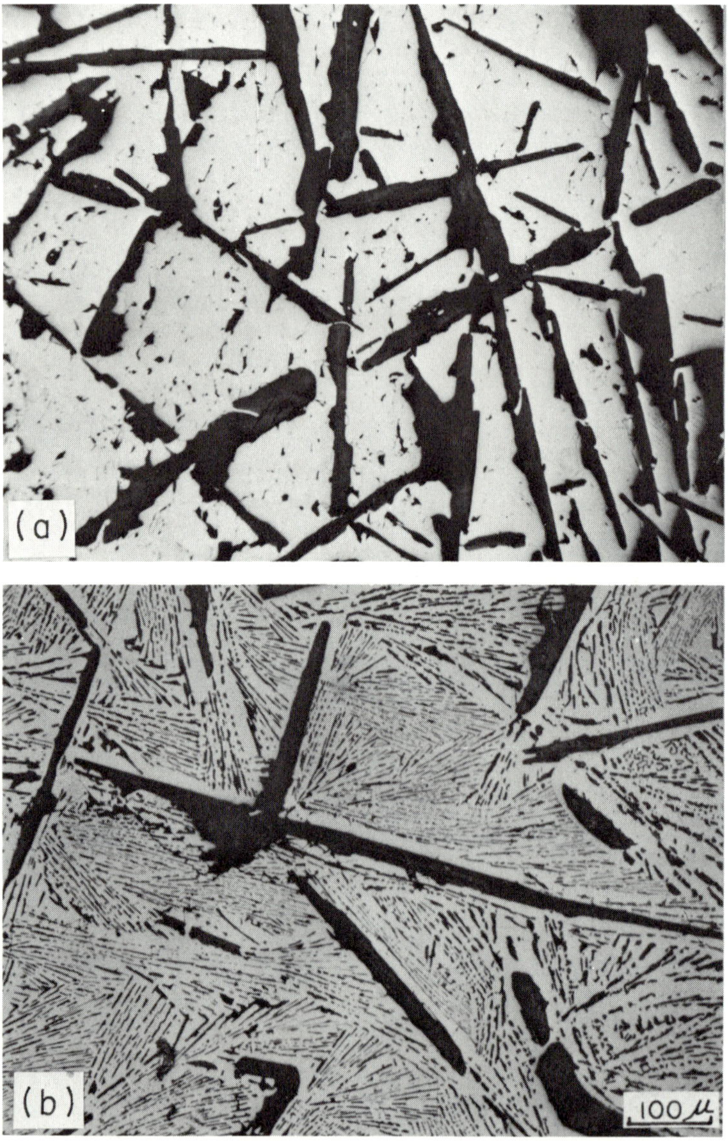

Fig. 22. Microstructure of (a) fusion-cast chromium carbide–graphite (peritectic system, 66% Cr–34% C) compared to (b) one of fusion-cast titanium carbide–graphite (eutectic system, 60% Ti–40% C).

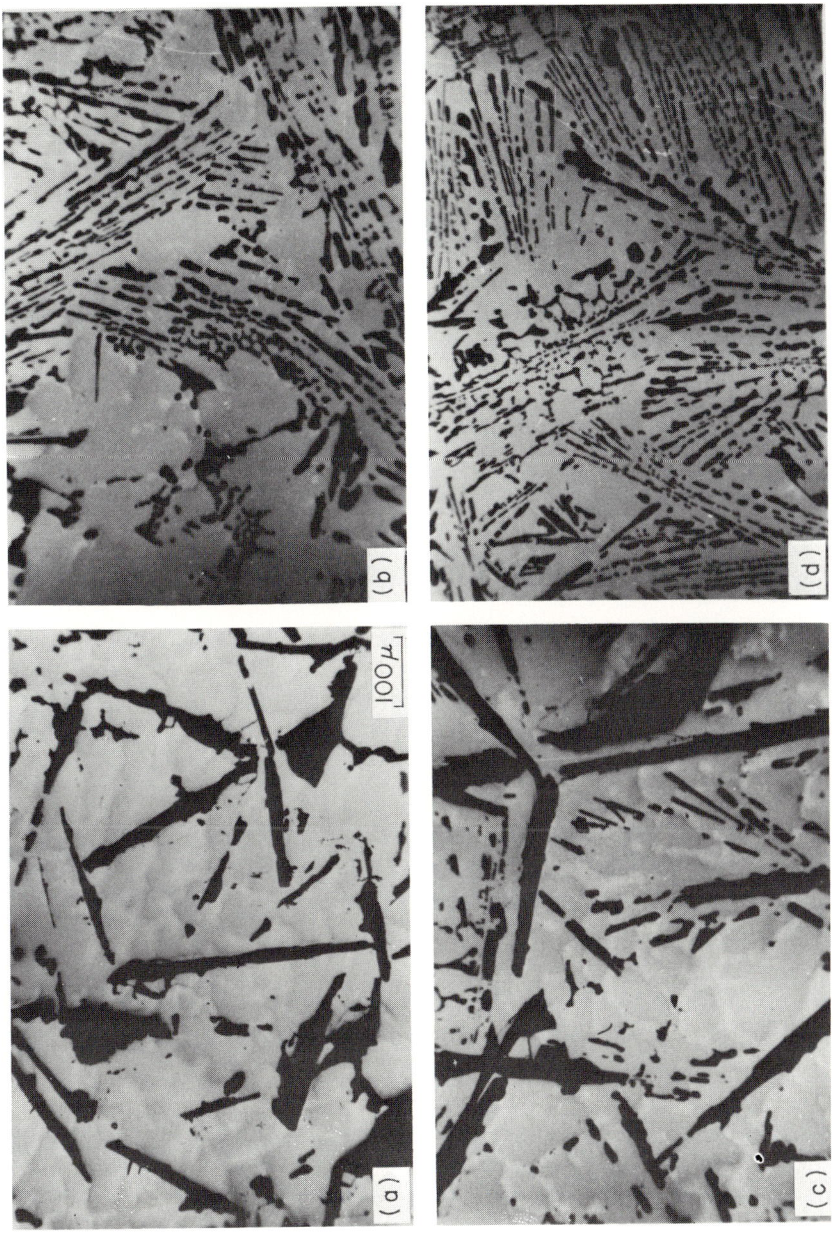

Fig. 23. Photomicrographs of fusion-cast titanium–chromium–carbon composition. (a) 2.8% Cr retained; (b) 4.6% Cr retained, (c) 9.9% Cr retained, (d) 18.0% Cr retained.

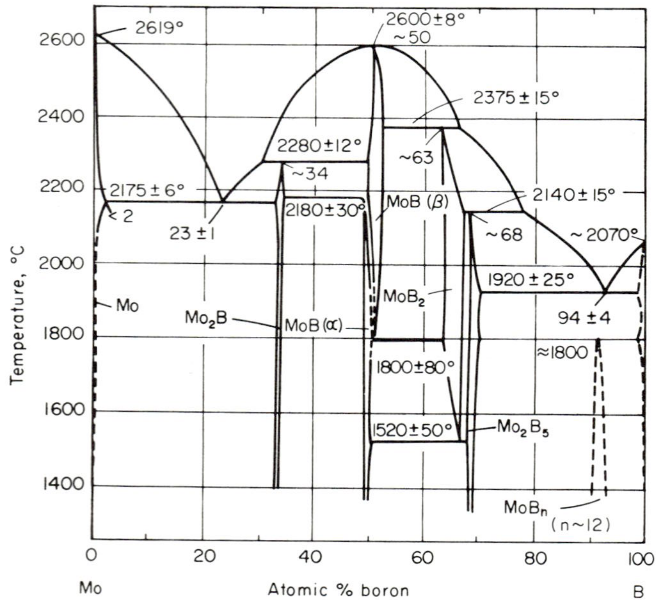

Fig. 24. Constitution diagram of molybdenum–boron system (89.7 wt% Mo, 10.3 wt% B) (Ruby and St Windisch, 1965).

Fig. 25. Constitution diagram of molybdenum–carbon system (88.9 wt% Mo, 11.1 wt % C) (Rudy *et al.*, 1964).

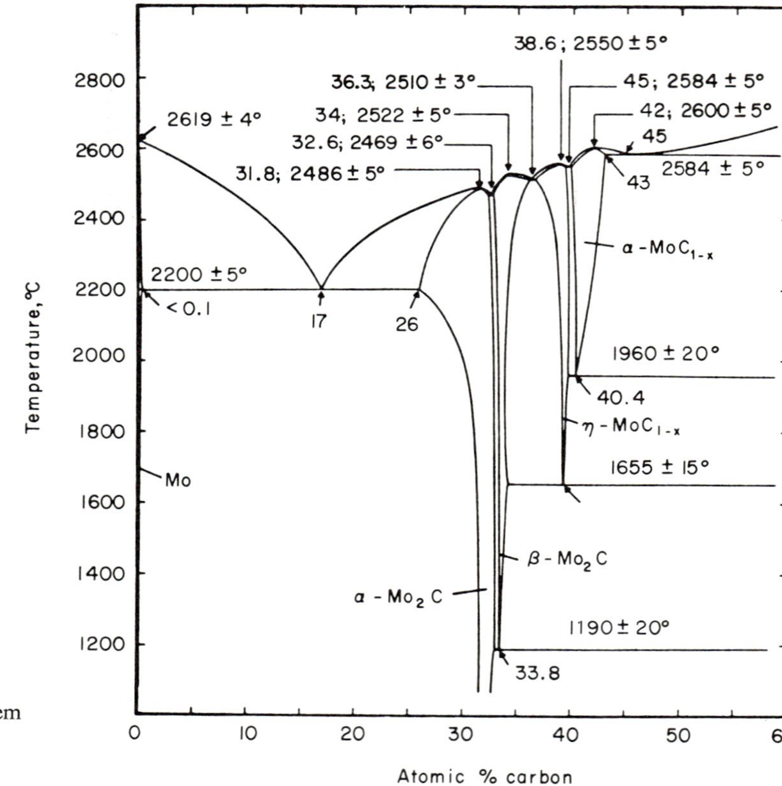

FIG. 25.

V. SYSTEMS WHERE NONEQUILIBRIUM PHASE ASSEMBLAGES ARE PREVALENT

As mentioned, systems that contain incongruent melting phases often result in nonequilibrium phase assemblages upon solidification, even in very fluid systems. This is more pronounced when there is a series of incongruent phases such as the molybdenum–boron system (Rudy and St. Windisch, 1965) (Fig. 24). The tendency to form nonequilibrium phases is even more enhanced if solid-state phase changes can also take place, which is shown in the constitution diagram of molybdenum–carbon (Rudy *et al.*, 1964) (Fig. 25). Often, when solid-state transformations occur, they are not completed when the ingot has cooled to room temperature. If equilibrium is desired, they should be reheated long enough for the transformation to take place. Figure 26 shows photomicrographs of Mo–B–C fusion-cast samples that tend to have nonequilibrium phase assemblages.

The tungsten–boron–carbon system also forms nonequilibrium fusion-cast alloys. Figure 27 is the constitution diagram of W–C (Rudy *et al.*, 1965a, b) which shows that incongruent melting phases and solid-state transformations occur. Figure 28 shows photomicrographs of polished sections of samples in the tungsten–boron–carbon system. Several have nonequilibrium phase assemblages. Since the phases are nonequilibrium, the microstructure can be readily modified by heat treatment.

VI. SYSTEMS CONTAINING LIQUID IMMISCIBILITY

It is possible to obtain a fine-grained structure in fusion-cast ceramics when immiscible liquids form upon melting. The melt separates into two liquids, which, upon solidification, often result in a structure composed of blebs of one phase surrounded by another phase. Sometimes the phases are interconnected. Figure 29 is a photomicrograph of a 50% chromic oxide–50% silica fusion-cast ceramic that was melted in an electric arc furnace with low voltage. Three immiscibility liquids formed—one high in silica, one high in Cr_2O_3, and one high in chromium. Upon solidification, a relatively fine-grained, fused ceramic was formed which is composed of cristobalite, Cr_2O_3, chromium, and residual glass.

VII. CONCLUSIONS

This chapter shows that a knowledge of the phase diagram of a system is very useful in understanding and predicting what microstructure will result in a fusion-cast refractory system. Systems that are essentially monophase often

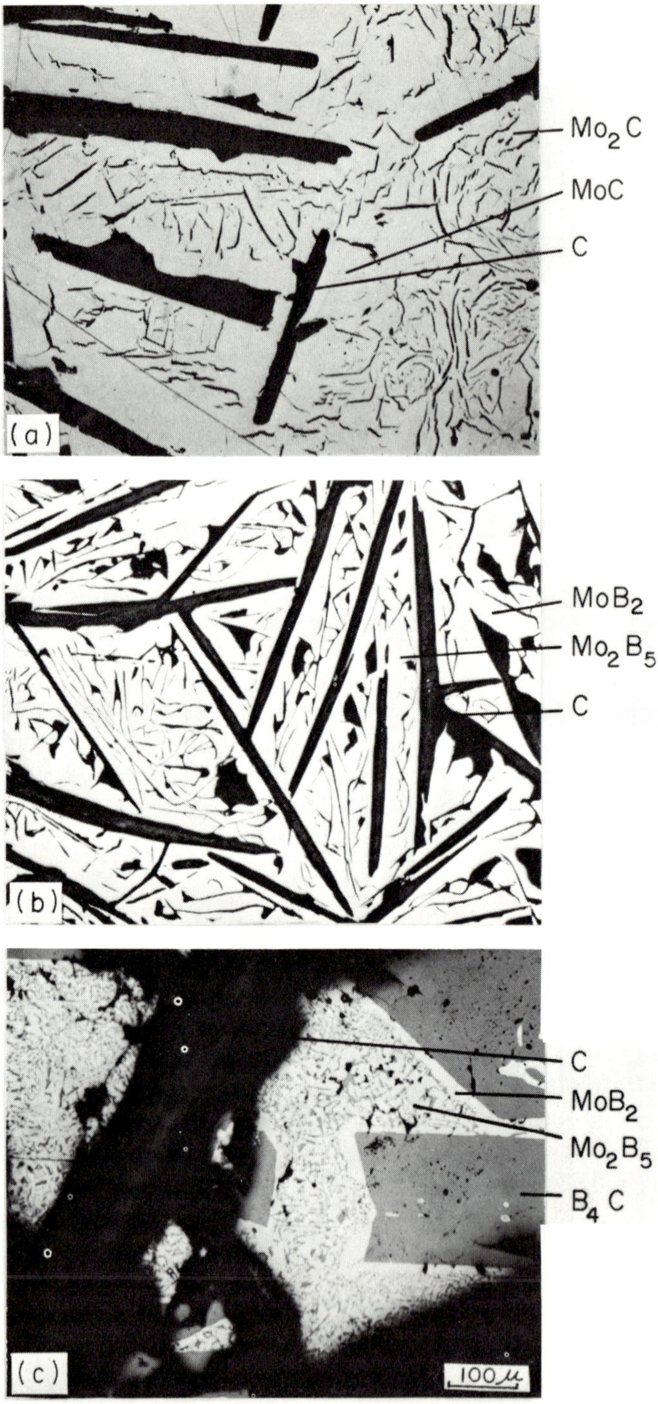

Fig. 26. Fusion-cast microstructures of the molybdenum–boron–carbon system. (a) 84% Mo–16% C, (b) 65% Mo–17% B–18% C, (c) 36% Mo–21% B– 43% C.

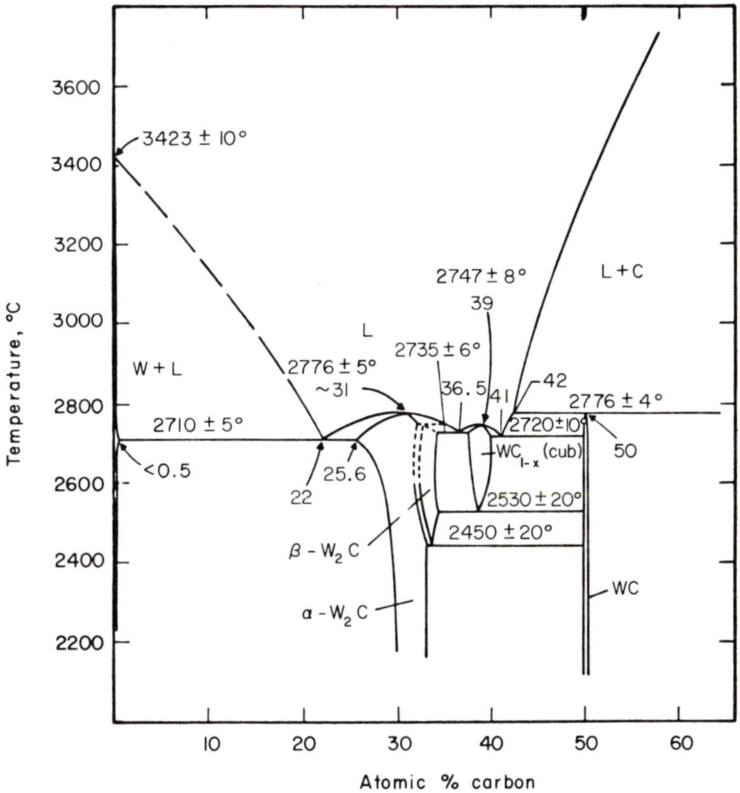

Fig. 27. Constitution diagram of tungsten–carbon system (93.77 wt % W, 6.23 wt % C) (Rudy *et al.*, 1965b).

crystallize in a columnar fusion with the lower-melting impurities surrounding the refractory phase partially to completely. Systems that have partial solid solution and a single eutectic can be manipulated to get a variety of microstructures, depending primarily on composition, solidification rate, and subsequent heat treatment. The existence of two refractory phases causes low-melting impurities to become isolated; whereas, when only one refractory phase is present, the impurities spread out along the grain boundaries between the refractory crystals. The addition of components that form a second phase appears to decrease size and orientation of grains in the primary phase. Eutectic structures are extremely important. They permit the formation of a very fine-grained structure. Crystallization of a eutectic can be controlled so that it is oriented in a particular direction. In some ceramic eutectic compositions, a structure can be made that consists of parallel needles or rods of one phase embedded in the more plentiful host phase.

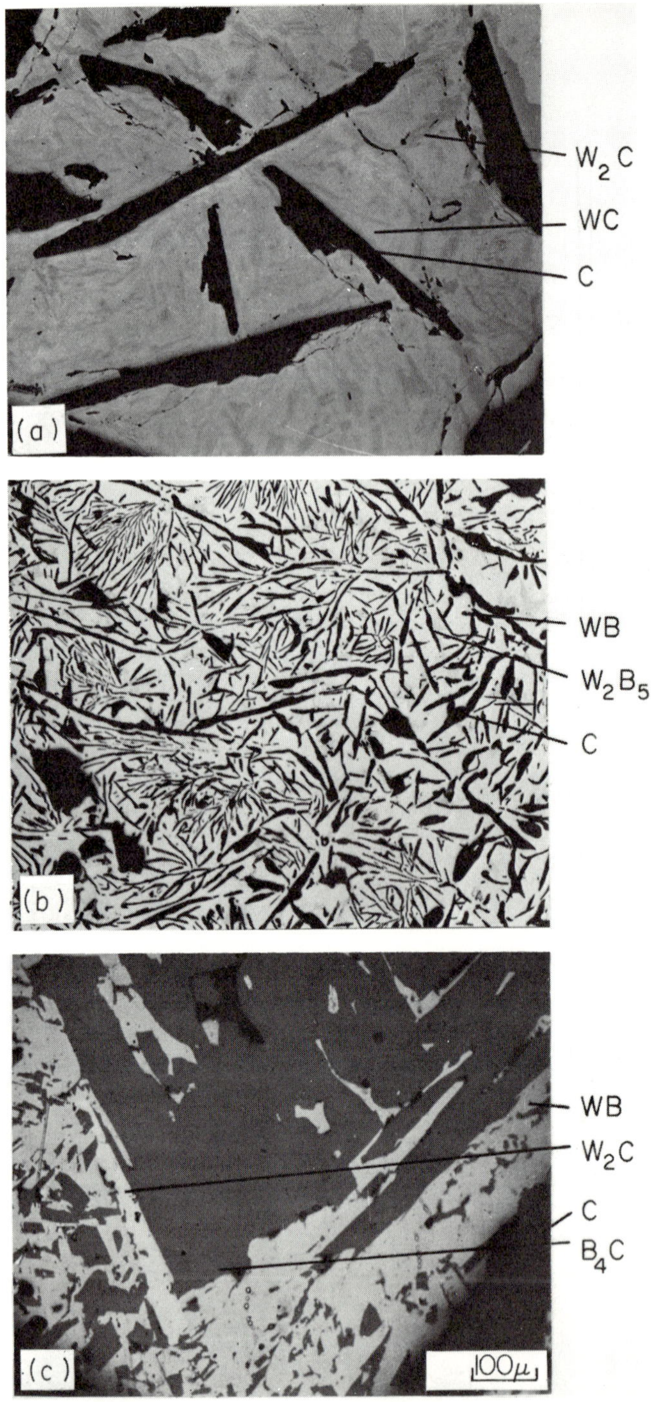

Fig. 28. Fusion-cast microstructures of the tungsten–boron–carbon system. (a) 87% W–13% C, (b) 83% W–6% B–11% C, (c) 49% W–13% B–38% C.

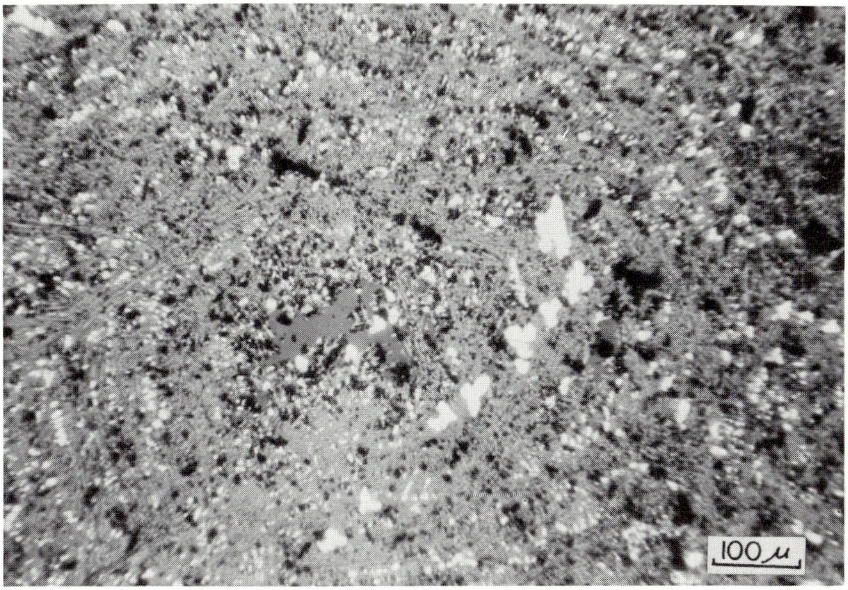

Fig. 29. Photomicrograph of 50% Cr_2O_3–50% SiO_2 fusion-cast ceramic.

Systems containing incongruent melting phases often contain a nonequilibrium phase assemblage. This occurs when the primary phase is not completely reabsorbed to form the incongruent melt phase. This causes the residual liquid to become enriched in low-melting components, which results in an excessive number of phases. The nonequilibrium phase assemblage consists of high- and low-melting phases.

Systems that contain either many incongruently melting phases or more solid transformations often depart far from the equilibrium phase assemblage. This is true even in systems that have very low viscosities when molten.

Systems that contain immiscible liquid phases permit the formation of a relatively fine-grained structure. Upon solidification, a fine dispersion of one phase in another phase occurs.

Without a eutectic, the matrix phase or phases are usually coarser-grained than when the matrix consists primarily of eutectic phases.

References

Alper, A. M., McNally, R. N., Ribbe, P. H., and Doman, R. C. (1962). *J. Am. Ceram. Soc.* **45**, 263.

Alper, A. M., McNally, R. N., Doman, R. C., and Keihn, F. G. (1964). *J. Am. Ceram. Soc.* **47**, 30.

BLOOM, D. S., and GRANT, N. S. (1950). *Trans. Am. Inst. Met. Engrs.* **188**, 41.
DOMAN, R. C., BARR, J. B., MCNALLY, R. N., and ALPER, A. M. (1963). *J. Am. Ceram. Soc.* **46**, 313.
JACKSON, K. A., HUNT, J. D., UHLMANN, D. R., and SEWARD, T. P., III (1966). *Trans. AIME* **236**, 49.
KISTLER, S. S., and RUE, C. V. (1964). U.S. Patent 3,156,545.
LEWIS, R. M. (1968). Personal communication.
MARSHALL, D. W., and ROSCHUK, S. J. (1965). U.S. Patent 3,181,939.
RUDY, E., and ST WINDISCH (1965). AFML-TR-65-2, Part I, Vol. II.
RUDY, E., ST WINDISCH, and CHANG, Y. A. (1964). AFML-TR-65-2, Part I, Vol. I.
RUDY, E., HARMON, D. P., and BRUKL, C. E. (1965a). AFML-TR-65-2, Part I, Vol. II.
RUDY, E., ST WINDISCH, and HOFFMAN, J. R. (1965b). AFML-TR-65-2, Part I, Vol. VI.
SARA, R. V. (1965). *J. Am. Ceram. Soc.* **48**, 243.
SARA, R. V., LOWELL, C. E., and DOLLOFF, R. T. (1963). WADD-TR-60-143, Part IV.

V

Application of the Phase Rule to Cement Chemistry

F. P. GLASSER

DEPARTMENT OF CHEMISTRY, UNIVERSITY OF ABERDEEN
OLD ABERDEEN, SCOTLAND

I. INTRODUCTION

The applications of the phase rule to studies of the chemistry of cements have been very helpful in systematizing a vast number of data. Many of these data can be interrelated in this way, and hence made more meaningful. The phase-rule approach to chemistry does not necessarily mean that an *a priori* assumption must be made that the system in question is in a state of equilibrium. This is a common, but unjustified, criticism that is often made of the phase-rule approach. As will be shown, nonequilibrium phenomena are frequently encountered in cement chemistry and these phenomena may also be treated using phase data obtained under equilibrium conditions. In the present chapter we shall review very briefly some of the procedures used for producing cements, discuss some of the available phase-equilibrium data, and apply them to specific situations. No attempt will be made, nor would it be possible in the available space, to discuss all the phase-equilibrium data. Therefore, the studies referred to by no means exhaust the important systematic observations that have accumulated over the past eight decades. An attempt will be made to select typical examples showing how the actual phase compositions of cements, both at equilibrium and nonequilibrium, can be assessed from the available data, and to discuss the present state of knowledge concerning the important hydration stage. Many of the anhydrous systems of special interest to the clinkering process have been very accurately studied. This may be contrasted with the relevant hydrous systems where, phase-"equilibrium" data scarcely exist for the low-temperature range, which is, of course, the range that is most relevant to cement hydration. Nevertheless, the concept of describing both the individual phases and also the coexisting phase assemblages has proved valuable. It has enabled workers to correlate data regarding the changes that are observed in the phases assemblages with variations in the other physical parameters of the system, such as changing bulk compositions, solution composition, and the temperatures and pressures used. This is, of course, an approach that is implicit in the phase-rule treatment. Examples will now be given of its application to cement chemistry.

II. CALCIUM ALUMINATE CEMENTS

A. Composition and Production of the Anhydrous Cement

Two types of calcium aluminate cements are produced commercially. The first, and commercially the most important type, has a typical bulk chemical composition (in weight percent): CaO, 36–40; Al_2O_3, 40–50; SiO_2 4–9; iron oxide, 1–15; TiO_2, 1–3; alkali, 0–1; and MgO, 0–2. The composition is variable only over rather narrower limits than would be suggested by the spread of values given above; thus, the sum of the CaO plus Al_2O_3 content is usually 80–90% of the total. This type of cement will be referred to as "normal-aluminous cement." The second type of cement, high-alumina cement, has a much higher Al_2O_3 content; typically, 72–82% Al_2O_3, and 15–27% CaO. Iron oxide, silica, and other impurities are kept low and often total only 1–4%.

Numerous combinations of raw materials have been patented for use in producing normal aluminous cements, such as phosphorous furnace slags, gypsum-plus-bauxite mixtures, etc., but most plants operate on mixtures of limestone and bauxite. Crude bauxites may be used for normal aluminous cements, but the high-alumina cements are usually prepared from purified bauxite. The appropriate combination of raw materials is usually melted in a furnace; the resulting melt is tapped off continuously, and the molten cement is finally chilled in air. Normal aluminous cements are usually chilled by pouring the melt into small iron moulds; each cast ingot usually weighs several kilos or more. The cast ingots are finally crushed and ground to cement-powder fineness. From the chemical analyses, it may be anticipated that the high-alumina cements will contain virtually all calcium aluminates, and that the normal aluminous cements will contain a high proportion of calcium aluminates. Therefore, we will first examine the $CaO–Al_2O_3$ system before presenting the more complex phase diagrams that are appropriate to describing the effects of iron oxide and the minor components.

B. The $CaO–Al_2O_3$ System

This system was first studied by Shepherd *et al.* (1909). With some modifications, their phase diagram is considered essentially correct. A modified version is shown in Fig. 1. Five binary phases: $3\,CaO{\cdot}Al_2O_3$, $12\,CaO{\cdot}7\,Al_2O_3$, $CaO{\cdot}Al_2O_3$, $CaO{\cdot}2\,Al_2O_3$, and $CaO{\cdot}6\,Al_2O_3$ are known. Liquidus temperatures drop rapidly upon addition of Al_2O_3 to CaO. Thus, $3\,CaO{\cdot}Al_2O_3$ melts incongruently to CaO and liquid at 1535°C. The minimum melting compositions are the eutectics between $12\,CaO{\cdot}7\,Al_2O_3$ and either $3\,CaO{\cdot}Al_2O_3$ or $CaO{\cdot}Al_2O_3$; these are located at 1400° and 1395°C, respectively. It is noteworthy that phase relations in this area are conveniently studied by the classical quenching techniques because the calcium aluminate liquids quench readily to yield glasses. These nonsilicate glasses form because the aluminate liquids have random-network structures built up of (AlO_4) tetrahedra, analogous to the random network structures based on (SiO_4) tetrahedra that are encountered in silicate liquids and glasses.

There is, however, some controversy regarding the stability of the $12\,CaO{\cdot}7\,Al_2O_3$ phase. At temperatures at or below ~1100°C, "$C_{12}A_7$" absorbs water from the ambient laboratory atmosphere to form a hydrate

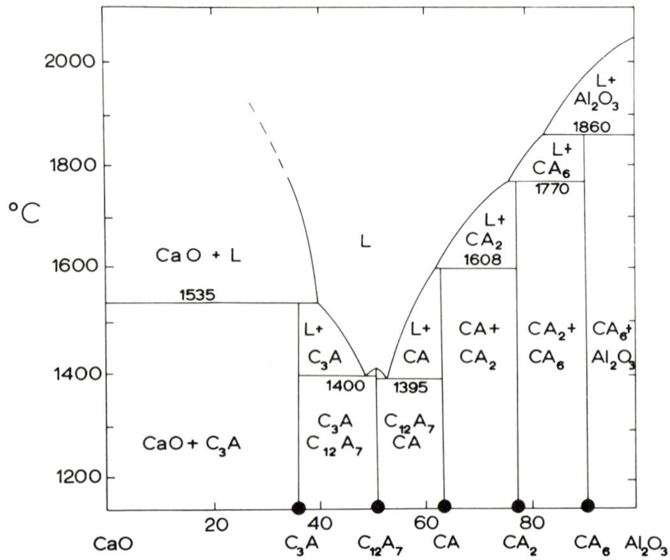

Fig. 1. Phase diagram for the system CaO–Al_2O_3. C, CaO; A, Al_2O_3; L, liquid. Compositions are in weight percentage. The congruent melting point of $C_{12}A_7$ is 1435°C.

that has the probable composition $11\,CaO \cdot 7\,Al_2O_3 \cdot Ca(OH)_2$ (Jeevaratnam *et al.*, 1964). The hydration $\rightleftarrows$ dehydration reaction may be reversed by raising or lowering the temperature; Roy and Roy (1962) have described this phase as an example of a high temperature "zeolite." Jeevaratnam *et al.* analyzed the "dehydrated" phase for total (CaO plus Al_2O_3). They concluded that, at temperatures above about 1200°C but below the melting point, the crystalline phase was truly anhydrous. By inference, therefore, it is a stable phase in the CaO–Al_2O_3 system. However, Welch (1964) and Nurse *et al.* (1965) have shown that if rigorously dry starting materials are used and if these are reacted in a moisture-free atmosphere, the $C_{12}A_7$ composition yields mainly (C_3A + CA) at temperatures above 1200°C. In their view, the $C_{12}A_7$ phase has a stability field in the CaO–Al_2O_3–H_2O system, but not in the CaO–Al_2O_3 system; that is $C_{12}A_7$ contains structural "water" at all temperature. In the present chapter, $C_{12}A_7$ is shown as a phase in the CaO–Al_2O_3 system, although doubts persist both about its thermodynamic stability, and also, its chemical composition.

Beyond the $C_{12}A_7$ eutectic with monocalcium aluminate, liquidus temperatures begin to rise rapidly with increasing alumina content. $CaO \cdot Al_2O_3$ melts incongruently to $CaO \cdot 2\,Al_2O_3$ at 1608°C; $CaO \cdot 2\,Al_2O_3$ melts incongruently to $CaO \cdot 6\,Al_2O_3$ at ~1790°C, and $CaO \cdot 6\,Al_2O_3$ melts incongruently to Al_2O_3 and liquid at ~1860°C. It can be seen that normal aluminous cements having CaO/Al_2O_3 ratios between 0.9 and 1.2, will have lower solidus and liquidus temperatures than the high-alumina cements, which have CaO/Al_2O_3 ratios between 1.8 and 2.5. The high-alumina cements are, consequently, more valuable as a refractory material; however, their higher raw-materials cost and the higher melting temperatures required add to the expense of their production.

Normal aluminous cements having CaO/Al_2O_3 ratios of less than 1.0 might be expected to contain $C_{12}A_7$ and CA as the principal aluminous phases: however, it will be shown that the effect of other oxides, especially iron oxide, tends to reduce the quantitative importance of $C_{12}A_7$. This is a satisfactory state of affairs, because $C_{12}A_7$ is an undesirable constituent of aluminous cements: it hydrates very rapidly and causes "flash setting." This results in partial hardening of the wet concrete before it can be placed. The monoaluminate, on the other hand, has very desirable hydraulic properties.

The high-alumina cements, depending on their CaO/Al_2O_3 ratio, contain mainly CA plus CA_2, or with increasing Al_2O_3 content, CA_2 plus CA_6. Both CA_2 and CA_6 are relatively inert to hydration in neutral solutions ($pH = 7$) but CA_2 will hydrate readily in alkaline solution. High-alumina cements containing CA_2 can be mixed with alkaline solutions ($pH \geq 12$) and poured or cast in the same manner as portland cements. Thus, CA and CA_2 are the most desirable phases in the anhydrous clinker.

C. The Iron-Oxide-Containing Systems

Quantitatively, the third most important component in normal aluminous cements is iron oxide. Unlike portland cement clinkers that contain iron mainly in the ferric state, aluminous cements may exhibit wide variation in their Fe^{3+}/Fe^{2+} ratios. The proportion of ferrous iron in aluminous cements is increased by (a) melting the cement under strongly reducing conditions, (b) by increasing the melting temperature as well as by (c) increasing the acidity of the melt: e.g., by adding such acidic oxides as SiO_2, TiO_2, etc. The Fe^{3+}/Fe^{2+} ratio that is present in the fused cement tends to be preserved during the subsequent chilling stage. If a melt is chilled in air some oxidation of the ferrous iron occurs, but in the time usually allowed for chilling, the extent of this oxidation is quantitatively insignificant. Thus, the melts are chilled under conditions that approximate to maintaining a constant oxygen composition of the condensed phases. Unfortunately, we have only limited information available on the system $CaO–Al_2O_3–FeO–Fe_2O_3$ at low levels

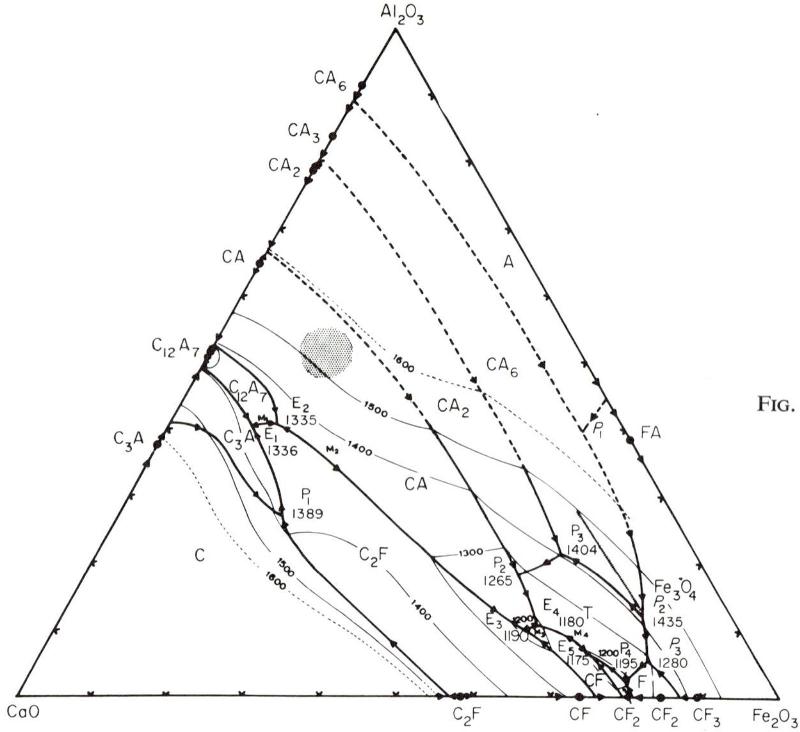

Al2O3
CA6
CA3
CA2
CA
C12A7
C3A
C
A
CA6
CA2
CA
C12A7
C3A
E2 1335
E1 1336
P1 1389
C2F
P1
FA
P3 1404
P2 1265
Fe3O4
P2 1435
E4 1180 T
E3 1190
E5 1175
P4 1195
P3 1280
CF
F
1600
1500
1400
1300
1200
CaO
C2F
CF
CF2
CF2
CF3
Fe2O3

FIG. 2a

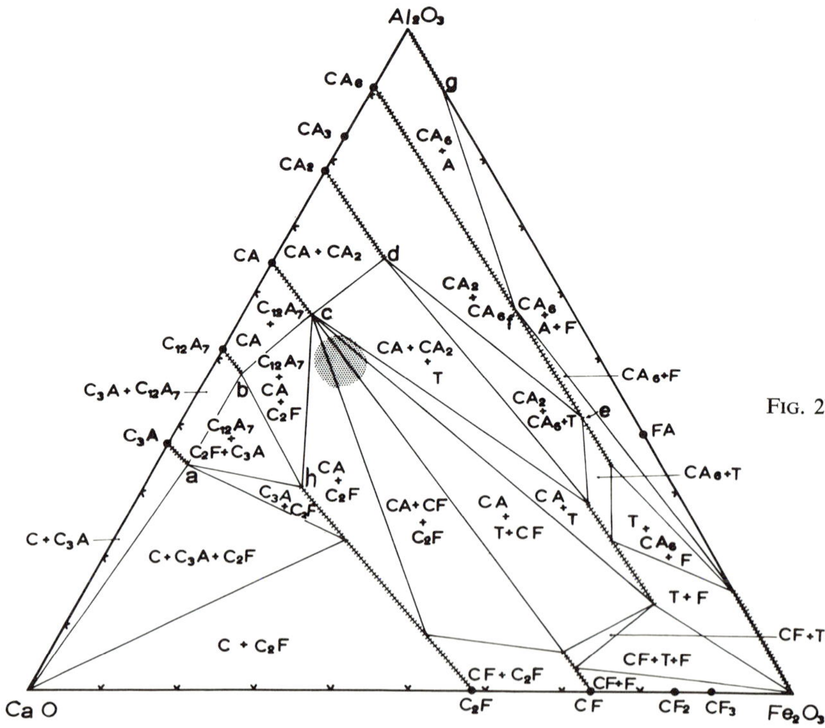

Al2O3
CA6
CA3
CA2
CA
C12A7
C3A
g
d
c
b
a
h
e
f
CA6 + A
CA + CA2
CA2 + CA6
CA6 + A + F
CA6 + F
C12A7 + CA
CA + CA2 + T
C3A + C12A7
C12A7 + CA + C2F
CA2 + CA6 + T
C12A7 + C2F + C3A
FA
CA + C2F
CA6 + T
C3A + C2F
CA + CF + C2F
CA + T + CF
CA + T
T + CA6 + F
C + C3A
C + C3A + C2F
T + F
C + C2F
CF + T
CF + T + F
CF + C2F
CF + F
Ca O
C2F
CF
CF2
CF3
Fe2O3

FIG. 2b

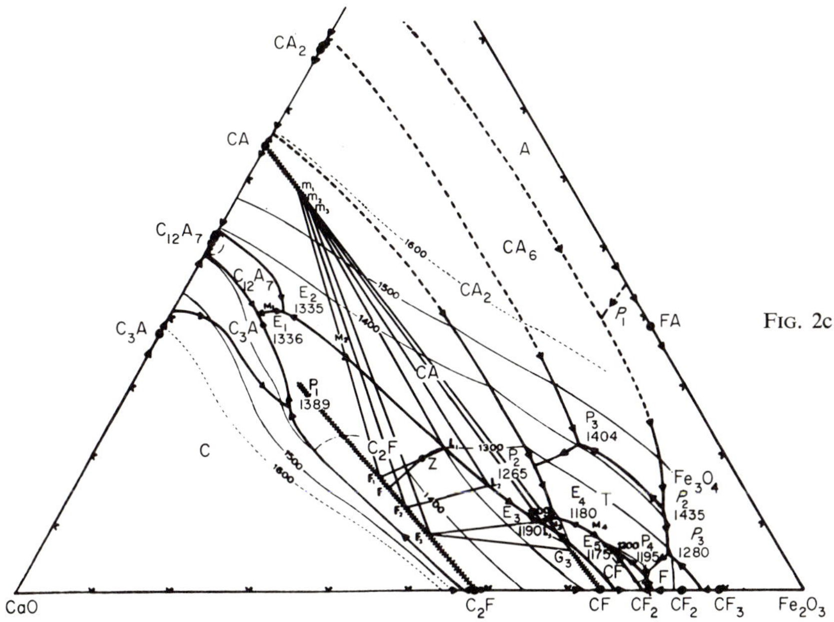

FIG. 2c

Fig. 2. Phase diagrams for the system CaO–Al_2O_3–"Fe_2O_3" showing (a) the liquidus surface at $P_{O_2} = 0.21$ atm, (b) subsolidus isothermal section at 1145°C, and (c) path of crystallization of a selected composition, Z. The crosshatching on joins indicates the limits of solid solutions. The shaded areas in Figs. 2a and 2b indicate the approximate composition of aluminous cements. Other abbreviations as in Fig. 1; also, F=Fe_2O_3.

of oxygen pressure, but rather more data are available at somewhat higher oxygen pressures, for example at $P_{O_2} = 0.21$ and 1.0 atm. Figure 2 summarizes the data obtained at $P_{O_2} = 0.21$ atm. and is based on the studies of Dayal and Glasser (1967). Phase relations in aluminous cements prepared under "oxidative" conditions may be described using these data.

Examination of Fig. 2 shows that relatively little iron substitution can occur in the three calcium aluminates: C_3A, $C_{12}A_7$, and CA. The maximum limits of solid solution can be estimated from the subsolidus isothermal section, which is also shown in Fig. 2. At subsolidus temperatures, all bulk compositions containing iron oxide in excess of the saturation limit of the calcium aluminates must develop a separate iron-bearing phase. This saturation limit of Fe^{3+} in the calcium aluminates is rather irregular in shape, and extends from CaO along the broken line that passes through points a–b–c–d–e–f–g (Fig. 2b). The bulk compositions of normal aluminous cements, shown by the shaded area, lie mainly on the high-iron side of this saturation line. The excess iron oxide must, therefore, form an additional phase or phases. At subsolidus temperatures and assuming equilibrium, the resulting phase distributions may be predicted from the subsolidus triangle or two-phase area that contains the bulk composition of interest. In this region of the diagram the appropriate tie lines radiate from monocalcium aluminate, or from solid-solution compositions very close to monocalcium aluminate. In addition, many of the compatibility regions involve as the principal iron-bearing phase, a solid solution based on $2\,CaO{\cdot}Fe_2O_3$. This solid solution is often referred to as the "ferrite" phase.* It is an important constituent of both aluminous and portland cement clinkers. Laboratory studies show that the ferrite solid solutions extend from $2\,CaO{\cdot}Fe_2O_3$ toward a hypothetical end member, $2\,CaO{\cdot}Al_2O_3$. Just below solidus temperatures, the limiting Al-rich solid solution contains approximately 69 mole % of this "$2\,CaO{\cdot}Al_2O_3$" end member. The importance of the ferrite phase in cement clinkers may be indicated by two observations. First, because the ferrite phase may contain relatively little iron, it may tie up a considerable proportion of the aluminum, not as calcium aluminates, but as "ferrite," which differs markedly from the aluminates in its hydration properties. Second, the location and existence of the thermodynamically stable two-phase region (CA ss + ferrite ss) shows why $C_{12}A_7$ is quantitatively unimportant in those normal aluminous cements having a CaO/Al_2O_3 molar ratio of slightly less than 1.0. The limiting compositions, over the area of interest, which can contain $C_{12}A_7$ at subsolidus temperatures, lie along the line c–h, Fig. 2a. This line, located at relatively low iron oxide contents, acts as a bar to the appearance of $C_{12}A_7$; thus $C_{12}A_7$ will not be encountered except in a restricted range of ternary compositions. Furthermore, the rather small expression of the $C_{12}A_7$ primary phase field at liquidus temperatures means that it is unlikely to appear even as a transient phase during either the equilibrium or fractional crystallization of normal aluminous cement liquids. Quantitative petrographic examination of normal aluminous cement clinkers shows that $C_{12}A_7$ is normally rare or absent, in accord with these predictions.

The constitution of the ferrite phase and its Al^{3+}-substituted solid solutions have been the subject of many investigations. Ideas that have been advanced in the literature and that are now believed to be incorrect include: first, that the solid solution extends from $2\,CaO{\cdot}Fe_2O_3$ toward $12\,CaO{\cdot}7\,Al_2O_3$, and second, that the solid solutions extends from $2\,CaO{\cdot}Fe_2O_3$ toward "$2\,CaO{\cdot}Al_2O_3$" not along a single line, but along a ternary band, or "swelled straight line" (Newkirk and Thwait, 1958). The solid solutions are now known to extend from $2\,CaO{\cdot}Fe_2O_3$ toward "$2\,CaO{\cdot}Al_2O_3$" and the mechanism of the solid solution is replacement of Fe^{3+} by Al^{3+}. The

* Not to be confused with the usage of the term *ferrite* which, in the electronics industry, is used to denote phases such as iron-containing spinels, etc.

solid solutions are essentially stoichiometric, and thus have a constant $CaO/Al_2O_3 + Fe_2O_3$ ratio. Nevertheless, replacement of Fe^{3+} by Al^{3+} in this series does not proceed as might be expected of an ideal solid-solution series. $CaO{\cdot}2\,Fe_2O_3$ itself affords two crystallographically distinct sites for the trivalent ions, and preferential replacement of one set of Fe^{3+} atoms by Al^{3+} leads to Al/Fe ordering at the 1:1 mole ratio. This is the "brownmillerite" or C_4AF composition, which some have held to be both a separate compound and also, the high aluminum limit of solid solution. We now know that the solid solutions extending from $CaO{\cdot}2\,Fe_2O_3$ can sweep past the brownmillerite composition to higher Al/Fe ratios. Furthermore, the cation ordering that occurs at the 1:1 ratio does not create a singular point at the liquidus surface. The expected singular point would mark the intersection of solidus and liquidus surfaces at the C_4AF isopleth. This intersection would be indicative of compound formation: since it is not observed, brownmillerite is not a discreet phase of the C_2F–C_2A join.

Another phase that might be encountered in aluminous cements is the "ternary" phase. Its composition lies on the join extending between the compositions: $3\,CaO{\cdot}Fe_2O_3$–$3\,CaO{\cdot}Al_2O_3$. Both end members are hypothetical, in the sense that the end members do not exist as stable phases. However, a range of solid solutions of intermediate Al/Fe ratios are thermodynamically stable to liquidus temperatures. This phase was first recognized by Tavasci (1937) and subsequently described more fully by Lister and Glasser (1967).

The products of the equilibrium crystallization of normal aluminous cement compositions may be predicted from the subsolidus diagram, Fig. 2b. However, it is likely that the cooling of compositions that were originally liquid will be sufficiently rapid so that nonequilibrium conditions will obtain. The fractional crystallization path of one typical composition, designated Z, is shown in Fig. 2c. The first crystalline phase to appear from the liquid is a dicalcium ferrite solid solution having the composition represented by point F. With falling temperature, the liquid composition shifts along a curved path. This path is outlined by the appropriate fractionation curve which is shown in Fig. 2c as a dot–dash curve Z–L_1. When the liquid composition intersects the univariant boundary curve at L_1, the ferrite solid solution that is crystallizing from this liquid has the composition F_1. The intersection of this fractionation curve with the univariant boundary curve occurs at about 1300°C: at this temperature a trace of aluminate solid solution whose composition is given by m_1 appears: this solid solution coexists with liquid L_1 and a ferrite, F_1, giving rise to a three-phase triangle having as vertices: F_1, L_1 and m_1. As heat is withdrawn the temperature continues to fall and, because all three of the condensed phases have variable compositions, the position of the three-phase triangle continues to shift (although the apex which indicates the composition of the aluminate solid solution changes position only slightly), until at 1190°C and maximum heat content, the eutectic liquid, L_3, is now crystallizing out solid phases having the compositions m_3 and F_3. At this temperature, the remaining liquid crystallizes yielding monocalcium ferrite as an additional crystalline phase.

The products that are obtained by equilibrium and fractional crystallization of the same composition may be contrasted as follows; in the latter case, liquid will generally be present to a much lower temperature: moreover, complete crystallization may yield an extra phase or phases which would not be present in the former case. In this context, it should be noted that the nature of the second or third crystalline phase to appear during fractional crystallization is very sensitive to small shifts in the position of the bulk composition. Examination of casts of commercial normal aluminous cements shows that considerable fractionation does in fact occur during the cooling stage. The crystals of the early formed phases, such as CA_{ss}, are either internally zoned or exhibit progressive compositional variation from one crystal to another, the latter-formed crystals being higher in Fe^{3+}. Another general effect of fractionation is to increase the potential quantity of iron-rich liquid available for crystallization at an invariant point; at equilibrium, most aluminous cement compositions would freeze completely before reaching an invariant point. The invariant points appropriate to the final freezing of fractionally crystallized aluminous cements cluster close together; all these points lie at relatively high ($CaO + Fe_2O_3$) contents and low temperatures (1180°–1265°C). These liquids, equilibrated in an air atmosphere, contain essentially all ferric iron; analyses give Fe^{3+}/Fe^{2+} ratios of approximately 20:1. The latter stages of the fractional crystallization of aluminous cement compositions would not be predicted accurately by this diagram owing, in part, to the much lower Fe^{3+}/Fe^{2+} ratios that are characteristic of aluminous cement liquids. In practice, the Fe^{2+} tends to accumulate in the residual liquid. Additional phase-equilibrium data on the quaternary system $CaO–Al_2O_3–FeO–Fe_2O_3$ obtained at lower partial oxygen pressures would, therefore, be helpful in predicting the effects of iron as Fe^{2+} rather than Fe^{3+}.

Fig. 3. Phase diagrams for the system $CaO–Al_2O_3–SiO_2$. (a) The liquidus surface (after Muan and Osborn, 1960). (b) Projected bulk composition of typical portland cement clinkers: one composition, designated X, is discussed in the text. The range of high alumina cement compositions includes the shaded area adjacent to the $CaO{\cdot}2\,Al_2O_3$ composition, and the range of aluminous cement compositions, recalculated to 100%, is shown as the shaded area adjacent to $CaO{\cdot}Al_2O_3$. The composition A is discussed in the text. (c) The 1500°C isothermal section of the $CaO–Al_2O_3–SiO_2$ system. The composition X also shown in Fig. 3b and discussed in the text is shown. At 1500°C the liquid composition which is in equilibrium with $3\,CaO{\cdot}SiO_2$ and $2\,CaO{\cdot}SiO_2$ is shown by L_X.

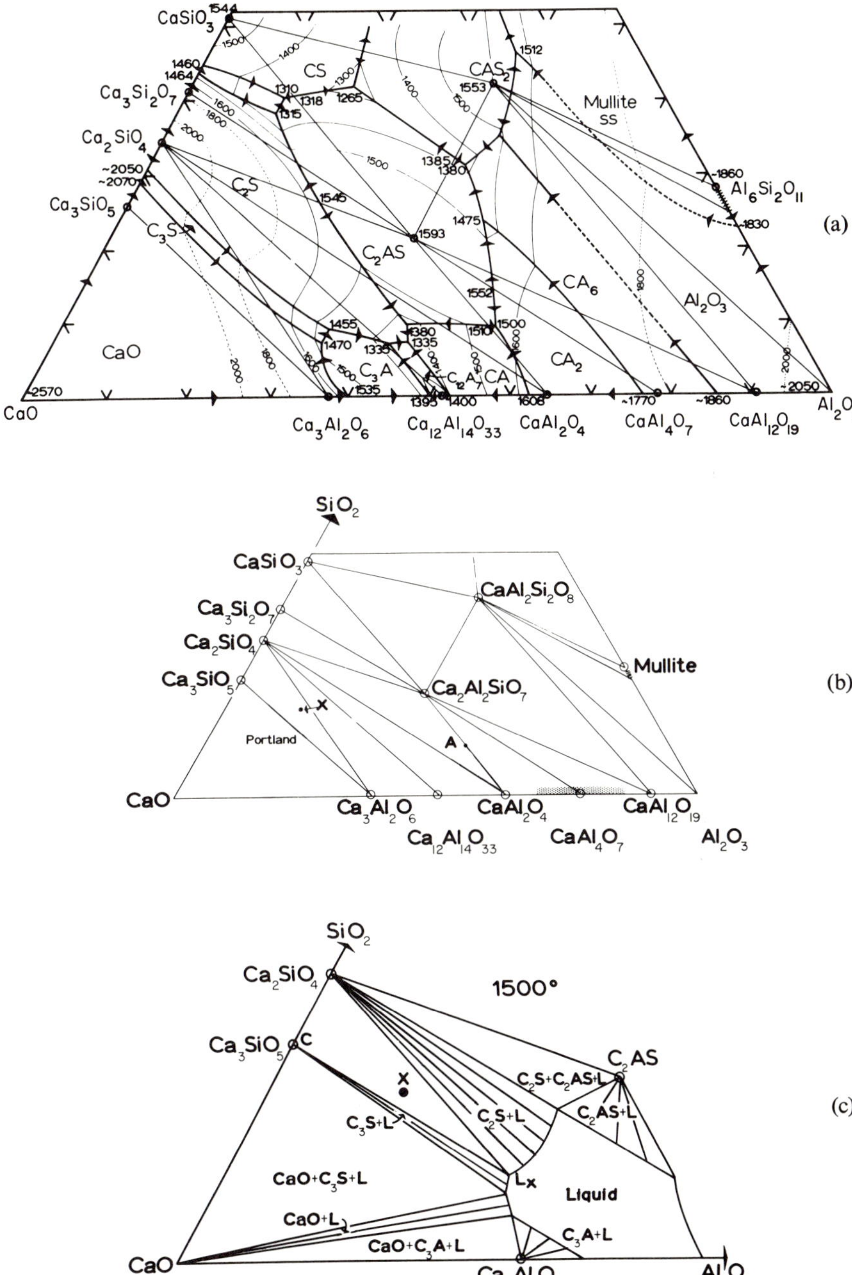

(a)
CaSiO3
1544
Ca3Si2O7
Ca2SiO4
Ca3SiO5
CS
CAS2
Mullite ss
C2S
C3S
C2AS
CA6
Al2O3
CaO
C3A
C12A7
CA
CA2
Al6Si2O11
CaO
Ca3Al2O6
Ca12Al14O33
CaAl2O4
CaAl4O7
CaAl12O19
Al2O3
~2570
~2050
~2070
1460
1464
1310
1318
1265
1315
1512
1553
1385
1380
~1860
~1830
1545
1593
1475
1552
1455
1470
1380
1335
1335
1510
1500
1535
1395
1400
1608
~1770
~1860
~2050
(b)
SiO2
CaSiO3
Ca3Si2O7
Ca2SiO4
Ca3SiO5
CaAl2Si2O8
Mullite
Ca2Al2SiO7
X
Portland
A
CaO
Ca3Al2O6
Ca12Al14O33
CaAl2O4
CaAl4O7
CaAl12O19
Al2O3
(c)
SiO2
Ca2SiO4
1500°
Ca3SiO5
C
C2AS
X
C2S+C2AS+L
C2AS+L
C2S+L
C3S+L
CaO+C3S+L
Lx
Liquid
CaO+L
C3A+L
CaO+C3A+L
CaO
Ca3Al2O6
Al2O3

FIG. 3.

D. The $CaO–Al_2O_3–SiO_2$ System

This system is shown in Fig. 3. As aluminous cements contain only low percentages of SiO_2, we will examine only the low-silica portion of this system and reserve general discussion of the phase relations until Section III, on portland cements. It is necessary to keep the silica contents of aluminous cements low. The need for this arises as follows: the diagram shows that a join exists between $CaO{\cdot}Al_2O_3$ and a ternary phase, gehlenite, $2\,CaO{\cdot}Al_2O_3{\cdot}SiO_2$. Gehlenite is one of the principal crystalline silica-containing phases in normal aluminous cements. Note that SiO_2 does not form solid solutions with any of the calcium aluminates, therefore no SiO_2 can be absorbed in the aluminates by a solid solution mechanism. This lack of solid solution, coupled with the low silica content of gehlenite itself, would cause a rapid rise in the proportion of gehlenite present in the cement as the SiO_2 content of the bulk composition rises. As an example, take the composition A, Fig. 3b, which lies at the intersection of the 10 wt % SiO_2 isopleth with the join extending from gehlenite to monocalcium aluminate. At subsolidus temperatures, the equilibrium phase assemblage would contain almost 50% gehlenite. Because gehlenite has poor hydraulic properties, this phase composition would produce an unsatisfactory cement, and the usual commercial practise is therefore to limit the SiO_2 content to much less than 10%; this is shown by the location of the stippled region marking the normal range of compositions. The bulk CaO/Al_2O_3 ratio could also be changed to avoid the formation of gehlenite. Thus, if it were necessary to use raw materials having higher silica contents, the CaO/Al_2O_3 ratio might be adjusted so that the bulk composition would be shifted along the same isopleth to its intersection with the join C_2S–CA. This would avoid gehlenite formation, but the resulting C_2S, while more hydraulic than C_2AS, also has a long setting time compared with that of the monoaluminate. Thus, formation of an excessive quantity of a crystalline silicate phase is not, in general, desirable in aluminous cements. Yet another possibility is that part of the SiO_2 present in the bulk composition may appear, not as a crystalline silicate phase, but as an aluminosilicate glass. The glass-forming properties of calcium aluminate liquids have been mentioned: silica enhances the resistance of these aluminate glasses to devitrification. It is probable that some liquid survives as a glass in normal aluminous cements, particularly in those zones of the casts that have been chilled most rapidly. Microscopic observation of this glass is handicapped both because of its dark color (the glass also contains considerable iron oxide) and also because it tends to be interstitial to the crystalline phases. Therefore, it is uncertain as to what proportion of the liquid is preserved as a glass. In any event, the devitrification of higher silica liquids also has other undesirable side effects. One phase that may thus be produced in quantity is the "pleochroic" phase (*vide infra*). Pleochroite is an undesirable constituent of the clinker. For all these reasons, the silica content of the cements is generally kept as low as possible.

E. The CaO–Al_2O_3–TiO_2 System

Equilibrium relations in this system may be used to show the application of phase diagrams to predicting the effects of minor constituents on the phase compositions of aluminous cements. Depending on the source of the bauxite used in aluminous cement production, considerable TiO_2 may be present. Phase relations in this system are shown in Fig. 4. The region of interest to aluminous cements is in compositions that lie close to the CaO–Al_2O_3 binary edge. Titanium is very insoluble in calcium aluminate liquids,

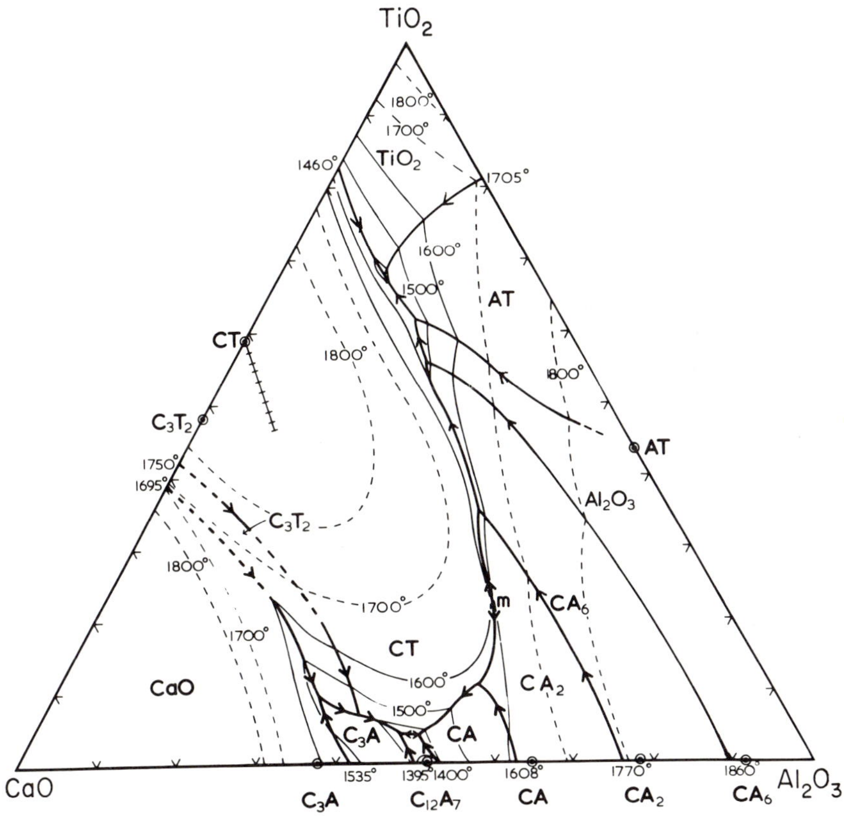

Fig. 4. Phase diagrams for the system CaO–Al_2O_3–TiO_2.

crystallizing out as perovskite (ideally, $CaTiO_3$). The pervoskite primary phase field extends to within a few weight percent of the $CaO–Al_2O_3$ binary edge, particularly in the vicinity of the $12\ CaO{\cdot}7\ Al_2O_3$ primary phase field. Upon entering the pervoskite primary phase field, liquidus temperatures rise very rapidly. Although no Ti^{4+} could be substituted in the calcium aluminates, it was found that the perovskite phase could take considerable Al^{3+} into solid solution; this is shown in Fig. 4 by the crosshatched line, whose length indicates the range of perovskite solid solutions. At subsolidus temperatures, perovskite is compatible with most of the calcium aluminates: the high lattice energy of the perovskite phase stabilizes phase assemblages containing perovskite in preference to other, alternative assemblages. Thus it would be predicted that all calcium aluminate cements would probably contain TiO_2 as perovskite. Perovskite has been reported as occurring in commercial normal aluminous cements containing more than a few percent TiO_2.

F. The Pleochroite Phase

Another interesting phase which sometimes occurs in aluminous cements is the "pleochroite" phase. The name is derived from its characteristic microscopic appearance; it occurs as bundles of needles or subparallel fibrous aggregates. These have parallel extinction and usually exhibit marked pleochroisim. The deepest pleochroic color in thin section is usually blue or violet. The reasons for the formation of the pleochroic phase are uncertain and its composition is known only approximately. Chemically, its appearance in normal aluminous cements has been correlated with two factors: high $(CaO + SiO_2)$ contents, and high FeO/Fe_2O_3 ratios. The appearance of this phase is of considerable commercial importance. Pleochroite has poor hydraulic properties: moreover, when it appears, the quantity of monocalcium aluminate present is very markedly reduced. Thus pleochroite-containing clinkers are associated with cements that have low set strengths and long setting times. Two suggestions have been made concerning the composition and origin of pleochroite that are of interest from the phase-rule standpoint. First, Rankin and Wright (1915), in their study of the $CaO–Al_2O_3–SiO_2$ system, noted that certain $CaO–Al_2O_3$ liquids sometimes gave a fibrous phase, which was formed during quenching runs. This phase has the composition $5\ CaO{\cdot}3\ Al_2O_3$; more recently, it was studied by Aruja (1957), who confirmed that this composition was correct. Its crystallography is similar to that of pleochroite extracted from aluminous cements. One suggestion, therefore, is that pleochroite is a thermodynamically metastable phase formed by partial devitrification of the liquid. In iron-containing aluminous cements, pleochroite would presumably occur as an iron-bearing solid solution. The other suggestion was first made by Parker (1952), who thought that a thermodynamically stable quaternary compound existed in the system $CaO–MgO–Al_2O_3–SiO_2$. This quaternary phase was assigned the approximate formula $6\ CaO{\cdot}4\ Al_2O_3{\cdot}MgO{\cdot}SiO_2$. Pleochroite was thought to be the ferrous–ferric iron-substituted solid solution based on the quaternary phase; it would thus have the formula $6\ CaO{\cdot}4(Al,Fe^{3+})_2O_3.(Mg,Fe^{2+})O{\cdot}SiO_2$. It is not now believed that a quaternary compound exists in the $CaO{\cdot}MgO{\cdot}Al_2O_3{\cdot}SiO_2$ system: certainly, intensive phase-equilibrium studies of this quaternary system have failed to disclose its existence. However, the picture has been complicated by the discovery of two new ternary phases in the system $CaO–MgO–Al_2O_3$ (Welch, 1964). These phases have the approximate compositions $7\ CaO{\cdot}MgO{\cdot}5\ Al_2O_3$ and $3\ CaO{\cdot}MgO{\cdot}2\ Al_2O_3$, and they appear to have a definite range of thermodynamic stability in the $CaO–MgO–Al_2O_3$ system. The crystallography of these phases shows that a simple relation exists between the unit-cell dimensions of these $CaO–MgO–SiO_2$ phases, as well as those of both unstable $5\ CaO{\cdot}3\ Al_2O_3$ and to pleochroite. Thus, although the "quaternary phase" is discredited, it is still possible that pleochroite is essentially identical to, or closely related to, some compound that is thermodynamically stable. Presently, it is not known if the crystallographic relationships outlined above are fortuitous or if they are indicative of a structural similarity. Further studies on the nature of pleochroite itself and on the $CaO–Al_2O_3$–iron oxide–SiO_2 system may help in fixing the composition and range of thermodynamic stability, if any, of this important phase.

III. PORTLAND CEMENTS

A. Preparation of the Cement

The chemical and phase compositions of anhydrous portland cements are somewhat more complex than those of the aluminous cements described in the previous section. Portland cements contain CaO, Al_2O_3, SiO_2, and Fe_2O_3 as major chemical components: i.e., those components that are present to the extent of 5 mole % or more. In addition, substantial quantities of Na_2O, K_2O, MgO, SO_4^{2-}, and PO_4^{3-} may be present. Before proceeding to examine the relevant phase diagrams, it would be appropriate to review briefly the production of portland cements in order to see what type of information might usefully be obtained from the phase diagrams.

The first stage in the manufacturing process is the selection of raw materials to give the correct bulk composition. Often, no single deposit or quarry will yield materials having the desired composition, and several types of rock, such as limestone and argillaceous shale, must be ground and intimately mixed. Next, the feed is introduced into the cool end of a gently inclined shaft kiln. The entire kiln is rotated slowly so that the feed moves on down the incline toward the hot end, or "burning zone," of the kiln. As the feed heats up, a complex series of reactions occurs. Those that take place at comparatively low temperatures, such as the loss of CO_2 from the carbonates, loss of H_2O from the clay minerals, etc., will not be discussed further. In the hot zone, a temperature of 1350°–1500°C is maintained. This temperature is always kept sufficiently high to allow a liquid phase to develop. The liquid phase hastens reaction, because material transport proceeds more rapidly through a liquid phase than would be the case in an all-solid-state reaction. Therefore, rather thorough reaction and recrystallization of the starting materials takes place in the burning zone. The temperature of this zone and the duration of the "burn" are chosen carefully. On the one hand, the temperature must not be allowed to rise to the point where the cement becomes fluid, or the resulting melt will attack the refractory lining of the kiln. On the other hand, the time and temperature conditions must permit a reasonably close approach to the equilibrium distribution of phases, and this state must be achieved rapidly enough to permit reasonable fuel economy. A usual compromise is thus to keep the proportion of liquid relatively low, and the physical texture of the resulting fired product is aptly described as "clinker." The clinker is a thoroughly indurated mass of crystalline phases that are held together by a ceramic bond between adjacent crystals as well as by an interstitial liquid phase.

The clinker is discharged continuously from the hot zone of the kiln and is chilled in an air blast. Finally, the cooled clinker is ground to produce cement powder. It is this cement powder that, when mixed with water and appropriate aggregates, hydrates to form cement or concrete. Thus the process is conveniently divided into two stages: first, manufacture of the anhydrous cement clinker and second, the hydration stage. Phase diagrams appropriate to each stage will now be considered, starting with the anhydrous clinker.

B. Phase Composition of the Clinker

The simplest phase assemblage found in normal clinkers usually comprises at least five or six phases: tricalcium silicate (C_3S); tricalcium aluminate (C_3A); an iron-containing "ferrite" phase, having the approximate composition C_4AF ($F = Fe_2O_3$), a silicate glass of variable composition; and possibly, free lime, CaO. In many low-iron-oxide cements, the sum of only three of the chemical components ($CaO + Al_2O_3 + SiO_2$) is equal to or greater than 90 wt % of the total. Therefore, it would be appropriate to consider the $CaO–Al_2O_3–SiO_2$ system first. One of the limiting binary systems, $CaO–Al_2O_3$, has already been discussed: to account for the quantitatively most important phases, we need also to examine the $CaO–SiO_2$ system.

C. The $CaO–SiO_2$ System

A phase diagram of the $CaO–CaSiO_3$ portion of the $CaO–SiO_2$ system is shown in Fig. 5. In the high-lime compositions, which are of interest to cement chemistry, two binary phases—Ca_3SiO_5 and Ca_2SiO_4—are encountered. The polymorphism of these compounds will be described in detail. At the liquidus, Ca_3SiO_5 melts incongruently to CaO and liquid at 2070°C;

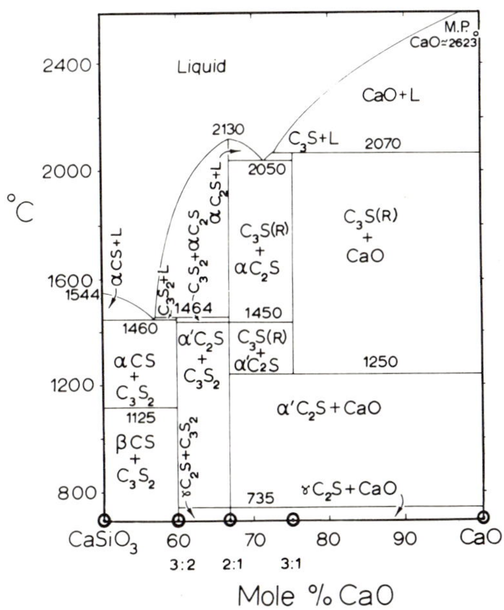

Fig. 5. Phase diagram for a portion of the system $CaO–SiO_2$. See text for a discussion of the polymorphism of C_3S (tricalcium silicate): $C_3S(R)$ indicates the rhombohedral phase.

Ca_2SiO_4 melts congruently at 2130°C. In more acid compositions two additional phases are encountered, $Ca_3Si_2O_7$ (rankinite) and $CaSiO_3$ (pseudowollastonite, or α-$CaSiO_3$). Rankinite melts incongruently to Ca_2SiO_4 and liquid at 1467°C; $CaSiO_3$ melts congruently at 1544°C. $CaSiO_3$ also has another polymorph (wollastonite, β-$CaSiO_3$), which is stable below approximately 1125°C. Rankinite and $CaSiO_3$ are not ordinarily encountered in portland cements.

1. Ca_2SiO_4

The polymorphism of Ca_2SiO_4 is shown in Fig. 6. Four crystalline phases are known: three of these, α, α', and γ-Ca_2SiO_4, have a range of thermodynamic stability at 1 bar pressure: the fourth phase, β-Ca_2SiO_4,

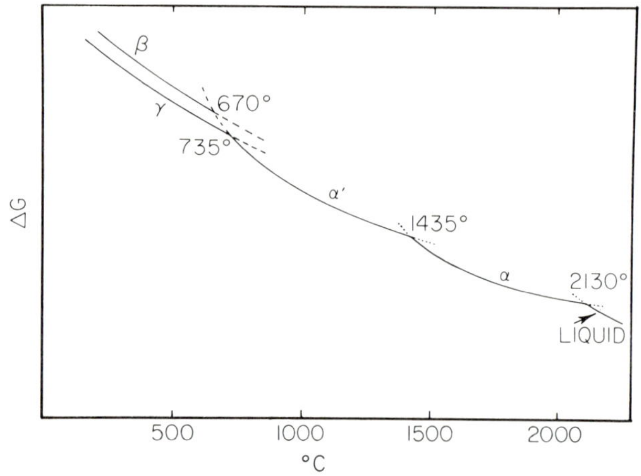

Fig. 6. Relative free energies of the polymorphs of Ca_2SiO_4 as a function of temperature.

is monotropic. The phase stable between 1435°C and the melting point is designated α-Ca_2SiO_4. This phase has a close structural relationship to the α' phase, which is stable over a range of temperatures below 1435°C. The inversion between the two phases is rapid and normally nonquenchable. With equilibrium cooling, α'-Ca_2SiO_4 inverts to γ-Ca_2SiO_4 at ~735°C. This inversion is, however, not rapid. The solid requires a good deal of structural reorganization to pass from the α' phase to the much more open, olivine-type, γ-Ca_2SiO_4 phase. Therefore, the α' phase tends to persist metastably below the equilibrium temperature: this possibility is shown in Fig. 6 by the metastable prolongation of the free-energy curve for the α' phase to temperatures below ~735°C. The metastable persistence of the α' phase does not continue to temperatures much below ~670°C, for at this temperature, another phase transition occurs, and the α'-Ca_2SiO_4 inverts to the monotropic β-Ca_2SiO_4. This reaction is characteristically rapid. Once β-Ca_2SiO_4 has formed, its conversion to the more stable γ phase is possible in theory, at any temperature; the use of a free-energy–temperature diagram shows this clearly, because both stable and metastable states can be represented. At any temperature, the stable phase in Fig. 6 is that which has the lowest free energy. However, when another component is added, e.g., CaO or SiO_2, only the lowest-free-energy portion of Fig. 6 would usually be shown: this can be seen by comparing Fig. 6 with the relevant portion of Fig. 4.

In practice, the kinetics of the $\beta \rightarrow \gamma$ reaction are rather sluggish, especially at ambient temperatures. Once initiated, however, the conversion of β- to γ-Ca_2SiO_4 tends to be autocatalytic. All polymorphic transitions must, in theory, be accompanied by changes attending the transformation of one phase to another. The volume change of the $\beta \rightarrow \gamma$ reaction is especially large. If the $\beta \rightarrow \gamma$ reaction occurs after the cement is partially hydrated, the resulting volume increase cannot be accommodated and the strength of the set cement is much reduced, often with spectacular results. Several solutions have been proposed for this serious technological problem. First, modern portland cements are often proportioned so as to contain a high ratio of C_3S/C_2S. One solution is thus to reduce greatly the quantity of Ca_2SiO_4 that is present, or even eliminate it. Its complete elimination is often economically not feasible or necessary. It might seem that, if Ca_2SiO_4 must be present, it would be more desirable to have it present as the γ phase. This is not the case, as the β phase is more hydraulic (e.g., hydrates more readily to form a strong bond with the aggregate) than is the γ phase. Provided that ways can be found to dependably prevent or inhibit the $\beta \rightarrow \gamma$ inversion, β-Ca_2SiO_4 may be a relatively desirable constituent. A number of factors that cannot be represented on a phase diagram are important in influencing the kinetics of the $\beta \rightarrow \gamma$ reaction. For example, the previous thermal and mechanical history of the Ca_2SiO_4 has an important bearing on the low-temperature sequence of inversions. Crystals that have passed through the $\alpha \rightarrow \alpha'$ inversion tend to convert more readily to γ-Ca_2SiO_4 upon cooling than those that have never been taken above the α–α' inversion temperature. Slow clinker cooling rates, particularly in the 600°–750°C range, favor the attainment of equilibrium, and hence formation of γ-Ca_2SiO_4.

The discussion has thus far been confined to pure Ca_2SiO_4. However, impurities also play a major role in shaping the sequence of low temperature inversions. A wide range of oxides are capable of forming at least a limited range of solid solutions with Ca_2SiO_4 at clinkering temperatures. Oxides whose solubility in the α or α' phases is equal to, or exceeds 0.5 mole %, include Na_2O, K_2O, MgO, FeO, Fe_2O_3, Al_2O_3, Mn_2O_3, Cr_2O_3, CrO_4^{3-}, VO_4^{3-}, and PO_4^{3-}. The temperature of the α–α' inversion is very sensitive to the presence of these oxides; in fact, Newman and Wells (1946) have used the changes in the α–α' inversion temperature with changing bulk composition to measure the equilibrium solubility of each of these oxides in the α-Ca_2SiO_4 phases. The effects of various substituents may be broadly classified as follows. First, there are substituents that form only a very limited range of solid solutions with both the α and α' phases. This case is shown schematically in Fig. 7a. Substituents in this class include Al^{3+}, Fe^{3+}, and Mn^{3+}.

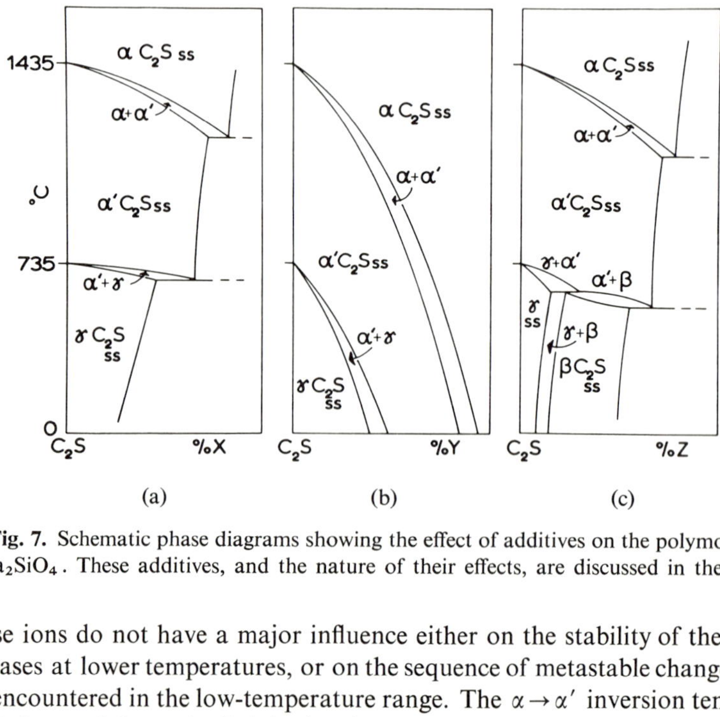

Fig. 7. Schematic phase diagrams showing the effect of additives on the polymorphism of Ca_2SiO_4. These additives, and the nature of their effects, are discussed in the text.

These ions do not have a major influence either on the stability of the β and γ phases at lower temperatures, or on the sequence of metastable changes that are encountered in the low-temperature range. The $\alpha \rightarrow \alpha'$ inversion temperature is lowered, but only slightly, by these substituents: the quantitative extent of solid solution is, at most, 1–2 mole % in the temperature range of interest, about 1430°C. Second, there are those substituents that form extensive solid solutions with either, or both, α- and α'-Ca_2SiO_4. The phosphorous-substituted Ca_2SiO_4 solid solutions are an example. The effect of this type of solubility is shown schematically in Fig. 7b: in all cases that have been studied, the solubility of the added ion is greater in the high form than in the low. This causes the inversion temperature to fall with increasing solid solution. The two phase region: (α + α' solid solutions) sweeps downwards, and the $\alpha' \rightarrow \gamma$

inversion temperature (which likewise becomes a two phase region) is lowered below room temperature. This means that the metastable $\alpha' \rightarrow \beta$ inversion temperature is probably depressed as well, and that only α or α', or a mixture of α and α' phases are likely to be encountered, provided that a sufficient quantity of substituent has been used. In practice the necessary quantity is only a few mole percent P_2O_5. Last are those substituents whose effects are not fully understood. B_2O_3 is an example of such a substituent. Ca_2SiO_4 solid solutions containing B^{3+}, even in amounts as small as 0.2–0.5 mole %, always give β-Ca_2SiO_4 upon either rapid quenching or slow cooling to room temperature. It is not certain if the B^{3+} merely prevents the β phases from converting to the γ phase, or if the relative order of stability of the β and γ phases are reversed by B^{3+} substitution. In the former case, γ-Ca_2SiO_4 would still be the thermodynamically stable phase at low temperatures, and the barriers to actually obtaining it during the cooling stage would be purely kinetic. However, in the latter case, which is shown schematically in Fig. 7c, the relative stabilities of β and γ phases are shown as having been reversed by increasing substitution of B^{3+}. It is not known which of these two possibilities is the correct one.

2. Ca_3SiO_5

Tricalcium silicate is stable from its incongruent melting at 2070°C to approximately 1250°C. The existence of this lower-temperature stability limit is perhaps an unusual feature, but one that is appearing in many oxide systems as the low-temperature phase relations are studied more thoroughly. Figure 5 shows that the products of dissociation of Ca_3SiO_5 are CaO and Ca_3SiO_5. The rate of the dissociation reaction is comparatively slow, even at these high temperatures, so that even moderately rapid cooling of industrial clinkers is sufficient to prevent dissociation of the Ca_3SiO_5 present. If clinkers are left to cool slowly, the Ca_3SiO_5 may begin to decompose: this trouble has occasionally been encountered in industrial products. Upon cooling below 1250°C, Ca_3SiO_5 is thus thermodynamically metastable. A number of polymorphic transformations are encountered during the cooling: these cannot be represented on the equilibrium diagram, but may be conveniently represented on a free-energy-versus-temperature plot, similar to that which was used to represent the sequence of phase changes in Ca_2SiO_4 (Fig. 5). These data for Ca_3SiO_5 are shown in Fig. 8, which is based largely on the recent study of Bigaré *et al.* (1967). The form stable at high temperatures, designated (R), has rhombohedral symmetry: basically, the polymorphic inversions encountered at lower temperatures are caused by structural distortions of the high-symmetry R phase. With decreasing temperatures, two monoclinic phases, designated M_{II} and M_{I}, are encountered, and

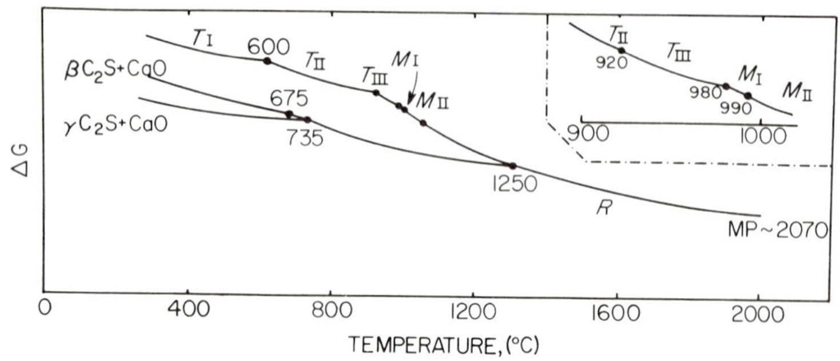

Fig. 8. Relative free energies of the polymorphs of Ca_3SiO_5 as a function of temperature.

then three triclinic phases: T_{III}, T_{II}, and T_I. For pure Ca_3SiO_5, these inversions are all rapidly reversible both on heating and cooling, and hence cannot be quenched. They must be detected by some direct method: in the study cited above both high-temperature powder x-ray photographs and D.T.A. were used. It is interesting to note that not all the phase transformations were detectable by *both* x-rays and DTA. The enthalpy change accompanying the $M_{II} \rightleftarrows R$ inversion is too small to be detected by DTA; on the other hand, the $T_{II} \rightleftarrows T_{III}$ transformation has a reasonably large enthalpy change, but the existence of the transformation could not be detected in high-temperature powder photographs. Small amounts, about 1–2% of alkali, MgO and Al_2O_3, have a marked influence on the sequence of polymorphic inversions; the experimental evidence is reviewed by Bigaré *et al.* (1967).

D. The $CaO–Al_2O_3–SiO_2$ System

The low-silica regions of the $CaO–Al_2O_3–SiO_2$ system are shown in Fig. 3. Insets to the diagram show the range of typical anhydrous cement compositions (center) and an isothermal section through the system at 1500°C (bottom). If the disposition of typical portland cement compositions is examined with respect to the subsolidus compatiability triangles, it can be seen that the range of compositions appropriate to portland cement production lies largely within the three-phase triangle that has as apices: C_3S, C_2S, and C_3A. Assuming, therefore, that the clinker is approximately at equilibrium during the cooling stage, the coexistence of the quantitatively most important phases would be accounted for. It is, however, advisable to follow a given composition through a thermal cycle simulating the clinkering process and predict the sequence of phase changes that would be encountered. A composition representative of a portland cement and designated X, is shown on the 1500°C isothermal section. At this temperature, it lies within the triangle: C_2S–C_3S–liquid. The liquid composition is fixed at L_X, and the liquid is thus enriched in Al_2O_3 relative to the bulk composition. If desired, the relative proportions of the three phases could be found by an appropriate geometrical construction. If the phase assemblage present at 1500°C is cooled sufficiently rapidly to room temperature, the phase distribution present at the higher temperature may be preserved for subsequent examination. The liquid phase will generally be found to have been preserved as a glass. However, these low-silica glasses are not very persistent, and the glass may crystallize readily during the quenching stage. Therefore, another case that may be treated is one where the liquid originally present at the high temperature crystallizes, but where the crystallizing liquid fails to react with the crystals already present. This corresponds to fractional crystallization. A simple analogy in a ternary aqueous salt system would be as follows: first, equilibrate a bulk composition under such conditions that an aqueous phase is in equilibrium with two crystalline phases. Centrifuge off the aqueous solution, and again crystallize the solution phase in part or in whole as, for example, by lowering the temperature. The process is easily visualized in an aqueous system of this sort, where an actual physical separation of the aqueous phase is made, but in the CaO–Al_2O_3–SiO_2 systems and indeed, in many other silicate systems, it is not necessary to make a physical separation to have fractionation processes operative. Fractionation processes become important because the rate-limiting step in the attainment of equilibrium is usually the rate at which material transport can occur across a crystal–liquid interface. Because this transport occurs only slowly, the crystalline phases are effectively prevented from reacting either with each other or with the liquid. As an example of this, consider the crystallization of the liquid L_X. This is the liquid composition that was generated in the previous example by partial fusion of a typical clinker. If the liquid crystallizes independently; that is, without reaction with the crystalline phases present at the clinkering temperature, the liquid phase will yield an assemblage containing C_2S, C_3A, and $C_{12}A_7$. This is because the liquid composition, unlike the bulk composition from which it originated, lies in the composition triangle C_2S–C_3A–$C_{12}A_7$. Thus, if we assume that the clinker composition designated X in Fig. 3c is first equilibrated at 1500°C, producing C_3S, C_2S, and liquid, and that in the subsequent cooling cycle the liquid crystallizes independently, we might obtain as many as five phases at room temperature. These phases would be C_3S, C_2S, and some residual glass, as well as the products of partially crystallizing the glass: namely, more C_3S, C_2S, and as the fifth phase, $C_{12}A_7$. Textural studies are often helpful in inferring the order of crystallization of phases and these textures become easier to interpret when the thermal history of the specimen is known and when the appropriate phase-equilibrium data are also known. In following crystallization paths in clinker compositions, the important role of the thermal history in generating complex phase assemblages, containing as many or indeed, sometimes more than the maximum permitted by the phase rule should not be overlooked.

E. The CaO–MgO–Al_2O_3–SiO_2 System

This system is one of the more completely studied quaternary silicate systems. As may be appreciated, the system represents not only a complex experimental problem, but also, it becomes more difficult to depict the data that have been obtained. Two main schemes of subdividing the experimental studies have been used. The first is to locate the composition planes within the quaternary system that represent ternary systems. A number of these subsystems are found that cut through the quaternary volume. An example of such a plane is the ternary system $CaMgSi_2O_6$–$CaAl_2Si_2O_8$–SiO_2 (diopside–anorthite–silica), which was studied by Osborn and Tait (1952). Although this system contains essential quantities of all four oxides, it is still a ternary system, and this simplifies the interpretation and presentation of results. Another method that has been used is to study planes of constant composition of one of the components. These planes have the appearance of ternary systems but, because the content of one component is arbitrarily fixed, these planes are isopleths and are not true ternary systems. Thus, neither crystallization paths nor the temperature or location of invariant points may, in general, be predicted from any one section. The sections may be used to give liquidus temperatures and primary phases. However, a series of these planes may be used to predict the location, nature, and temperature of invariant points whose compositions lie between any two planes. This type of analysis is facilitated if the series of planes are taken parallel to each other (e.g., differ by a fixed percentage of the same component). The scheme of invariant points may also be summarized in skeletal form; provided that one is not required to show the geometrical relationships between the invariant points accurately, only two dimensions are required for such a summary. For example, the scheme of phase relations in the Ca_2SiO_4–12 $CaO \cdot 7\ Al_2O_3$–CaO corner of the CaO–Al_2O_3–SiO_2 system can be represented with geometric accuracy as in Fig. 3a or schematically, as scheme (I) where E indicates a eutectic, P a peritectic, C a congruent melting point and $\rightarrow$ indicates the direction of falling temperature. These studies of isoplethal planes and true ternary subsystems are complimentary, a fact that is often overlooked. In a quaternary system, the intersection of a ternary plane with an arbitrarily chosen plane will generate a line element in common between the two planes; this will represent a series of compositions

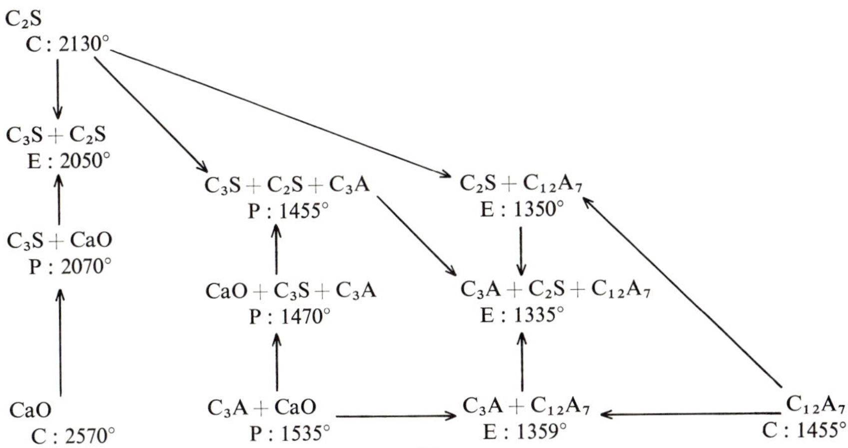

(I)

that are common to the two interesting planes. All features plotted on the one section, for example, liquidus isotherms, boundary curves, isofracts, etc., must meet the corresponding feature plotted on the other plane without discontinuity at their mutual intersection.

The experimental studies of the more basic parts of the CaO–MgO–Al_2O_3–SiO_2 system show that the MgO phase has a very large primary phase volume. The liquid phase that is present in portland cement clinkers may, therefore, be expected to become saturated in MgO even at comparatively low MgO contents. This is important for two reasons: first, because it may be difficult to keep the MgO content of raw materials low and second, because the presence of free MgO in the clinker is undesirable. Any free MgO that is present in the clinker tends to hydrate only after the cement is set and this hydration causes an undesirable volume expansion. McMurdie and Insley (1936) provide important phase-equilibrium data bearing on this problem: they determined the position of the divariant surface representing the limits of the MgO primary phase volume in the high-CaO portions of the CaO–MgO–Al_2O_3–SiO_2 system. Because the position of this geometrically complex surface was approximately parallel to the CaO–Al_2O_3–SiO_2 face of the composition tetrahedra, they chose the method of plotting that is shown in Fig. 9. In this case the contours represent not isotherms, but the weight percentages of MgO that are necessary to saturate the liquids in periclase (MgO), as well as in the phase that is shown on the diagram. Thus, the liquid composition in equilibrium with CaO, C_3A, C_2S, and MgO would contain slightly less than 6% MgO. The solubility of MgO in the liquid phase that is present in commercial clinker is, of course, influenced by the concentration of other components (principally

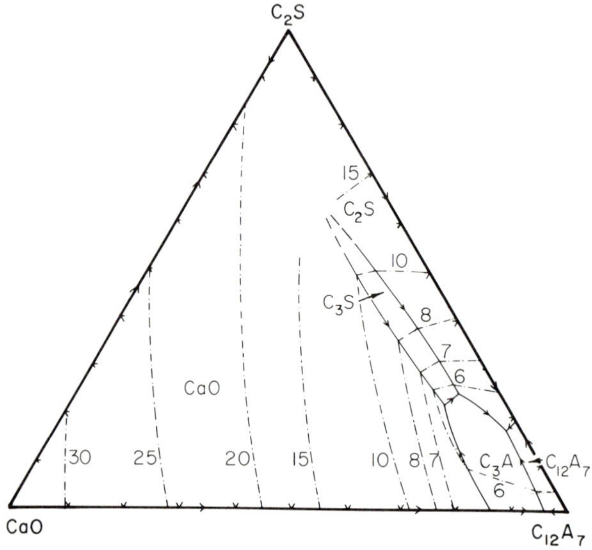

Fig. 9. Phase diagram for part of the system CaO–MgO–Al_2O_3–SiO_2. Projection of the MgO saturation contours on the CaO–Al_2O_3–SiO_2 face. These contours are marked in dot-dash lines. Percentages are in weight percent MgO.

alkali) in the liquid, as well as by solid solution of some Mg^{2+} in the crystalline phases, e.g., in C_3S. However, the general conclusion that would be made from these data—namely, that relatively small MgO contents would result in the appearance of free periclase in the clinker—is confirmed in practice.

Other types of representation have also been used to demonstrate the application of these phase-equilibrium data to cements. Thus, the primary phase volume of C_3S in the CaO–MgO–Al_2O_3–SiO_2 system has been mapped. This volume extends into the quaternary system as an irregular wedge-shaped volume; its shape and disposition have an important bearing on the relative proportions of crystalline phases that are at equilibrium with liquid during the clinkering stage and upon the paths of crystallization, either equilibrium or fractional, of the liquid phase.

One remaining problem, for which phase-equilibrium studies have not yet yielded a definite answer, is the existence and stability of a quaternary phase in the CaO–MgO–Al_2O_3–SiO_2 system. Midgely (1968), reviewing earlier studies and providing new data, presents evidence that a quaternary phase exists in cement clinkers and that it has the formula (given in terms of the unit cell content): $Ca_{35}Al_{53}Mg_4Si_5O_{128}$. On the other hand, extensive phase-equilibrium studies in the $CaO–MgO–Al_2O_3–SiO_2$ system have not yet disclosed a stability region for this phase. This apparent contradiction may be tentatively explained in several ways: for example, it may be that this phase has a lower limit of stability, in which case it would not appear in the subsolidus quaternary phase volumes; alternatively it might be that it is metastable under all conditions and arises only as a metastable devitrification product of a range of liquid compositions.

F. The $CaO–Al_2O_3–Fe_2O_3–SiO_2$ System

One other system that assumes considerable importance with reference to the production of portland clinkers is the $CaO–Al_2O_3–Fe_2O_3–SiO_2$ system. References to the numerous experimental studies in this system will be found in Bogue (1955), Taylor (1964), and in other standard collections of phasc diagrams. The actual phase diagrams will not be reproduced here. There is, however, an additional method of treating phase-equilibrium data that is commonly used in cement chemistry. This is the "Bogue calculation." The object of the calculation is to calculate the phase composition of a cement clinker from its chemical composition. As will be appreciated, the quantitative determination of all the phases present in a complex mixture, such as a clinker, remains a difficult task, although many attempts have been made to devise a quantitative phase-analysis scheme. Because the hydration properties of a cement vary with its phase composition and because the chemical and phase compositions cannot be equated directly, calculations of this sort still have a useful place. One phase that is, however, readily determined quantitatively is free CaO. Therefore these calculations usually start not only with a bulk chemical analysis, but also with a free-lime determination. The calculation has been devised to take into account the phase-rule relationships insofar as they are known. The bulk composition of a normal portland cement, projected into the $CaO–MgO–Al_2O_3–SiO_2$ system, falls in the composition volume that has as its apices: $3\,CaO{\cdot}SiO_2$, $2\,CaO{\cdot}SiO_2$, $3\,CaO{\cdot}Al_2O_3$ and a ferrite solid solution having the approximate composition $4\,CaO{\cdot}Fe_2O_3{\cdot}Al_2O_3$. These are therefore selected as the "normative" phases. Any approach toward equilibrium at or just below the solidus temperature must result in the development of these phases. Nevertheless, the calculation need not assume that equilibrium has been attained. This is implicit in the first step of the calculation, for having converted the chemical analysis to mole percentage, the quantity of free lime (as found by direct analysis), is subtracted from the "total CaO." Next, each mole of Fe_2O_3 is combined with four moles of CaO and one of Al_2O_3 to form $4\,CaO{\cdot}Fe_2O_3{\cdot}Al_2O_3$. The remaining Al_2O_3 is recalculated as $3\,CaO{\cdot}Al_2O_3$. Now, the remaining CaO is combined with SiO_2 to form x moles of $3\,CaO{\cdot}SiO_2$ and y moles of $2\,CaO{\cdot}SiO_2$. The values of x and y are determined from the equations

$$3x + 2y = m\mathrm{CaO}$$

$$x + y = n\mathrm{SiO_2}$$

where m and n are the number of moles of CaO and SiO_2, respectively, in the analysis. Therefore two more useful relationships may be derived: first, that $x = m - 2n$ and second, that $y = 3n - m$.

Up to this point, the quantitative calculation of the phases present does not differ greatly in principle from the usual geometrical method of calculation except that departures from equilibrium (caused by incomplete reaction of the CaO) can be taken into account. However, the calculation may be further modified in two ways, first, to take into account other types of departure from the equilibrium phase compositions, and second, to allow for additional phases that might be formed by the minor components. As an example of how the calculations may be modified to allow for the minor components that appear in the analysis, take the case of SO_3 and MgO. Most SO_3-containing clinkers are observed to contain $CaSO_4$ as the main sulfur-bearing phase. Therefore the SO_3 in the analysis is usually combined with CaO, and both this CaO, which is combined as $CaSO_4$, as well as the free lime, are subtracted from the total CaO as the first stage in the recalculation. MgO is usually treated as MgO, rather than combining it. This is not strictly accurate from the standpoint of the known phase equilibrium relations, because as has been shown, Mg^{2+} is soluble to some extent both in the crystalline phases, especially Ca_2SiO_4 and Ca_3SiO_5, as well as in the glassy phase. However, as free MgO is deleterious to the cement, it is usually calculated in this way so as to set a maximum upper limit to the periclase content that could be present.

Departures from equilibrium are treated in various ways. One type of departure, the presence of free lime, has already been mentioned. Much of the free CaO that is usually found to be present is not at equilibrium with the bulk composition. This nonequilibrium lime content can arise in several ways; for example, during the initial firing, coarser lumps of $CaCO_3$ that have escaped grinding will decompose to yield free CaO, and this lime recrystallizes, becoming rather unreactive. The CaO tends to become coated with tricalcium aluminate during the clinkering stage; this tricalcium aluminate is itself in equilibrium with the liquid phase, but because the C_3A coats the CaO grains, it forms a barrier to further reaction between CaO and liquid. Only slow diffusion through this barrier allows the CaO to react further with either the liquid or with other crystalline phases. The CaO is thus said to be "protected" against reaction. In many compositions, it would be predicted that, at equilibrium, free CaO would be developed at clinkering temperatures but would subsequently disappear by peritectic reaction with liquid during the cooling stage. Again, this CaO tends to persist, and is often found to occur in the cooled clinker as a "protected" phase. Fortunately, the free-CaO test selectively determines all the CaO present as lime, irrespective of its origin or textural relationships. Another possibility is that the liquid phase that is present at clinkering temperatures may not crystallize, but may persist in whole or in part as a glass. The composition and proportion of this glass may be estimated, the former from a knowledge of the liquid composition (as derived from phase equilibrium studies) which is in equilibrium with the appropriate crystalline assemblage; the latter, from a knowledge of the bulk composition and the liquid composition. An actual physical separation of the glass phase is difficult, but the available analytical data suggest that the glass has a composition close to that which would be predicted from the relevant phase equilibria studies. These analyses of the glass compositions have been made either by direct high-temperature studies, where the liquid phase was centrifuged off from the coexisting crystalline phases, or by some indirect procedure. With the advent of electron-probe analyzers, it should be possible to accumulate much new information on the compositions of clinker liquids. This will be especially helpful in assessing the distribution of the minor constituents between coexisting crystalline and liquid phases. Alternatively, the proportion of liquid present (but not its composition) may be determined by petrographic thin-section methods. Once the proportion and composition of the liquid phase is known, the scheme of calculation that is employed may be further modified in various ways, to allow for total or partial independent crystallization of the glass phase.

IV. HYDRATION OF CEMENTS

A. The CaO–Al_2O_3–H_2O System

The calcium aluminates are only sparingly soluble in water. Consequently, when a calcium aluminate cement is placed in contact with sufficient water to ensure the presence of an aqueous phase, the hydration proceeds in two ways: first by solution of some of the anhydrous material and subsequent redeposition of solid hydrates, and second, by direct hydration of the solid phases. The former process is termed "solution hydration"; the latter, "paste hydration." Both processes play an important role in the hydration and setting of commercial cements. However, it is experimentally an easier task to study the equilibria through an examination of the solution-hydration processes.

In the presence of an excess of solid material, supersaturated aqueous solutions of the calcium aluminates are readily obtained. Solutions may be prepared that are supersaturated with respect to lime, to alumina, or to both lime and alumina. The subsequent behavior of these solutions plays an important role in understanding the hydration stages. Upon standing, a clear, supersaturated solution will eventually begin to crystallize out a hydrate phase. The hydrate phase that appears will not necessarily be the equilibrium phase and, in fact, both stable and metastable crystallization phenomena have been intensively studied in the $CaO–Al_2O_3–H_2O$ system. Each *crystalline* hydrate phase has a definite solubility curve; the position of this curve varies as a function of both temperature and solution composition. It is important to study these curves and locate their positions on an isothermal section and to identify the crystalline phase coexisting with solution along these curves for two reasons: first, because reaction may proceed so slowly that the true equilibrium phases can be identified only by determining which crystalline phase has, for a given temperature and CaO/Al_2O_3 ratio, the lowest solubility. Second, the nonequilibrium crystallization processes are important in their own right. Many of the metastable phases that have been encountered in laboratory solution hydration experiments are also encountered during the hydration of large masses of paste, as for example in the setting of a large concrete mass. Many of these metastable phases and phase assemblages will, once formed, persist apparently indefinitely. The kinetics of the transformation of the metastable states to the stable phase or phases are found to occur over a wide range of rates; in some cases, transformation rates are so slow that one might legitimately refer to many of the metastable solubility curves as quasiequilibrium solubility curves.

A series of these solubility curves are shown for the hydrate phases in the $CaO–Al_2O_3–H_2O$ system, in Fig. 10. These figures are not plotted on the triangular representation that has been used for anhydrous ternary systems. The reason for this is because the solids are only sparingly soluble in the solution phase, even allowing for supersaturation effects. Therefore, on the usual ternary representation the solution compositions would all cluster very close to the H_2O apex. With the method of representation used in Fig. 10, the range of solution compositions are spread out. The contours of solution compositions that have equal CaO/Al_2O_3 ratios are not shown on the diagram, but may be pictured as a series of straight lines radiating from the left-hand corner of the diagram. One other shortcoming of this type of diagram is that it does not allow the composition of the solid phases coexisting with liquid to be represented.

The significant solution compositions are those which are in univariant equilibrium with a crystalline phase; these solution compositions are usually determined in two ways. The equilibrium or quasiequilibrium may be

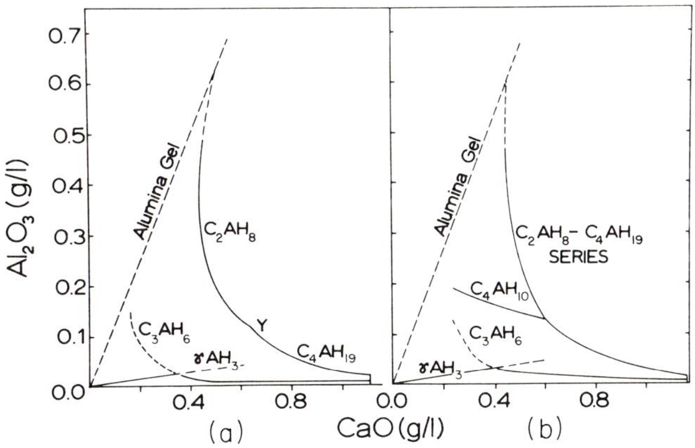

Fig. 10. Solubility curves in the $CaO–Al_2O_3–H_2O$ system at (a) 25°C, and (b) 21°C. An invariant point is placed at a point designated *Y* (25°C section). This is based on evidence that C_4AH_{19} and C_2AH_8 are two different phases. This feature is not shown at 21°C, because Jones (1959) believed that C_4AH_{19} and C_2AH_8 were end members of a solid solution series. See text for a discussion of this and other features of the diagram.

approached either from supersaturation, or from undersaturation, with respect to the same crystalline phase. In the former case, a supersaturated solution is first prepared and then allowed to crystallize at a constant temperature. Portions of the filtered liquid are analyzed from time to time until the solution composition remains constant. The analysis of the solution phase may be done using conventional chemical techniques. Usually, both CaO and Al_2O_3 are determined. In the latter case, where a point on the solubility curve is to be determined starting from undersaturation, distilled water is placed in contact with an excess of the appropriate crystalline phase, and the mixture is again held at the same constant temperature. Again, the solution composition is followed as a function of time, until no further changes are observed. The method presupposes that both the undersaturated and supersaturated solutions have, at the end of the experiment, reached an equilibrium or quasiequilibrium with the same crystalline phase; adequate means must therefore exist whereby the solid phase (or phases) can be characterized. Provided that these conditions are fulfilled, and that both solution compositions finish at the same CaO and Al_2O_3 concentrations, another point on a given saturated curve will have been located. Not only may other points be located on the same curve, but a whole family of these solubility curves can be obtained, each involving a univariant equilibrium between a given crystalline phase and a range of solution compositions. The family of equilibrium and quasiequilibrium solubility curves are shown in Fig. 10. The thermodynamically stable equilibrium may be found for any given CaO/Al_2O_3 ratio by projecting that ratio across the diagrams and noting its intersections with the various solubility curves. The curve with the lowest solubility represents the stable equilibrium for that ratio. Thus, at both 21° and 25°C, the only thermodynamically stable equilibria between solutions of all possible CaO/Al_2O_3 ratios and solid are those involving the following solid phases: alumina gel, γ-$Al_2O_3 \cdot 3\,H_2O$ and $3\,CaO \cdot Al_2O_3 \cdot 6\,H_2O$. The nature of the alumina gel "phase" deserves some comment: it is poorly crystallized, and may be almost colloidal in nature. It is not surprising, therefore, that its exact solubility varies markedly as a function of its crystallinity. The position of the solubility curve for alumina gel cannot be specified exactly; however, it is certainly shown schematically in its correct position relative to the other aluminate hydrates in both these diagrams. Phases such as the C_4AH_{10} hydrate that are much better crystallized than alumina gel and thus have better defined solubility curves are nevertheless thermodynamically metastable over the entire range of CaO/Al_2O_3 ratios, at both 21° and 25°C.

There are some differences between the two isothermal sections despite the rather small temperature interval between them. The phases that are shown as having the highest solubilities at both 21° and 25°C are C_2AH_8 and C_4AH_{19}. These two phases are structurally related, and it is not certain if they are best regarded as a single phase that has a variable composition or if they are, in fact, two discrete phases. The 21° and 25°C isothermal sections have been studied by different investigators (Jones, 1959; Percival and Taylor 1959), and their respective interpretations of the solubility data on C_2AH_8 and C_2AH_{19} have been shown. The solubility data in the case of the 21°C isothermal section are shown as falling on two separate curves, therby producing a quasi-invariant point between C_2AH_8, C_4AH_{19}, and liquid. If the solubility data were good enough to establish the existence of a break at the point in question, this would constitute sufficient proof that C_2AH_8 and C_4AH_{19} were separate phases. In fact, the accuracy of the local tion of the solubility curves is not good enough to prove the existence or nonexistence of this invariant point. The existence of one, or possibly two higher hydrates is thus seen as representing a difference in viewpoint as to what constitutes a "phase." Another, more important, difference between the two isothermal sections is the appearance of the C_4AH_{10} phase at 21°C, but not at 25°C. This is an example of a quasiequilibrium that can be followed over the temperature interval from 1° to 21°C, but which definitely terminates between 21° and 25°C. All investigators agree that the C_4AH_{10} phase is not formed at higher temperatures.

All the aluminate hydrate phases that are produced by paste hydration are also encountered in solution hydration. Thus, paste hydration does not generally produce any new phases, but the phases having a relatively low degree of hydration do tend to be quantitatively much more important.

Examples are also known where the higher hydrates (which tend to be formed via the solution phase that is present during the initial setting stages of a cement), slowly convert to lower hydrates. This process is sometimes observed to occur spontaneously. Production of the thermodynamically stable C_3AH_6 hydrogarnet phase tends to be favored by these conversion reactions involving the higher hydrates. The factors governing the kinetics of these solid-solid transformations in set cements are not fully understood. It is nevertheless an important concern because the conversion may give rise to textural and volumetric changes that may lead to some loss of strength in set cements. Further studies of these slow reactions would be desirable. These differences which have been noted between the products resulting from paste or solution hydration become less important at higher temperatures. Above the normal boiling point of the solution phase, hydration experiments must, of course, be carried out under pressure. The results of such studies are of more than academic interest, as cement products are increasingly set under pressure, e.g., as in "autoclaved" cement products, or else where set cements are used in higher-pressure higher-temperature service environments. In order to study the equilibria, it is necessary to construct special apparatus to sample the liquid phase present at temperatures above 100°C; ingenious filtration devices have been constructed to allow separation of the pressurized aqueous phase from the coexisting solids, in order that both phases may be analyzed separately. The solution phase may remain homogeneous, but the solids obtained are still wet with mother liquor, and hence contaminated. Provided that the solubility of the solids in the aqueous phase is sufficiently low that the bulk composition of the solid portion is not shifted significantly, it may not be necessary to have a total analysis of the solids in order to establish the end point of a tie line. With increasing temperatures, however, the solubility of the solids generally increases, and it may be desirable to either limit the total volume of solution phase or else to analyze both solid and liquid. The analysis of the solid phase(s) can be interpreted using the method of wet residues. This procedure is commonly used in salt-water systems where one of the phases is homogeneous and the other is a mixture consisting of solid plus solution. It is fortunate that at elevated temperatures, the differences between paste hydration and solution hydration tend to become less important and also, that the approach to equilibrium becomes more rapid.

Experimental studies on the $CaO–Al_2O_3–H_2O$ system have also been made at very much higher pressures, e.g., 10^3–10^4 bars. These data are usually presented in two ways. First, they may be represented by $P–T$ curves either for a given reaction or for a number of separate reactions. In this way, for example, the $P–T$ curve for the dissociation of $Ca(OH)_2$ to ($CaO+$ vapor) has been followed to the critical point at which $Ca(OH)_2$ melts at approximately 1000 bars pressure and 840°C (Harker *et al.*, 1962). An example of a *P–T* curve which is of more direct relevance to cement chemistry is the curve representing the dehydration of the hydrogarnet phase, C_3AH_6 (Majumdar and Roy, 1956); the equilibrium curve: $3[3\ CaO{\cdot}Al_2O_3{\cdot}6\ H_2O] = [4\ CaO{\cdot}3\ Al_2O_3{\cdot}3\ H_2O] + 5\ Ca(OH)_2 + 10\ H_2O$ shifts to higher temperatures with increasing pressure. As is usually observed, the pressure dependence of the reaction temperature is most marked at low pressures. Second, the data may be presented as a series of isobaric sections.

These hydrothermal studies have produced a number of results that are also applicable to less extreme pressure–temperature conditions of hydration. For example, many of the crystalline phases that are encountered at lower pressures can also be found to occur at higher pressures. The more extreme conditions of synthesis often favor better-crystallized products, and it becomes easier to characterize the individual phases with the aid of the better-crystallized materials. Also, it is easier to recognize the equilibrium phase distribution at higher temperatures. Extrapolation of these data back to lower temperatures and pressures is helpful in proving that the conclusions as to the direction of the equilibria that were first obtained from low-temperature solubility studies, are, in fact, correct. In many cases, useful thermochemical data, such as heats of formation and reaction, can also be obtained from the *P–T* stability data.

B. The $CaO–Al_2O_3–H_2O–CO_2$ and $CaO–Al_2O_3–H_2O–SO_3$ System

The effect of impurity on the phase relations in the $CaO–Al_2O_3–H_2O$ system is considerable. Even the presence of small quantities of carbonate and sulfate substantially alter the phase relationships that are observed in the absence of these components. CO_2 and SO_3 are, of course, two impurities which are apt to be found in cements. Carbon dioxide uptake from the atmosphere or from CO_2-saturated solutions occurs quite rapidly. This carbon dioxide uptake constitutes a well-known experimental difficulty in preparing the calcium aluminate hydrates on a laboratory scale, free from CO_2 contamination. Phase-rule studies of the system $CaO–Al_2O_3–H_2O–CO_2$ are, however, still very incomplete. The various polymorphs of $CaCO_3$ occur: in addition, several ternary crystalline phases have been prepared and characterized. Carlson and Berman (1960) reported that many preparations gave a phase having the formula $C_3A{\cdot}CaCO_3{\cdot}11\ H_2O$. An ettringite-like phase having the formula $C_3A{\cdot}3\ CaO_3{\cdot}nH_2O$ is also known: reported values for n range from 31 to 32.

Phase relations in the $CaO–Al_2O_3–H_2O–SO_3$ system have been examined more thoroughly; the studies of Jones (1944) and by D'Ans and Eick (1953) are important. Principal features of these studies were the discovery of a compound $C_3A \cdot 3\ CaSO_4 \cdot nH_2O$ ($n \simeq 32$) and the discovery that many of the known calcium aluminate hydrates can form limited or complete solid solutions with the corresponding sulfate-containing end member. The first of these compounds, $C_3A \cdot 3\ CaSO_4 \cdot 32\ H_2O$, was shown to be identical with the natural mineral ettringite. Solid solutions were also encountered between the hydrated and sulfated aluminates: e.g., between C_4AH_{13} and the sulfate-containing end member $C_4A(SO_4)H_{12}$ (also written as $C_3A \cdot CaSO_4 \cdot 12\ H_2O$). This latter solid solution series was reported to be complete: several nomographs have been reported that show the variation in the properties of the solid solutions. These nomographs usually show the variation in a specific property, e.g., *d* spacings of a strong x-ray reflection, mean refractive index, etc., as a function of solid-solution composition.

The experimental studies of the $CaO–Al_2O_3–H_2O–SO_3$ system have been further complicated by the existence of relatively long-lived metastable states, as well as by the general difficulties inherent in studying a four-component system. The general approach that has been used in these studies

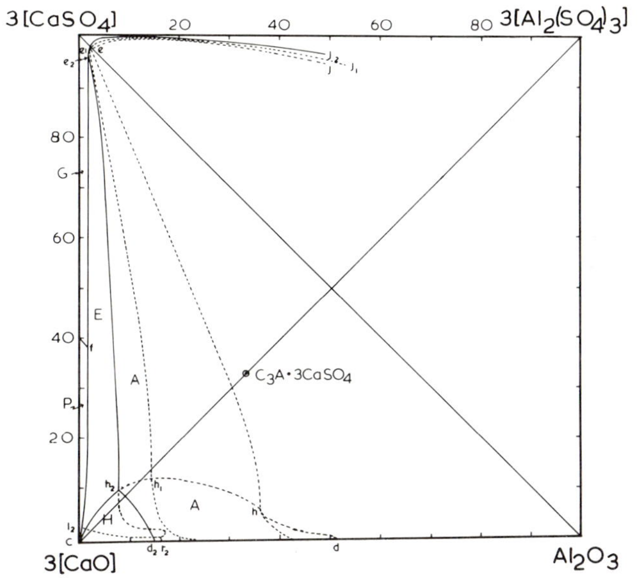

Fig. 11. Some solubility data on the system $CaO–Al_2O_3–H_2O–SO_3$. Metastable solubility curves are shown by continuous dashed lines, stable curves by lines. The invariant points (stable and metastable) are lettered using the same symbols as Jones (1944) except that lower-case letters are used. The projected primary phase fields are shown by upper-case letters: P, portlandite [$Ca(OH)_2$]; G, gypsum ($CaSO_4 \cdot 2\ H_2O$); E, ettringite ($3\ CaO \cdot Al_2O_3 \cdot 3\ CaSO_4 \cdot 32\ H_2O$); A, alumina trihydrate (usually bayerite and gibbsite); H, hydrogarnet ($3\ CaO \cdot Al_2O_3 \cdot 6\ H_2O$).

is similar to that which has been described in connection with the CaO–Al_2O_3–H_2O system. Jones has chosen to represent his data on a two-dimensional plot of a type that follows logically from an extension of that used to represent the data on the CaO–Al_2O_3–H_2O system. It will be recalled that this method was chosen in order to represent the solution compositions coexisting with various phases, and also, to allow representation of both equilibrium and quasiequilibrium solubility curves. Figure 11 shows a typical example of the method as applied to the sulfate-containing system. The basis of the two-dimensional projection shown in the figure is a square pyramid. The four corners of the square base of the pyramid are used for the reciprocal-salt end members: CaO, Al_2O_3, $CaSO_4$, and $Al_2(SO_4)_3$: thus, compositions are recalculated to two oxide and two sulfate end members. The apex of the pyramid, which is shown in the projection at the intersection of the two square diagonals, is H_2O. The compositions of the saturated solutions that are in equilibrium with various two-phase crystalline assemblages are determined experimentally, and these results are projected onto the base. The method of projection is as follows: each solution composition is actually represented by a point lying within the pyramidal volume. A line is directed from the H_2O apex through this point, and extended toward the base. The intersection of this line with the base gives the projected location of the solution compositions and it is these projected compositions that are shown in Fig. 11. Provided that the results shown all pertain to an isothermal section and that they represent the equilibrium between two crystalline phases and solution, the series of projected points for any one equilibrium will trace out the projection of a univariant curve.

D'Ans and Eick, on the other hand, use a triaxial, geometrically orthogonal plot. The solution compositions, as determined by chemical analysis, are recalculated to the end members CaO, Al_2O_3, and $CaSO_4$. Each component is represented along one vertical edge of a triangular prism. Again, only isothermal results can be shown on any one diagram. This method of representation has the advantage of not only showing the location of univariant curves but also, the relationships between families of univariant curves that delineate a primary phase volume are brought out. Thus, for example, the mode of presentation shows with particular clarity the large range of solution compositions from which ettringite will crystallize as a primary phase. It should be emphasized that, despite the differences used in representation, the results of both sets of studies are in good agreement. However, as more data are obtained, the type of representation that is employed will undoubtedly undergo further changes, as more experimental details (such as the exact compositions of solid solutions in equilibrium with liquids) become known, and it is wished to represent them on phase diagrams.

C. The $CaO–SiO_2–H_2O$ System

Our knowledge of phase relations in the system $CaO–SiO_2–H_2O$, particularly at the temperatures and pressures usually encountered in cement hydration, is still largely in the formative stage. One principal experimental difficulty has been in the characterization of the individual phases that are found to occur. Many of the hydration phases appear in the form of extremely small crystallites: moreover, the crystallites often have poor internal order and may contain an appreciable fraction of their water content as absorbed, rather than structural, water. In fact, one school of cement chemists has advocated that the hydration products that are obtained at or near room temperature should be treated as colloids rather than as crystalline phases. During the early stages of cement hydration, this view is probably correct and the hydration products that are obtained do not constitute a "phase" or "phases" in the classical sense of the word. However, the time factor is also important in discussing the crystallinity of the hydration products. Most portland cements continue to gain in strength with time, indicating that the setting reactions continue over a period of months or longer. After several weeks or months, it is found by direct examination that the colloidal material gives way, in large part, to materials that are definitely crystalline.

Space does not permit a critical discussion of the experimental techniques which have been used in the identification of the individual phases: infrared spectroscopy, electron microscopy and diffraction, DTA and TGA, petrographic microscopy, etc., have all played important roles. This characterization of the individual phases is not yet complete—a number of problems still remain—but already, many data have accumulated on the nature of the crystalline products that may be obtained reproducibly, and of the probable coexisting phase assemblages.

One method of studying the phase relations is to correlate the phases that are formed as a function of three variables: CaO/SiO_2 starting ratio, temperature and pressure. These early attempts to systematise knowledge on the $CaO–SiO_2–H_2O$ system result in what might be termed a "mode-of-occurrence" diagram. A typical mode-of-occurrence diagram is shown in Fig. 12. Its rationale is as follows. The phases that are formed by reacting a given mixture are known to be very sensitive both to the CaO/SiO_2 ratio of the starting material and also to temperature. However, although the CaO/SiO_2 ratio of the starting material is known, the exact composition of the individual phase(s) produced were not always known at this stage of the investigation. The phase distribution is also found to vary only slightly with pressure over the range that has been most accessible to experimental studies, typically, 0–500 bars H_2O pressure. Therefore, the mode-of-occurrence diagram can be treated as being essentially isobaric, provided that only

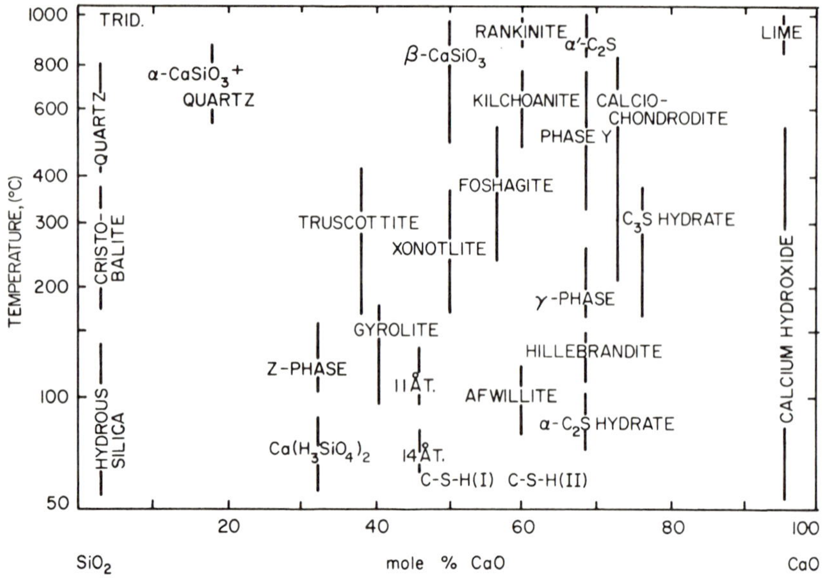

Fig. 12. Occurrence and approximate compositions of the principal phases in the system $CaO–SiO_2–H_2O$. T, Tobermorite; C–S–H, $CaO–SiO_2·H_2O$ phases: these have an indefinite composition, phase (II) being richer in CaO than phase (I).

these data obtained at low pressures are shown. Only the CaO/SiO_2 ratio of the starting material and the temperature of the run need, therefore, be specified in order to place the occurrence of a given phase in its appropriate position on the diagram. This means that a given phase can be located on such a diagram even if its exact CaO/SiO_2 ratio is not yet known, or its water content not yet fully established. Inspection of the diagram shows that a large number of calcium silicate hydrates have been identified in this way and that products having a crystalline nature can be obtained even at room temperatures. Many of the phases encountered have also been found to occur in nature, hence the frequent use of mineralogical names. One weakness of this type of representation is that it gives little information as to which sets of phases are compatible. Therefore, as more actual phase-equilibrium data become available, and as the composition of the individual phases becomes better known, it becomes increasingly desirable to use a more informative method of presentation.

Pressure–temperature curves are now available for the formation and decomposition reactions of most of the individual phases. A series of such *P–T* curves for some of the phases in the lime–silica–water system are shown

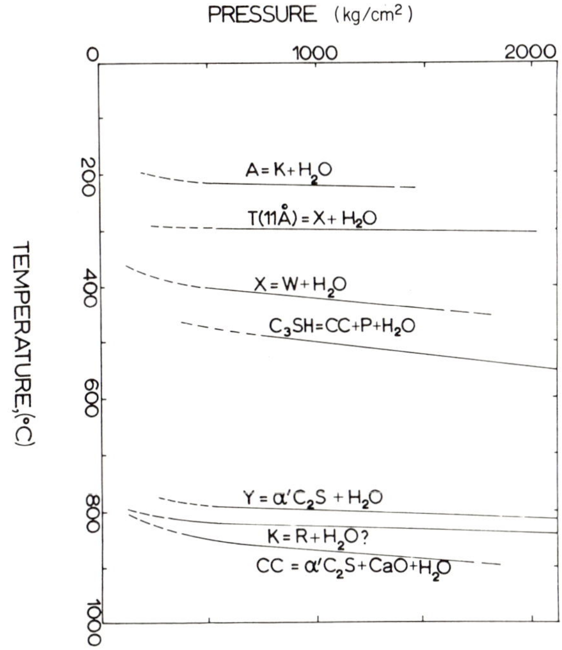

Fig. 13. Some pressure–temperature data for phases in the system $CaO–SiO_2–H_2O$. A, afwillite; K, kilchoanite; T(11 Å), 11 Å tobermorite; X, xonotlite; W, wollastonite (β–$CaSiO_3$); C_3SH, tricalcium silicate hydrate; P, portlandite [$Ca(OH)_2$]; Cc, calcio-chondrodite; Y, phase Y; K, kilchoanite. The position of the curve for the dehydroxylation of kilchoanite is uncertain. It is probably schematically correct if kilchoanite contains some water. Rankinite, $Ca_3Si_2O_7$ is anhydrous.

in Fig. 13. As in the case of the $CaO–Al_2O_3–H_2O$ system, many of the reactions have been followed to temperatures and pressures higher than are usually encountered in cement hydration, but the same justifications attach to the value of the information thus obtained as in the previous cases (Section IV,A).

Once *P–T* data have been obtained for the individual reactions, a logical further step is to establish the compatibility relationships between coexisting phases. Results of these studies are best shown on a conventional ternary diagram, as in Fig. 14. This shows the 350°C isothermal section of the $CaO–SiO_2–H_2O$ system; the section is also approximately isobaric, and is meant to be representative of the low-pressure range. The solution phase lies very close in composition to pure H_2O. In the $CaO–Al_2O_3–H_2O$ system, it will be recalled, it was advantageous to be able to represent the solution

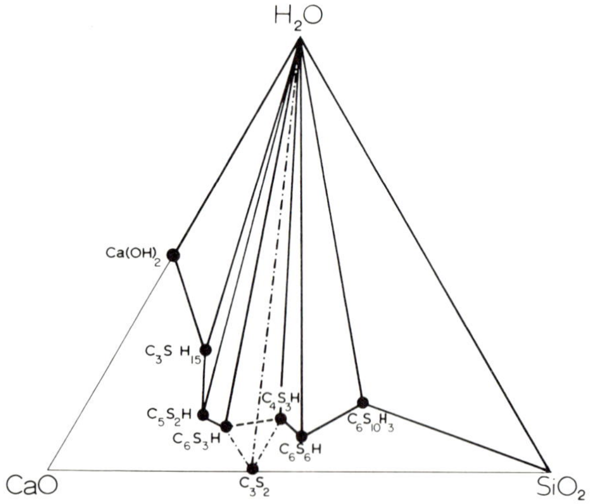

Fig. 14. The 350°C isothermal section of the CaO–SiO_2–H_2O system at low H_2O pressures (after Roy and Harker, 1962). Uncertain compatibility relations are shown by dash-dot lines.

compositions accurately. While solubility studies of the CaO–SiO_2–H_2O system have been undertaken along lines parallel to those of the CaO–Al_2O_3–H_2O system, they have generally proved less helpful in establishing the probable phase–equilibrium relationships. This is because the solution compositions are so close to pure H_2O that it is difficult to observe significant solubility differences between different phases. It is also more difficult to prepare supersaturated solutions of the calcium silicates, especially the more acidic silicates. Hence, the triangular representation is usually used, except in the few cases where data are available to represent solution compositions.

The relationship between an isothermal section and the mode-of-occurrence diagram, which was discussed earlier, is shown in Fig. 15. This figure shows part of a triaxial temperature-composition prism for the CaO–SiO_2–H_2O system. Each apex of the triangular base is used to represent one of the three components: CaO, SiO_2, and H_2O. Temperature is shown along the vertical edges, and the direction that extends upwards is used to show rising temperature. The 350°C isothermal section is shown (in perspective) intersecting the mode-of-occurrence diagram. It can be seen that, if the phase compositions are projected from the water vertex through the appropriate bulk composition on the isothermal section, they intersect the front face at the CaO/SiO_2 ratio appropriate to its position on the mode-of-occurrence diagram. In general, the isothermal section is the more useful method of

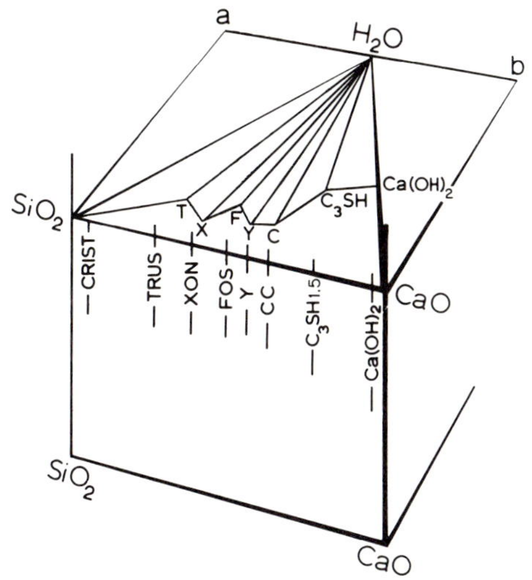

Fig. 15. Temperature–composition diagram for part of the system $CaO–SiO_2–H_2O$. This shows the relationship between an isothermal section (Fig. 14) and the mode-of-occurrence diagram, shown in more complete form in Fig. 12.

representation. Especially at the higher temperatures, where equilibrium is more readily established, isothermal sections can now be drawn with some confidence from the data available in the literature.

It would, however, be wrong to conclude that our knowledge of phase equilibria (or, indeed, of the quasiequilibria) in this system were complete. First, a number of new phases have recently been discovered (such as kilchoanite). Thus, the 350°C isothermal section shows that some doubt still exists as to the phase relations involving kilchoanite. Second, the identity of the crystalline products that are obtained, even after long runs, tend to reflect the nature of the starting materials which were used. Many of the experimental studies have used so-called "gel" starting materials. These are essentially coprecipitated hydrous oxides; as first prepared they are usually contaminated by nitrate or other foreign ions, which must be removed by gently igniting at about 400°C. During this treatment, the gel usually improves considerably in crystallinity. It was thought that gels represented the most reactive set of starting materials and would, therefore, yield the equilibrium products more rapidly than other, alternative, starting materials. This assumption is probably not universally true, and a real need exists for further studies of the products of reaction which are obtained from other calcium silicate starting materials, such as glasses or fine-grained crystalline oxide mixtures, etc.

D. Phase Relations in Other Hydrate Systems

Systematic phase-equilibrium studies of more complex hydrate systems, such as the CaO–Al_2O_3–SiO_2–H_2O system, are far from complete. As in the case of the CaO–SiO_2–H_2O system, more data are available at the high temperatures and pressures, especially for the more acid compositions, which are of interest to petrologists. Fewer data are available for the range of compositions reacted at the temperatures and pressures that obtain during cement hydration. Some studies may be cited as examples of the type of data that are presently available. Kalousek (1957) has shown that tobermorite can take in up to 4–7 wt % Al_2O_3 in solid solution. The partial or complete isomorphic replacement of Al by Si, or vice versa, is a common feature of silicate crystal chemistry, and it appears likely that the determination of the limits of solid solutions between the silicate and aluminate hydrate phases will make an important contribution to our knowledge of this system. As an example of another type of solid-solution series. Roy and Roy (1962) have studied phase relations on the "garnet join" of this system: this join extends between $3\,CaO{\cdot}Al_2O_3{\cdot}6\,H_2O$ (hydrogarnet) and $3\,CaO{\cdot}Al_2O_3{\cdot}3\,SiO_2$. This latter compound, the garnet grossularite, is not shown in the phase diagram for the CaO–Al_2O_3–SiO_2 system. At 1 bar pressure, the anhydrous phase is probably thermodynamically stable at temperatures below ~600°–800°C, but its formation is in general, not observed in anhydrous compositions because of the sluggishness of reaction at these temperatures. Roy and Roy (1962) were able to show that at temperatures below 275°–375°C, and at 1000–2000 bars pressure, an extensive range of solid solutions form along this join. The solid solutions start at hydrogarnet and extend to a limiting composition containing more than 60 mole % of the silicate end member. The solid solution mechanism involves replacement of $4OH^-$ by SiO_4^{4-}. This represents yet another, and unusual, type of solid-solution phenomenon that has been shown to exist in this system. At higher temperatures, the garnet join is not a binary system and the garnet solid solutions break down, yielding as important crystalline phases $C_4A_3H_3$, $Ca(OH)_2$, and the hexagonal polymorph of $CaO{\cdot}Al_2O_3{\cdot}2\,SiO_2$. The latter phase, sometimes incorrectly called "hexagonal anorthite" is probably metastable under all conditions.

Another anhydrous CaO–Al_2O_3–SiO_2 phase which is encountered in clinkers is gehlenite. Its hydration has been studied by zur Strassen and Strätling (1940). The principal hydrate phase obtained is the so-called "gehlenite hydrate," $2\,CaO{\cdot}Al_2O_3{\cdot}SiO_2{\cdot}8\,H_2O$. Inasmuch as this compound has also been obtained from a wide variety of starting materials (including kaolin plus lime water), by reaction of $3\,CaO{\cdot}SiO_2$, $3\,CaO{\cdot}Al_2O_3$, and H_2O and by hydration of a glass of the gehlenite composition, this hydrate probably plays an important role as a low-temperature reaction product in the $CaO–Al_2O_3–SiO_2–H_2O$ system.

Similar types of study have disclosed some information on the principal hydration products of the other important clinker phases, such as the ferrite phase. The iron-containing phases have proved more difficult to study, because the techniques that were used in the $CaO–Al_2O_3–H_2O$ system are not, in general, applicable to the ferrite systems. This is partly because it is difficult to prepare initially supersaturated ferrite solutions. Also, direct hydration of the crystalline phases proceeds very slowly at or just above room temperature.

References

ARUJA, E. (1957). *Acta Cryst.* **10**, 337.

BIGARÉ, M., GUINIER, A., MAZIÉRES, C., REGOURD, M., YANNAQUIS, N., EYSEL, W., HAHN, TH., and WOERMANN, E. (1967). *J. Am. Ceram. Soc.* **50**, 609.

BOGUE, R. H. (1955). "Chemistry of Portland Cement," 2nd ed. Reinhold, New York.

CARLSON, E. T., and BERMAN, H. A. (1960). *J. Res. Natl. Bur. Std.* **64A**, 333.

D'ANS, H., and EICK, H. (1953). *Zement-Kalk-Gips* **6**, 197.

DAYAL, R. R., and GLASSER, F. P. (1967). "Science of Ceramics" (G. H. Stewart, ed.), Vol. 3, pp. 191–214. Academic Press, New York.

HARKER, R. I., ROY, D. M., and TUTTLE, O. F. (1962). *J. Am. Ceram. Soc.* **45**, 471.

IMLACH, J., and GLASSER, F. P. (1968). *Trans. Brit. Ceram. Soc.* **67**, 581.

JEEVARATNAM, J., GLASSER, F. P., and GLASSER, L. S. D. (1964). *J. Am. Ceram. Soc.* **47**, 105.

JONES, F. E. (1944). *J. Phys. Chem.* **48**, 311.

JONES, F. E., and ROBERTS, M. H. (1959). "The System $CaO–Al_2O_3–H_2O$ at 25° C," D.S.I.R. (Building Research Station) Note E965, Her Majesty's Stationery Office, London.

KALOUSEK, G. L. (1957). *J. Am. Ceram. Soc.* **40**, 74.

LISTER, D., and GLASSER, F. P. (1967). *Trans. Brit. Ceram. Soc.* **66**, 293.

MCMURDIE, H. F., and INSLEY, H. (1936). *J. Res. Natl. Bur. Std.* **16**, 472; R.P. 884.

MAJUMDAR, A. J., and ROY, R. (1956). *J. Am. Ceram. Soc.* **39**, 434.

MIDGELY, H. G. (1968). *Trans. Brit. Ceram. Soc.* **67**, 1.

MUAN, A., and OSBORN, E. F. (1960). "Large Scale Phase Diagrams of Oxide Systems," Plate 1. American Ceramic Society, Columbus, Ohio.

NEWKIRK, T. F., and THWAITE, R. D. (1958). *J. Res. Natl. Bur. Std.* **61**, 233.

NEWMAN, E. S., and WELLS, L. S. (1946). *J. Res. Natl. Bur. Std.* **36**, 137.

NURSE, R. W., WELCH, J. H., and MAJUMDAR, A. J. (1965). *Trans. Brit. Ceram. Soc.* **64**, 323.

OSBORN, E. F., and TAIT, D. B. (1952). *Am. J. Sci.* (Bowen Volume), 413.

PARKER, T. W. (1952). *Symp. Chem. Cement, 3rd London,* p. 485.

PERCIVAL, A., and TAYLOR, H. F. W. (1959). *J. Chem. Soc.* (London) **1959**, 2629.

RANKIN, G. A., and WRIGHT, F. E. (1915), *Am. J. Sci.* **39**, 1.

ROY, D. M., and HARKER, R. I. (1962). "Chemistry of Cement" (*Proc. Intern. Symp., 4th, Washington, D.C.,* 1960) (Natl. Bur. Std., Monograph 43, U.S. Dept. of Commerce), p. 196.

Roy, D. M., and Roy, R. (1962). "Chemistry of Cement" (*Proc. Intern. Symp., 4th, Washington, D.C.*, 1960) (Natl. Bur. Std., Monograph 43, U.S. Dept. of Commerce), p. 307.

Shepherd, E. S., Rankin, G. A., and Wright, F. E. (1909). *Am. J. Sci.* [4] **28**, 293.

Tavasci, B. (1937). *Ann. Chim. Appl.* **17**, 505; *Tonindustr. Ztg.* **61**, 717, 729.

Taylor, H. F. W., ed. (1964). "The Chemistry of Cements," Vol. 1, Chapter 2, pp. 49–88 (by J. H. Welsh); Vol. 1, Chapter 5, p. 215. Academic Press, New York.

Welch, J. H. (1964). Private communication reported in: Majumdar A. J. (1964). *Trans. Brit. Ceram. Soc.* **63**, 347.

zur Strassen, H., and Strätling, W. (1940). *Z. Anorg. Chem.* **245**, 267.

VI

Phase Diagrams in Extraction Metallurgy

J. TAYLOR

DEPARTMENT OF EXTRACTION METALLURGY
UNIVERSITY OF STRATHCLYDE, GLASGOW, SCOTLAND

I. INTRODUCTION

The thermal equilibrium diagrams of the widest general interest to the extraction metallurgist are those of oxide systems that represent the slags so important to the chemistry of the classical smelting and refining operations. In this context slags formed in the melting of metals and alloys for casting may also be included in refining slags. Before discussing the more important oxide systems in relation to smelting and refining, a brief review of the general role of slags in these operations is desirable.

Essentially slag consists of the oxide impurities that must be separated from the molten metal or, in some cases the molten metal sulfide. The less dense slag floats on the surface of the metal and imposes a barrier between the metal and the ambient gas atmosphere. The scientific study of these processes is largely concerned with reactions at the phase boundaries, slag/metal and slag/gas, separation of the phases, and heat and mass transport through the phases, in particular the slag phase.

Smelting and refining slags are formed in five main ways.

(1) In smelting, which is an oxide reduction process,* those oxides with a much higher free energy of formation than the primary oxide undergo little reduction and constitute the slag. These oxides are commonly silica, alumina, lime, and magnesia, and they may be present as impurities in the ore or may be added. The first two are usually referred to as gangue and the latter as flux. During smelting, conditions within the charge change progressively, the temperature increases, and the oxygen potential may decrease (oxide smelting) or increase (matte smelting). There can, therefore, be considerable changes in composition between initial and final slag.

(2) It is theoretically impossible to achieve perfect separation between the main metal and oxide impurities. Impurity oxides must always be reduced to some extent and enter into the metal solution just as some of the main metal oxide must remain unreduced in the slag solution. Also, the metal oxide (oxygen) may be soluble to some extent in the metal phase. After removing the bulk of the oxide impurities during smelting, the process is continued in the fire-refining process. Since in this case the bulk of slag is much less, the oxygen potential may be raised much above that of the smelting process and impurities plus, of course, some primary metal oxidized out. The oxides, together with any oxides added as fluxes or refining agents, constitute the refining slag. In the case of refining, there is usually little change in temperature during the process but the oxygen potential gradually increases and this, together with the solution of oxide additions is chiefly responsible for changes in slag composition during refining.

(3) At the end of the oxidation refining process the concentration of oxygen in metal solution may build up to a significant amount. This is particularly so in the case of copper and iron. Deoxidation is effected by the addition of an element with a strong affinity for oxygen. In the case of iron (steel) this is usually silicon plus manganese and/or aluminum. The resulting oxides are precipitated as liquid (slag) or solid products.

(4) In some refining operations preformed slags are used. This permits more precise control and results in better quality products. An example of this type of practice is electroslag melting. As the emphasis on quality in steel making increases, this type of process seems certain to increase in importance.

(5) Smelting slags are also preformed in the sintering of ores prior to the

* Matte smelting may be considered as a reduction process with sulfur or sulfide as the reducing agent.

actual smelting operation. Partial fusion is an essential step in the mechanism of this type of sintering and control of the extent of fusion is the most important factor in the quality of the product.

The most general and important function of slags in metallurgical operation is to take part in slag/metal reactions. The typical slag/metal reaction for both smelting and refining processes is

$$\mathrm{MO(slag)} + \mathrm{I} = \mathrm{IO(slag)} + \mathrm{M} \tag{1}$$

where M and I are the main metal and impurity, respectively. In smelting the aim is usually to have the concentration of MO in the discard slag as low as possible (maximum recovery of metal) and in refining the aim is to have the concentration of I as low as possible (maximum purity of metal). In any chemical reaction the final concentrations are a function of the thermodynamics and kinetics of the process.

The equilibrium in reaction (1) is

$$K = \frac{f_1 \times \%\mathrm{IO} \times a\mathrm{M}}{f_2 \times \%\mathrm{MO} \times f_3 \times \%\mathrm{I}}$$

Since the concentration of the metal approaches 100%, the activity is not a variable. The activity coefficient of the impurity in metal solution may vary due to the composition of the crude metal but normally this composition cannot be varied at will by the metallurgist. The only independent variables within his control are the temperature, which affects the equilibrium constant, and slag composition, which affects the activities of the oxides in slag solution. These in turn are interdependent variables because of the usual requirement that slags should be liquid and any variations in slag composition and temperature must in the first place then be referred to the appropriate thermal equilibrium diagram. Temperature and slag composition are also important variables for the kinetics of slag/metal reaction. This follows since most of these reactions are controlled by mass transport across the boundary layer on the slag side of the slag/metal interface. Apart from agitation the important variables are the viscosity, density, and diffusion coefficient of the reactant in the slag. These in turn are a function of temperature and composition.

There is still further restriction on the temperature and composition of metallurgical slags. Processes involving slags are of necessity carried out in refractory-lined vessels, and slag attack is inevitable. Several promising new processes in the field of direct iron or steel making have foundered on this rock. Attack of refractories by slag is an exceedingly complex process, but the study must begin with the relevant thermal equilibrium diagram, i.e., that for the oxides of the refractory plus those present in the slag.

Very similar considerations apply to the transport of a reactant between a metal and gas phase through an intermediate slag phase. In addition to a slag/metal there is also a slag/gas reaction, and the latter is influenced by temperature and slag composition just as is the former. One small difference is that, where it is desirable to minimize transport, the metallurgist may aim to produce a semisolid rather than a liquid slag.

Deoxidation reactions are basically the same as reaction (1) but since we are concerned with oxygen in solution may be written

$$O + I = \mathrm{IO(deoxidation\ product)} \quad (2a)$$

In practice one may precipitate a substantially pure oxide or a mixture of oxides IO, I′O, and some of the primary metal oxide by a simple partition reaction

$$\mathrm{M} + O = \mathrm{MO(slag\ solution)} \quad (2b)$$

This mixture of oxides may well be molten, in which case it is a slag. Important aspects of deoxidation practice concern the elimination of the precipitate and the effect of residual particles on the properties of the metal. In both these respects crystalline solid particles and liquid droplets behave very differently. It is obviously, therefore, most important to know the conditions governing the formation of each type, and this requires data on the appropriate equilibrium diagrams.

In some ways thermal equilibrium diagrams might be considered most important in the study of slag formation during the sintering or smelting processes. Here we are concerned with physical effects only, the binding of particles together by partial fusion. In sintering this is necessary and desirable; in blast-furnace smelting it is unavoidable but can have serious practical effects by causing sticking and hanging of the descending stock. Although clearly the study of these processes must begin with the appropriate diagrams, in practice changes in temperature, oxygen pressure, and composition are so rapid that even if precise data on these changes were available, analysis of the phenomena would be very difficult. There is no question of equilibrium being attained, and therefore only the most qualitative deductions can be made from the equilibrium diagram.

One further important use of the thermal equilibrium diagram is in connection with the thermodynamics of the oxides in liquid slags. At liquidus boundaries the activity of the phase at saturation is unity and this then serves as an independent reference to confirm or extend experimentally determined activities. This has proved particularly useful when the phase in the saturated solution is an interoxide compound and an equilibrium of the type

$$x\mathrm{M'O} + y\mathrm{M''O} = x\mathrm{M'O} \cdot y\mathrm{M''O} \quad (3)$$

is established. Three independent sets of data can be related through this equilibrium, viz., the liquidus boundary, the activities of the oxides at this boundary, and the free energy of formation of the interoxide compound. It is obvious that through this relationship any one unknown can be calculated if values for the other are available, but so far it has been most widely used to extend our knowledge of oxide activities. This is largely because thermal equilibrium diagrams have been actively investigated for years, whereas the importance of the activities of oxides in slag solution has only been appreciated in comparatively recent times.

The application of thermal equilibrium data to metallurgical problems can best be elaborated by consideration of specific diagrams. Before proceeding to consider such specific cases, we can make one general comment. One common difficulty is that few metallurgical slags are simple binary or even ternary systems. Data on higher systems are scanty and, in the absence of such data, the metallurgist must frequently base his deductions on a simple system. Fortunately, in many important cases the oxides of the simplest system make up the greater proportion of the slag and the deductions are good approximations and, for comparative purposes, are invaluable.

II. OXIDE SYSTEMS

Unless otherwise stated, the equilibrium diagrams used in the subsequent discussion are taken from "Phase Diagrams for Ceramists," American Ceramic Society, Inc., 1964.

A. $CaO–Al_2O_3–SiO_2$ (plus MgO)

This system affords one of the clearest examples of the application of thermal equilibrium data to metallurgical problems. The diagram, certainly in its main essentials, is full and accurate, and the equilibria are not significantly affected by oxygen pressure. This represents a considerable simplification vis-à-vis those systems containing oxides of variable valency. Furthermore, the solution thermodynamics for the main reactions, i.e., those involving lime and silica have been fairly well determined.

Lime, alumina, and silica constitute the basic composition of slags arising from iron-smelting operations. By far the most important of such operations is the iron blast furnace. Blast-furnace slags frequently contain substantial amounts of magnesia, up to about 12%, as well as 5–6% of calcium sulfide, alkalies, and manganese oxide, but the effect of composition can be discussed

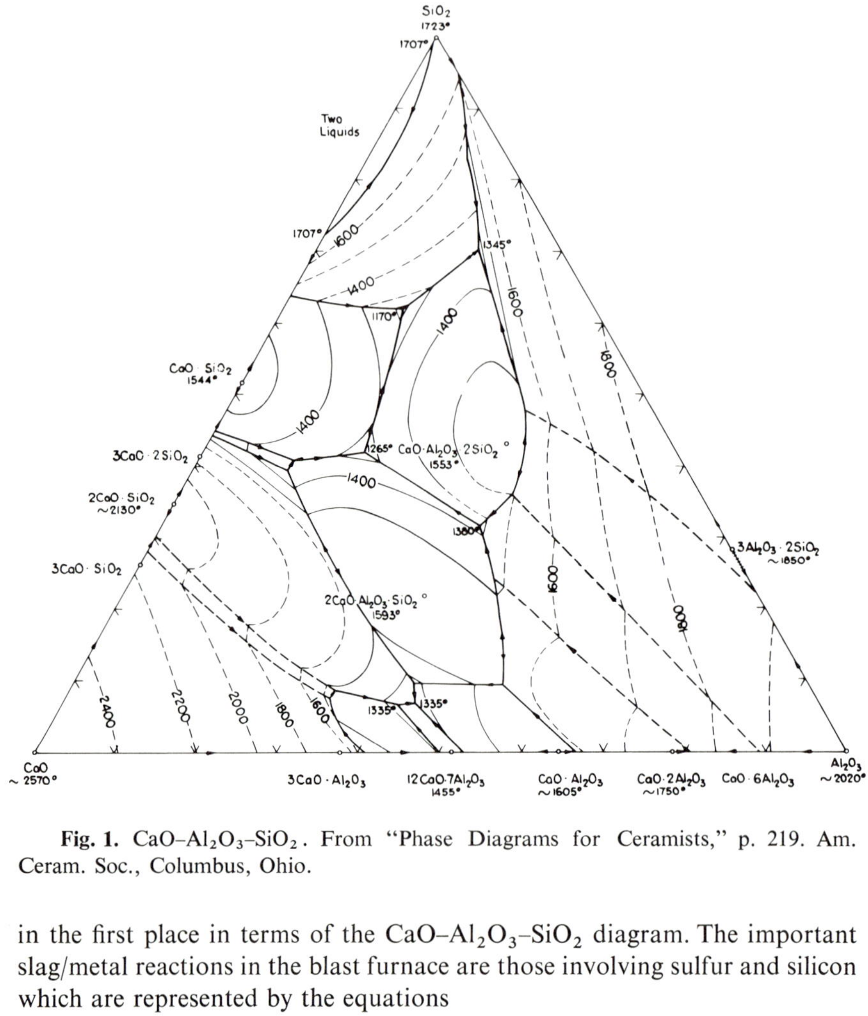

Fig. 1. CaO–Al_2O_3–SiO_2. From "Phase Diagrams for Ceramists," p. 219. Am. Ceram. Soc., Columbus, Ohio.

in the first place in terms of the CaO–Al_2O_3–SiO_2 diagram. The important slag/metal reactions in the blast furnace are those involving sulfur and silicon which are represented by the equations

$$S + \mathrm{CaO} + \mathrm{C} = \mathrm{CaS} + \mathrm{CO} \tag{4}$$

$$\mathrm{SiO_2} + 2\mathrm{C} = Si + 2\ \mathrm{CO} \tag{5}$$

It is usually necessary to have both sulfur and silicon in blast-furnace iron as low as possible. The two most important variables are temperature and slag composition and, since the temperature is controlled by the liquidus temperatue of the slag, these are in fact dependent variables. Both reactions (4) and (5) are displaced to the right with increase in temperature but, whereas the temperature coefficient for (4) is fairly small, that for (5) is very large; therefore lower temperatures are preferred. Increase in basicity will tend to increase the activity of lime and decrease that of silica, displacing both reactions in the desired direction.

It can readily be seen from Fig. 1 that maximum basicity and minimum temperature are conflicting requirements. The best compromise requires detailed information on the thermodynamics of reactions (4) and (5) in addition to the thermal equilibrium diagram. In default of data on both the activity of lime and calcium sulfide, the desulfurizing power of slags can be represented by the sulfur capacity C_S (Fincham and Richardson, 1954). On the basis of extensive experimental data, Bell (see Kalyanram *et al.*, 1960) has shown that C_S can adequately be described as a linear function of

$$\frac{\text{NCaO} + \text{N/2 MgO}}{\text{NSiO}_2 + \text{N/3 Al}_2\text{O}_3}$$

In Fig. 2 values of C_S, the activity of silica (Taylor, 1964) and liquidus isotherms are shown for a portion of the $CaO–Al_2O_3–SiO_2$ system appropriate to blast-furnace slags.

Figure 2 illustrates very clearly the importance of alumina content of slags. The best combination of desulfurizing power, silica activity, and liquidus temperature is obtained with slags of 11–15% alumina. As the alumina increases, for what are substantially the same values of sulfur capacity

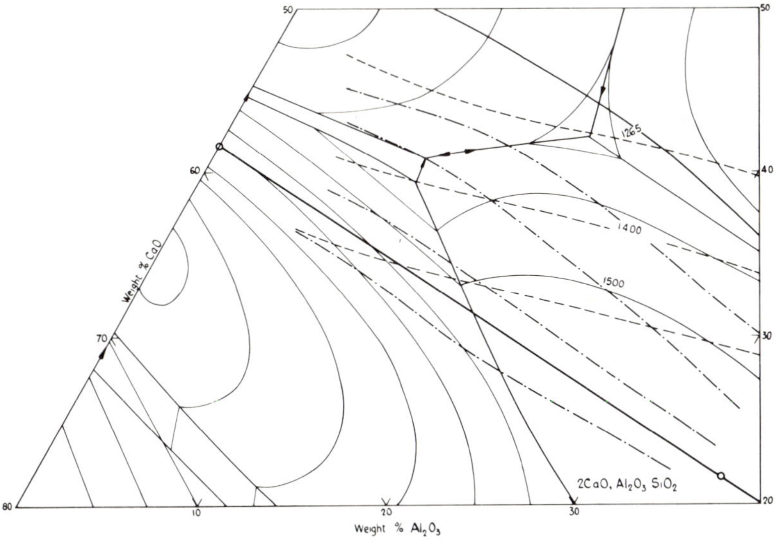

Fig. 2. Part of $CaO–Al_2O_3–SiO_2$ system showing ——: lines of constant sulfur capacity C_S, —·—: lines of constant a_{SiO_2}.

and silica activity, there is a marked increase in the liquidus temperature that begins to be quite serious above 22% or so alumina. The effect at low alumina contents, e.g., <10%, is not so clear cut. The liquidus temperature may or may not increase, depending upon the chosen value of C_S, but there is an increase in silica activity. Low-alumina slags, however, also suffer from a less obvious disadvantage and that is that the liquidus surface rises very steeply in this region. This means that composition control has to be very exact or the operator must allow a greater margin for error, i.e., a greater degree of superheat above the liquidus to allow for compositional variations.

To some extent the data represented in Fig. 2 have only rationalized and made more exact the experience of generations of blast furnace operators. On the other hand, extension of the range of slag compositions to include substantial proportions of magnesia has been based on the technical application of thermal equilibrium data. Osborn *et al.* (1954) critically examined

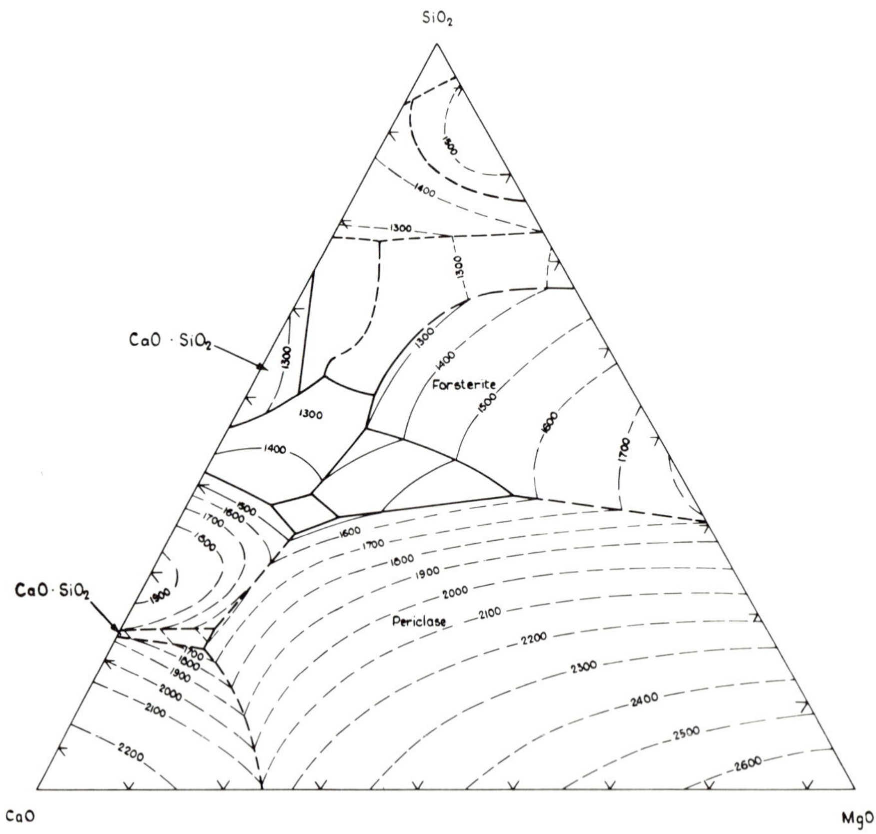

Fig. 3. CaO–Al_2O_3–SiO_2–MgO. Ternary section at 15% Al_2O_3. From Osborn, E. F., De Vries, R. C., Gee, K. H., and Kraner, H. M. (1954). *Trans. AIME* **200**, 33.

data on the CaO–Al_2O_3–SiO_2–MgO system from the point of view of the blast furnace. They concluded that for all alumina concentrations a better combination of desulfurizing power and liquidus temperature could be obtained by the addition of magnesia. In the original paper these conclusions were based on ternary sections at 5, 10, 15, 25, 30, and 35% alumina. The section at 15% alumina is shown in Fig. 3 as illustration. Small-scale reproduction of these sections is not very informative, but the data have proved so valuable to the blast-furnace industry that detailed tables of liquidus temperature versus composition have been compiled. The variables in the slag/metal reaction are so complex that the effect of magnesia can only be illustrated by typical examples. Osborn *et al.* suggested the best values of magnesia for different levels of alumina and these are the basis for the comparison with the magnesia-free slags in Table I.

TABLE I

A COMPARISON OF SLAGS OF THE SAME DESULFURIZING POWER WITH AND WITHOUT MAGNESIA

Slag composition (%)				Liquidus temperature	Silica activity
CaO	Al_2O_3	SiO_2	MgO	(°C)	$\times 10^2$
51	10	39		1500	6
40	10	36	14	1450	8
48.5	15	36.5		1455	5.5
38	15	34.5	12.5	1430	8
46.5	20	33.5		1510	4.5
37	20	32	11	1450	6
44	25	31		1545	3.5
37.5	25	29.5	8	1470	5

The slags are estimated to have the same desulfurizing power, and the effect of magnesia on the silica activity and liquidus temperature is shown. In all cases for the optimum magnesia content, the liquidus temperature is lowered and the silica activity raised. However, since the equilibrium silicon value in the iron increases or decreases by a factor of 3 for $\pm 50°C$ it can be seen that, despite the slightly higher silica activities, the tendency for silica reduction is decreased for the magnesia-containing slags. Development of the use of magnesia slags is an excellent example of the application of scientific data to a practical problem.

Attack of slag on refractories poses a problem in the iron blast furnace as in all high-temperature smelting operations. The possibility of such attack can be assessed from thermal equilibrium diagrams. Slag in contact with refractory can dissolve the refractory until saturation is reached. When this happens the composition of the slag moves along the tie line joining the two compositions until the liquidus for the operating temperature is reached. The corrosive effect due to thermodynamic considerations can then be estimated by applying the lever principle to the three compositions along the tie line, i.e., slag, refractory, and liquidus. Reference to Fig. 1 shows that alumina–silica refractory is liable to severe attack by blast-furnace slags at temperatures of 1500–1600°C. From a purely thermodynamic point of view, the corrosive effect should decrease as the composition moves from pure silica at one extreme to pure alumina at the other but, at best, the effect is severe. Similarly, from Fig. 3 it can be seen that magnesite offers no substantial improvement in this respect. It must be concluded, therefore, that none of the usual oxide refractories can be expected to withstand the action of blast-furnace slag at high temperatures, and this is recognized by heavily water cooling such refractories in the blast furnace. As it happens, this is an acceptable solution in this case since the furnace is stationary and very large, hence the engineering of water cooling is straightforward and the heat loss per unit of iron made is small. It is very different if the furnace is of the rotary type and of much more modest dimensions. Many new methods for the production of iron other than by the blast furnace have been tried, and all too often the magnitude of this problem of refractory attack has not been appreciated.

So far only slags containing substantial concentrations of silica have been considered. Reference to Fig. 1 shows that there is a low-melting region adjoining the $CaO–Al_2O_3$ binary. Slags from this range of compositions offer attractive possibilities. As might be expected the desulfurizing power is high and the silica activities are so low that the possibility of silica reduction is negligible. The problem, as can be seen from the diagram, is that the silica content is very critical because the liquidus temperature rises very steeply for concentrations of $\sim 10\%$. There are a number of iron-ore deposits in which the gangue is largely alumina that might be smelted with this type of slag. Unfortunately, the normal fuel, i.e., coke, always contains silica as a constituent of the ash and unless the silica present in the gangue of the ore is very low, the slag liquidus temperature is uneconomically high. In the few cases where this type of slag has been used, it appears to be directed as much to recovery of alumina or the production of high-aluminous cement as to the production of iron.

Lime alumina slags are also attracting attention in connection with steel making. Increasing demands for cleaner, high-quality steel are forcing the steel maker to consider the use of preformed slags as refining agents instead of, as in the past, compromising with slags that arise from the steel-making process. Because of the properties mentioned above, lime alumina slags look very attractive, all the more so if compositions that are almost saturated with calcium aluminate at 1500–1600°C are used. Premelted slags must be held in a refractory vessel under conditions very conducive to slag attack. As the diagram shows high alumina/silica refractories should in fact withstand very well the attack of slags of the suggested composition.

Fairly exhaustive studies of the solution thermodynamics of the CaO–Al_2O_3–SiO_2 system have been made by Rein and Chipman (1965), Sharma and Richardson (1961, 1962), and Kay and Taylor (1960) among others. In all these studies continual reference is made to the thermal equilibrium diagram. Saturation surfaces, eutectic troughs, or invariant points, together with the necessary free energy of formation of the many interoxide compounds, have been invaluable for calculating silica activities from lime activities or vice versa, alumina activities from lime activities, etc. The relationship between location of saturation lines, interoxide free-energy data, and oxide activities has been used to test the self-consistency of the data. Generally this is fairly satisfactory, although there are still some discrepancies in the CaO–Al_2O_3 binary and adjacent region.

This system is also useful in a slightly different connection. Dicalcium silicate undergoes a 10% volume change in going from the high–temperature to low-temperature form. Solid slags containing dicalcium silicate are therefore liable to disintegrate; so-called falling slags. This can be very serious for instance where blast-furnace slag is used as an aggregate for roads, concrete, etc. The liability to form dicalcium silicate can be assessed from the equilibrium diagram. This is a sufficient guide for the simple CaO–Al_2O_3–SiO_2 slag but not for those containing significant percentages of magnesia. The transformation in dicalcium silicate can be retarded or suppressed by a number of substances of which magnesia is one. Dicalcium silicate can be present in such slags, and the slags may still be volume stable.

B. CaO–SiO_2–Al_2O_3–FeO

The chief interest in this system lies in its relation to the process of slag formation in the iron blast furnace. Melting of the partially reduced burden takes place far down in the furnace. As fusion proceeds the burden goes through a plastic sticky stage, which leads to a large decrease in the permeability of the stock column and a corresponding increase in the resistance to gas flow. There is a considerable pressure drop, i.e., a large buoyancy effect, and this, combined with the sticky conditions, can lead to erratic stock movement and therefore poor and inefficient operation. It is to be expected that the shorter the range over which melting takes place the narrower will be the melting zone and the less the effect on stock movement.

It is known from the temperature distribution in the furnace that the temperature rise in the fusion zone is very rapid, also the ferrous oxide, which is a constituent of the primary slag, is reduced very quickly once the liquid slag comes into contact with the coke. Not only are the conditions difficult to specify precisely but equilibrium in the melting slag cannot be expected. The equilibrium diagram therefore is only a guide to the changes that take place; nevertheless, it is a very useful one. It may also be noted that, in any system, the nearer the approach to equilibrium, the narrower is the melting range.

There are two principal types of blast-furnace burden, in (a) the flux is charged as a separate constituent and in (b) the flux is incorporated into the charge before the furnace, usually as so-called fluxed sinter. Within each type there are many composition permutations depending on the proportion of gangue in the ore, composition of the gangue, and the coke ash that enters the slag during combustion in the tuyere zone. It is only possible, therefore, to consider typical examples.

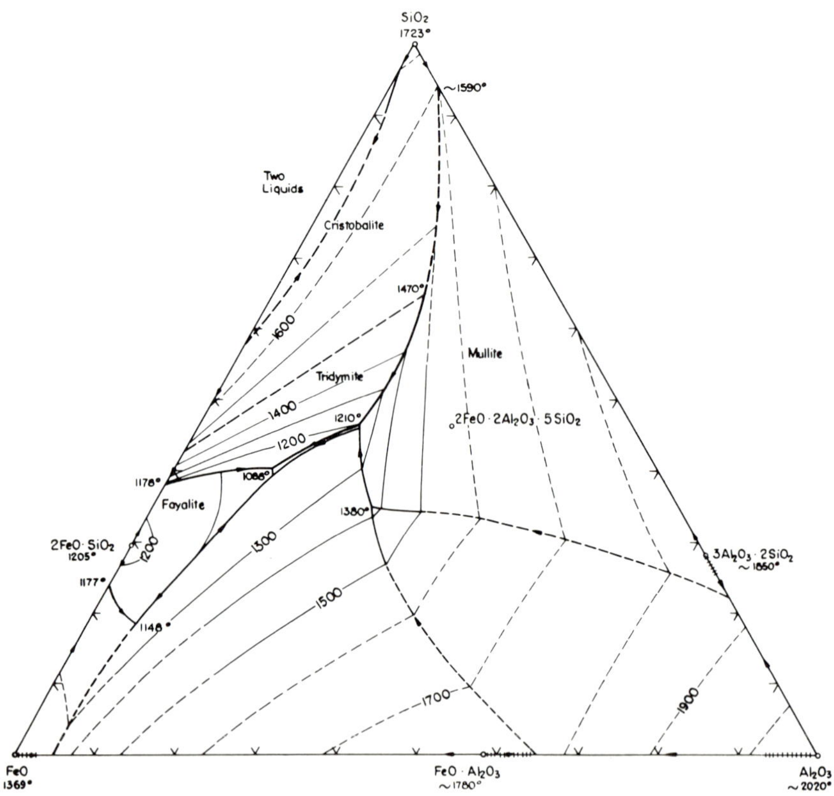

Fig. 4. $FeO–Al_2O_3–SiO_2$. From "Phase Diagrams For Ceramists," p. 241. Am. Ceram. Soc., Columbus, Ohio.

Consideration of type (a) is based on the $FeO–Al_2O_3–SiO_2$ system, Fig. 4. Slag formation begins with the mixture of unreduced ferrous oxide and the alumina and silica of the gangue. As seen from Fig. 4, melting begins at just under 1200°C, and the higher the ferrous oxide content the greater is the amount of liquid formed at low temperature; but regardless of the initial content, as the slag becomes liquid, the ferrous oxide content is reduced and the composition moves toward the $Al_2O_3–SiO_2$ binary in which the minimum liquidus temperature is 1595°C.

The most favorable melting conditions are those for which the slag composition moves along the eutectic trough leading toward the $Al_2O_3–SiO_2$ binary. This corresponds approximately to an Al_2O_3/SiO_2 ratio of 13, which on average would give an alumina content of 16–18% in the final slag (Taylor, 1962). The melting range will be extended for alumina concentrations on either side of this optimum, but more particularly on the high-alumina side. This picture is of course modified by the fact that the composition of the liquid slag is continually changing owing to absorption of lime. Nevertheless, it may be concluded that a high ferrous oxide content and a high Al_2O_3/SiO_2 ratio tend to give a long melting range. The former may be due either to poor reduction efficiency or low gangue content. In the latter case the lower total slag volume more than offsets the effect of the high ferrous oxide concentration. The former implies that inefficient furnace operation may contribute to poor physical conditions in the fusion zone, which in turn further impairs furnace efficiency.

Consideration of type (b), e.g., fluxed sinter, is based on the quaternary system. This has been only partially investigated, but two ternary sections are useful in the present context. These are those for $2CaO{\cdot}Al_2O_3{\cdot}SiO_2$–$CaO{\cdot}SiO_2$–FeO and $2CaO{\cdot}Al_2O_3{\cdot}SiO_2$–$2CaO{\cdot}SiO_2$–FeO (Fig. 5 a, b). If tie lines are drawn on Fig. 1 from gehlenite to the calcium meta- and orthosilicates, it is seen that typical blast-furnace slag compositions lie about midway between these two tie lines, and the more basic bosh slags rather nearer the gehlenite–orthosilicate tie. As a first approximation melting may be considered in terms of Fig. 5. Three alumina compositions representing typical high, medium, and low concentrations are shown on the base line of the silicates and tie lines drawn to the ferrous oxide corner. Typical ferrous oxide concentrations when melting begins should be 30–50%. It can be seen that for all three types melting should begin about 1200–1250°C and the amount of liquid formation between 1300° and 1400°C would seem to be very similar for the high and medium alumina contents. At the low-alumina composition, there is a considerable difference between the two diagrams, and it would be dangerous to draw conclusions for compositions intermediate between the two cases covered by the diagrams. If account is taken of the fact that reduction of iron oxide takes place as melting proceeds,

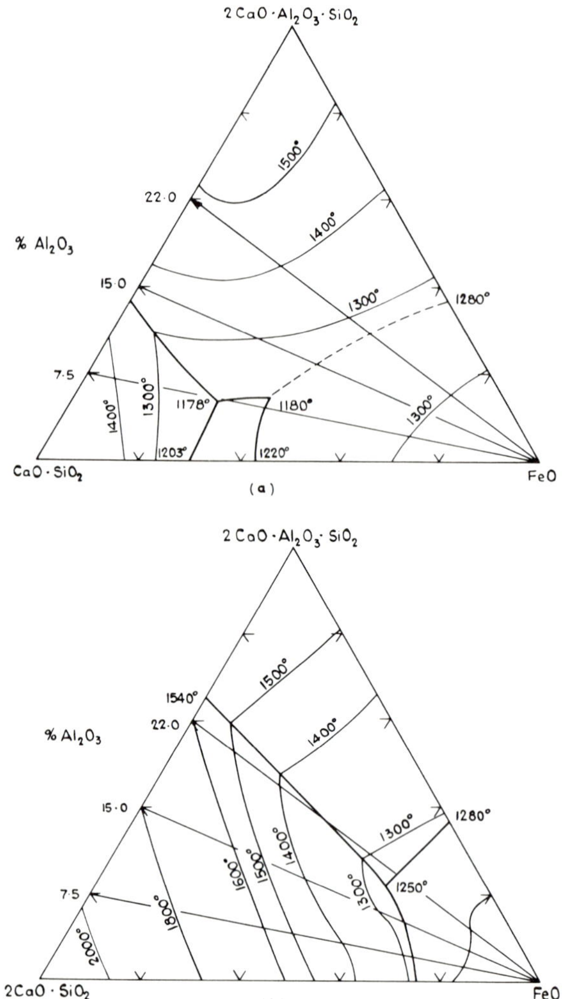

Fig. 5. (a) $2CaO{\cdot}Al_2O_3{\cdot}SiO_2$–$CaO{\cdot}SiO_2$–FeO. (b) $2CaO{\cdot}Al_2O_3{\cdot}SiO_2$–$2CaO{\cdot}SiO_2$–FeO. From A. Muan and E. F. Osborn, Yearbook Am. Iron Steel Inst. 1951, pp. 321–359.

then the final melting temperature should be that of the bosh slags. The calculated liquidus temperatures for the three examples are 1500°, 1450°, and 1500°C, respectively. The medium-alumina slag should therefore have the slightly shorter melting range, but alumina does not appear so important in this respect as for the case when lime is charged separately. What does seem certain is that in all cases the melting range for a prefluxed charge should be shorter, the more particularly since departure from equilibrium melting should be less in this case. This is an important conclusion for blast-furnace operation, implying as it does that, quite apart from other considerations, a prefluxed charge should give a smoother and, therefore, harder-driving furnace.

Up to a point the sintering of iron ore can be considered to be similar to the early stage of slag formation in the furnace. The main difference is

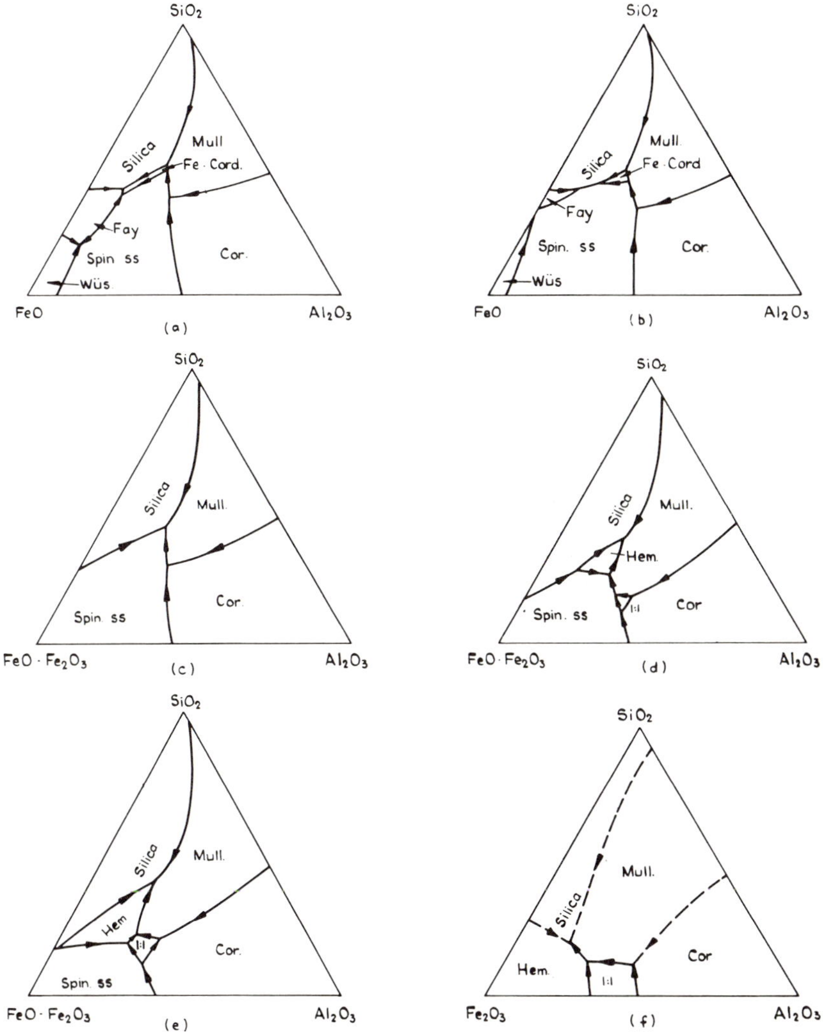

Fig. 6. $FeO(Fe_2O_3)$–Al_2O_3–SiO_2. (a)–(f) Effect of increasing oxygen pressure. From "Phase Diagrams for Ceramists," p. 76. Am. Ceram. Soc., Columbus, Ohio and Muan, A. (1957). *J. Am. Ceram. Soc.* **40**, 429.

that the oxygen pressure is high to begin with, decreases to a low value in the fusion zone, and then increases again. Iron oxide, which is normally ferric oxide as charged, is largely reduced to the ferrous state in the fusion zone and is subsequently reoxidized. These changes are extremely rapid as the time cycle is measured in minutes rather than hours and, despite the small particle size and intimate mixing, it is not to be expected that equilibrium will be attained. Sinter quality is closely allied to the final degree of oxidation, which is a function of the reduction during fusion and reoxidation during subsequent cooling. The effect of oxygen pressure in the FeO (Fe_2O_3)–Al_2O_3–SiO_2 system, which applies to simple iron ore sintering, has been very adequately described in a series of diagrams by Muan (1957), some of which are reproduced in Fig. 6. These show that, while at low oxygen pressures and under equilibrium conditions the iron should be present as fayalite, wüstite, and spinel (magnetite–hercynite) solid solution, with increasing pressures first the spinel phase grows and then breaks down to give hematite. Under operating conditions it is very probable that some of the highly siliceous slag formed by partial fusion will undercool and remain as a glass even in the finished product and the iron oxide in solution will oxidize much less easily than when present in a crystalline form.

The partial fusion of fluxed sinter, on the other hand, should be considered through Fig. 5. The liquid formed in the fusion zone again has a high ferrous oxide content but also contains lime as well as alumina and silica in solution. On resolidification the other oxides crystallize as compounds while the iron should largely crystallize as wüstite with a little olivine. Because of the less siliceous nature of the melt, crystallization is more probable and reoxidation to hematite kinetically more favorable. It should therefore be easier to have a high state of oxidation and good reducibility characteristics in fluxed sinter. While this deduction is substantially correct conditions in practice are not quite so straightforward. Not all the iron oxide is reduced to the ferrous state, even under the reducing conditions in the combustion zone. Calcium ferrites are found in these fluxed sinters, which implies that ferric oxide plays some part in the process. There are, however, few data on the CaO–Al_2O_3–SiO_2–Fe_2O_3 system and what there are, are not relevant to the present problems. Meanwhile, the precise role of ferric oxide on the fusion process must remain a subject of conjecture.

C. CaO–FeO–SiO_2

Just as iron smelting slags are based on a silica-alumina gangue with lime added as a flux, so nonferrous smelting slags are based on a silica–iron oxide gangue and lime as a flux. The appropriate equilibrium diagram is

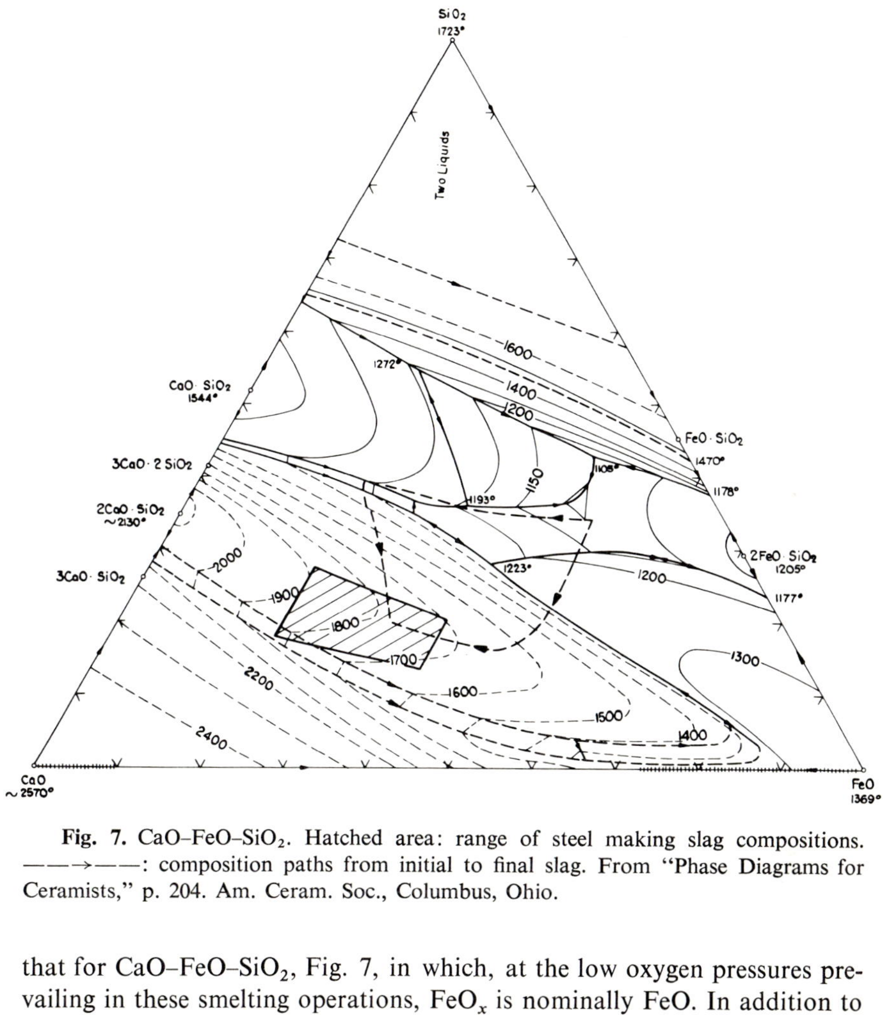

Fig. 7. CaO–FeO–SiO_2. Hatched area: range of steel making slag compositions. ——→——: composition paths from initial to final slag. From "Phase Diagrams for Ceramists," p. 204. Am. Ceram. Soc., Columbus, Ohio.

that for CaO–FeO–SiO_2, Fig. 7, in which, at the low oxygen pressures prevailing in these smelting operations, FeO_x is nominally FeO. In addition to the main oxides, these nonferrous smelting slags frequently contain significant proportions of alumina, arising from coke ash in blast-furnace smelting and metal oxides such as tin and zinc. Few data are available on these more complex systems, and the simpler systems are used as an approximate guide. The metals to be considered in this context are lead, tin, and zinc; copper could be included, but in this case, oxide smelting has been displaced by matte smelting.

In iron smelting the main factors were control of sulfur and silicon; iron loss in the slag was of no significance. These same factors plus volatilization and fume formation are common to nonferrous smelting, but the relative importance varies for the different metals. Metal loss in the discard slag both on technical and economic grounds is always an important consideration. It is bound up with the problem of reduction of a gangue oxide, ferrous oxide in this case through reaction (1), in this instance:

$$MO_{(slag)} + Fe = FeO_{(slag)} + M \tag{1'}$$

The simplest of the three cases is that of lead. The negative free energy of formation of lead oxide is much less than ferrous oxide so that the equilibrium in reaction (1′) is well over to the right. Furthermore, iron is virtually insoluble in liquid lead so that low values of lead oxide in slag solution can be obtained without reduction of iron oxide, even when the activity of the ferrous oxide is high. Similarly the solubility of sulfur in liquid lead is negligible and a desulfurizing slag is unnecessary. The chemical properties of the slag are therefore not important, it is only necessary to have a suitable low-melting composition. As can be seen from Fig. 7, there are low-melting slags in the ferrous oxide–silica binary but the limiting composition is $SiO_2/FeO = 40/60$ at smelting temperatures. For the higher ratios of silica to iron oxide normally present in the concentrates, a low-melting slag can be most economically achieved by the addition of 10–20% lime.

The main problem in tin smelting is that of metal recovery. Sulfur is not very important because the solubility in tin is negligible. Stannous sulfide can contribute to the fume problem because of its relatively high vapor pressure, but the sulfur content is usually low and it is not a major factor. The problem in metal recovery is that the standard free energies of stannous oxide and ferrous oxide at smelting temperatures are not very different and iron is fairly soluble in liquid tin. As the stannous oxide in the slag is reduced, so the concentration of iron in tin increases. The equilibrium concentrations can be influenced by temperature and slag composition. In the case of tin, the standard enthalpy change for reaction (1′) is positive but for the actual reaction conditions the partial molar heat of solution of iron in tin is also required. This should be positive since there is a large miscibility gap in the iron–tin system. Probably, therefore, the reaction enthalpy change is small and the temperature coefficient also small. With regard to the effect of slag composition, there are data on the activity of ferrous oxide in this system but not for stannous oxide, but empirical studies indicate that the ratio activity coefficient stannous oxide/activity coefficient ferrous oxide increases with lime concentration. A suitable slag composition would therefore have maximum lime and minimum iron oxide concentrations and a low liquidus temperature. The latter is desirable to minimize volatilization of stannous oxide. In terms of the $CaO–FeO–SiO_2$ diagram this would be about 25% CaO, 40% FeO, 35% SiO_2 depending upon the FeO/SiO_2 ratio in the concentrates. Industrial slags frequently contain 20% or more of other oxides, but if the composition is calculated onto a $CaO–FeO–SiO_2$ basis the above is a good guide.

Zinc presents slightly different problems as it is not a true smelting process. In retort distillation slag formation is to be avoided and, as Fig. 7 shows, that is simply a question of maintaining a high SiO_2/FeO ratio in the gangue. Blast-furnace smelting requires slag formation, and the problems are essentially the same as in lead and tin smelting. The metal recovery equilibrium in this case can be written

$$ZnO_{(slag)} + Fe = Zn(g) + FeO_{(slag)}$$

The temperature coefficient for this reaction is large, and increase in temperature favors zinc recovery. Sulfur is also a problem in the blast furnace process. Formation of zinc sulfide not only represents a loss in yield, but, because it has a high melting point and is insoluble both in slag and lead (which is always produced along with the zinc), it tends to give dirty, difficult conditions on the hearth. The first step in sulfur control is to increase the lime content of the slag. Reference to the diagram, however, shows that the possibilities in this direction are restricted by the high-melting dicalcium silicate field, which extends across the diagram almost to the iron oxide corner. Even operating at a rather higher temperature is not much help because the liquidus surface is so steep that, e.g., a rise of 100°C in liquidus temperature only increases the lime solubility by about 2%. The direction of these liquidus isotherms corresponds very nearly to a constant CaO/SiO_2 ratio, i.e., lime and silica can only be increased together. Sulfur-capacity data for this system are not known, but one would not expect rapid changes along the middle part of this dicalcium silicate liquidus surface, nor does the activity of ferrous oxide vary much in this region. It may be concluded that the optimum composition of slag should lie near the dicalcium silicate saturation surface, the exact position being largely determined by the iron oxide/silica ratio in the zinc calcine.

Basic steel-making slags are multicomponent, having four or five main and two or three minor components. Modern steel-making practice tends to separate the slags into two main groups, those of low phosphorus pentoxide content produced by the open-hearth, basic electric, and top-blown oxygen converter and those of high phosphorus pentoxide content produced in the basic Bessemer and oxygen-lime–blown converters, e.g., LDAC and OLP. The former group may be considered as a first approximation in terms of the $CaO–FeO–SiO_2$ system when the magnesia is included with the lime, manganese oxide with the iron oxide and phosphorus pentoxide with the silica.

Basic steel-making slags are usually characterized by the $CaO/SiO_2 + P_2O_5$ or V ratio. This is taken as an empirical measure of the basicity or lime activity, which requires to be high for dephosphorization and desulfurization. In practice the V ratio varies between about 2.2 and 3.5 and the iron oxide plus manganese oxide content between about 20% and 40%.

The corresponding range of compositions is indicated in Fig. 7, the main features of this range of compositions in the simple system are that (a) it is in the high-melting dicalcium silicate field, (b) the liquidus temperatures decrease somewhat as the V ratio increases, and (c) the liquidus temperature decreases with increase in the ferrous oxide concentration.

Although it is evident that final steel-making slags must be very close to the liquidus temperature, it is probable that they should be just liquid at the final temperature under equilibrium conditions. The most critical conditions should correspond to the manufacture of high-carbon steels when both the iron oxide content and the final temperature are low. A high manganese charge helps in this respect by increasing the manganese oxide content of the slag. The main problem with these slags, however, is the attainment of the final equilibrium composition. Slag formation takes place by the SiO_2–FeO_x–MnO produced by oxidation of the metal, taking into solution the lime that has been included in the charge. It is very easy for the particles of lime to become coated with the high-melting dicalcium silicate, and this retards the solution of lime to such an extent that such coated particles remain undissolved in the final slag. Lime not in solution is useless for chemical refining, and such slags are less basic than appears from the nominal composition.

The tendency to form this dicalcium silicate layer can be illustrated by two extreme cases of the way the slag composition may change as lime goes into solution. The two routes are shown in Fig. 7. In each case the starting point is an assumed high iron oxide (manganese oxide) of 50% and a low lime of 15% corresponding to an early stage in slag formation. By route (a) the dissolution of lime up to 40% is accompanied by a drop in iron oxide to 20% after which, in order to attain the final composition, the slag composition has to cross a ridge on the dicalcium silicate liquidus surface. Dicalcium silicate must form on the lime particles and can only dissolve by a very slow solid diffusion process. By route (b) the solution of lime takes place with the "iron oxide" concentration maintained at about 50% until it begins to approach the final composition. In this case no coating of dicalcium silicate need form. The latter case is an unreal one in the context of either the open hearth or the top blown oxygen converter but it illustrates the most favorable conditions for lime solution. High iron oxide is essential and, in both the above-named processes, a highly oxidized slag can be formed in the early stages before the carbon boil begins in earnest. Thereafter, the iron oxide is largely controlled by the carbon content of the metal. It should also be noted that in the highly oxidized slag the Fe^{3+}/Fe^{2+} ratio is greater, and this tends to lower the liquidus temperatures and also make the contours less sharp than they are in Fig. 7. This also applies to manganese oxide so that a high manganese charge, which also gives a high iron oxide content is also favorable to lime solution.

The best conditions for "making up" the refining slag may be summarized as (a) a high manganese charge or, more correctly, a high Mn/Si ratio as shown in the section on the $FeO–MnO–SiO_2$ system, (b) a highly oxidized early slag, (c) crushed and graded lime, (d) agitation in the early stage to promote rapid dissolution of the lime. Conversely, late addition of lime to correct for earlier miscalculation is poor practice as such lime must be difficult to dissolve. The importance of an early highly oxidized slag is illustrated by the comparison between the top-blown and bottom-blown converters. In the former the oxygen gradient is gas→slag→metal and the oxygen potential of the slag can be much higher than that of the metal, whereas, in bottom blowing the reverse is the case. This is reflected in the two processes. In the top-blown process, the lime is taken into solution early and dephosphorization can proceed alongside carbon oxidation. In the Bessemer, lime solution does not take place until carbon is almost eliminated and dephosphorization has to be effected in the so-called "after blow."

D. Fe–Si–O

In the oxide smelting processes, the oxygen pressure is so low that it is permissible to regard the iron oxide as in the ferrous state even though some ferric iron must always be present. This is not so in the matte smelting and converting of copper and nickel. For these slags both oxygen pressure and temperature are important variables, especially with regard to magnetite precipitation. The pertinent equilibrium diagram is shown in Fig. 8.

The oxygen pressure of the matte slag system at equilibrium can be calculated from the reaction

$$FeO + SO_2 = FeS + 3/2\ O_2 \tag{6}$$

for which

$$\log K = -25{,}800/T + 24.0/4.575 \tag{7}$$

If, for reverberatory smelting, $p_{SO_2} \approx 1$ atm, a_{FeO} in silica saturated slag is $\sim$0.4 and a reasonable value for $a_{FeS} = 0.2$ is assumed, then $p_{O_2} = 2 \times 10^{-7}$ and 6×10^{-8} for 1400°C and 1300° respectively. Reference to Fig. 8 shows that in neither case should the slag be saturated with regard to magnetite. Since magnetite does separate in the reverberatory furnace, equilibrium is not established. When the conditions of smelting are considered, this is not surprising. A rather heterogeneous mixture of oxides and sulfides is melted down, and the molten matte and slag then flow into the settling zone where placid conditions are maintained to encourage separation. Any magnetite in the calcines that is not reduced into the slag in the melting zone, because of

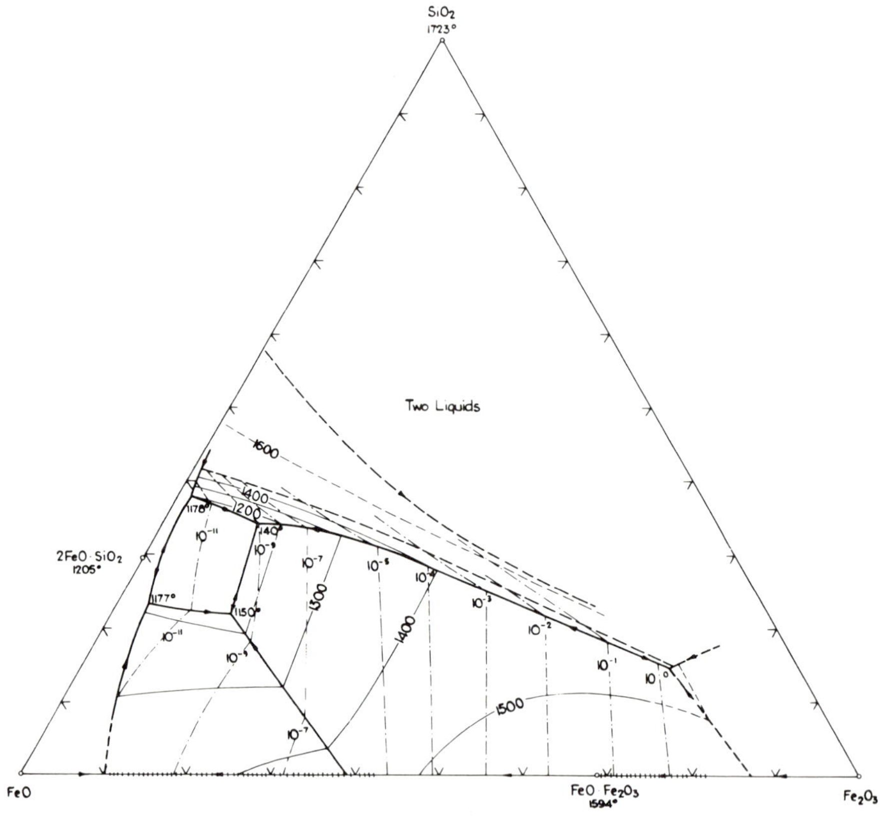

Fig. 8. FeO–SiO_2–Fe_2O_3. ———: line of equal oxygen pressure. From "Phase Diagrams for Ceramists," p. 60. Am. Ceram. Soc., Columbus, Ohio.

the high density, gradually sinks through the matte and builds up on the furnace hearth. Once magnetite enters the matte phase, it can only be reduced by the reaction

$$3\ Fe_3O_4 + FeS = 10\ FeO + SO_2 \tag{8}$$

In order that this reaction should proceed, p_{SO_2} must exceed 1 atm and this requires $a_{FeO} < 1$. Ferrous oxide partitions between slag and matte and at equilibrium $a_{FeO} \approx 0.4$ but, in the placid conditions, transfer from matte to slag must be very slow, the solubility of ferrous oxide is small and locally the concentration can readily increase so that reaction (8) is suppressed. In practice this difficulty is surmounted by addition of green ore that contains some silica so that slag can form at the reaction site. In order to avoid magnetite separation, reaction between slag and matte must take place before they separate, i.e., in the melting zone. The most important factor in this respect is temperature. Reference to Fig. 8 shows that at 1400°C and $p_{O_2} \times 2 = 10^{-7}$ atm the slag is further from magnetite saturation than at 1300°C, and $p_{O_2} = 6 \times 10^{-8}$ atm which, together with more rapid reaction rates, means that higher temperatures should promote solution of magnetite. The effect of grade of matte is rather uncertain. The less the ferrous sulfide content, the higher the equilibrium pressure but the less the iron oxide to be taken into solution. Since even very rich grades should produce an equilibrium slag well below magnetite saturation the amount of iron oxide is probably more important than the grade of matte. A further point brought out by the diagram is that, at any given operating temperature, the solubility of silica decreases with increase in oxygen potential, which may require the addition of extra flux such as lime.

In the converter similar considerations apply. Even at the end of the blow to white metal, the equilibrium partial pressure of oxygen should be 5×10^{-6} and 1×10^{-6} at 1400° and 1300°C, respectively. Either then, some of the iron sulfide is oxidized to magnetite and does not dissolve, as in the reverberatory, or the oxygen potential of the slag is raised well above the equilibrium value and magnetite crystallizes and settles out on the lining. Growth of magnetite is most pronounced toward the end of the blow when the oxygen potential and amount of slag are greatest. The next charge of matte is reducing with respect to magnetite and, encouraged by the highly turbulent conditions, magnetite is reduced and taken into solution. In the converter, therefore, the thickness of magnetite undergoes a cyclic variation and, provided this can be controlled, it serves a useful purpose in protecting the magnesite lining.

E. MgO–FeO–SiO_2

The interest of this system lies in the fact that magnesite is the most widely used refractory for contact with metallurgical slags. The tendency for reaction between a FeO–SiO_2 slag and magnesite refractory can be seen from Fig. 9. Magnesia and wüstite form a continuous series of solid solutions so the resistance to wüstite is excellent, but the trend of the liquidus lines shows that the resistance decreases with increase in silica content of the slag. However, even for a silica saturated slag the resistance should be fairly good at 1400°C but it decreases rapidly with temperature and slag attack at 1600°C can be expected to be fairly severe.

Industrial slags are not so simple; there may be substantial proportions of ferric oxide, lime, and manganese oxide singly or together. Only in the first case are data on the relevant system available. These show that as the

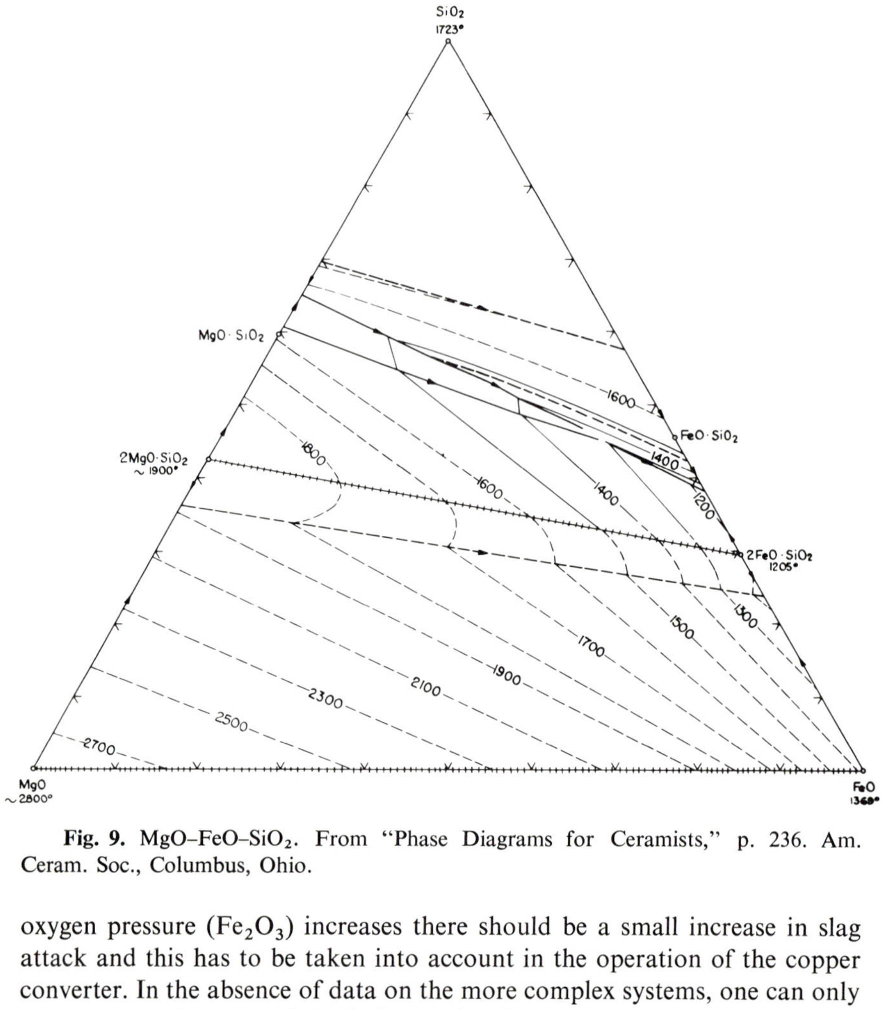

Fig. 9. MgO–FeO–SiO_2. From "Phase Diagrams for Ceramists," p. 236. Am. Ceram. Soc., Columbus, Ohio.

oxygen pressure (Fe_2O_3) increases there should be a small increase in slag attack and this has to be taken into account in the operation of the copper converter. In the absence of data on the more complex systems, one can only assume from the general similarity of the CaO–FeO–SiO_2 and MgO–FeO–SiO_2 systems on the one hand, and the MgO–MnO–SiO_2 and MgO–FeO–SiO_2 on the other that the effects of CaO, MgO, and FeO, MnO will be supplementary. The importance of this with respect to steel making is discussed in the following section.

F. FeO–MnO–SiO_2

The only feature of this system of real importance to the extraction metallurgist is the delimitation of the field of silica saturation at 1500°–1600°C. Slags of this system are the initial product of oxidation in the steelmaking process and also of the deoxidation of steel with silicon and manganese. The composition of the oxide product is determined by the reactions

$$\underline{Si} + 2\,\underline{O} = SiO_2 \tag{9}$$

$$\underline{Mn} + \underline{O} = MnO \tag{10}$$

$$Fe + \underline{O} = FeO \tag{11}$$

Combination of the first two equations leads to

$$K_{Si-Mn} = (a_{MnO}^2/a_{SiO2})(a_{Si}/a_{Mn}) \tag{12}$$

and for the latter two

$$K_{Mn} = (a_{MnO}/a_{FeO})(1/a_{Mn}) \tag{13}$$

From the thermodynamics of these reactions, the equilibrium composition of the product and the oxygen in solution can be calculated in terms of the manganese and silicon concentrations in the metal. An example of this for the most important temperature of 1600°C is shown in Fig. 10, in which is also shown the silica saturation boundary.

The most important application of this is to deoxidation. In the classic theory of deoxidation, separation of the product from the liquid steel has always been ascribed to Stokes-law-type flotation. It is now appreciated that in the separation of alumina particles produced by aluminum deoxidation surface energy forces play the major part, but flotation remains the more

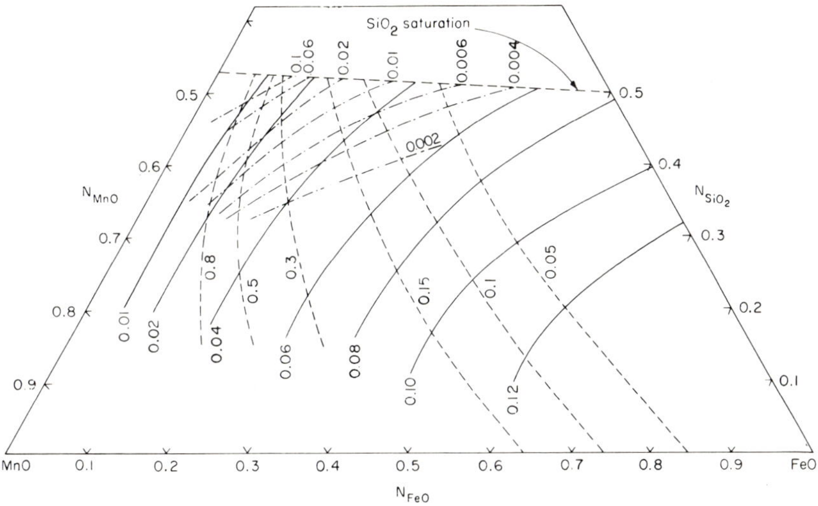

Fig. 10. FeO–MnO–SiO$_2$. Composition of steel in equilibrium with FeO–MnO-SiO$_2$ at 1600°C. —: % O, ---: % Mn, -·- : % Si. From Bell, H. B. (1963). *J. Iron Steel Inst. (London)* **201,** 116.

important mechanism for silicon–manganese deoxidation. In the latter case liquid particles are to be preferred, for not only have spherical drops a smaller Stokes diameter than irregular crystalline particles of equivalent mass, but agglomeration of liquid drops is more probable. The aim, therefore, is to produce a liquid product, i.e., one that is not silica saturated. From Fig. 10 it can be seen that for decreasing oxygen content in the metal the residual manganese and silicon must increase and the slag composition move toward the $MnO–SiO_2$ binary.

In acid steel making, as in deoxidation, the composition of the initial slag is determined by the Mn/Si ratio. A low ratio gives a semisolid slag that is in fact preferred in one type of acid Bessemer practice. A high ratio gives a fluid-reactive slag that is preferred in Swedish Bessemer practice. In each case, as refining proceeds and the oxygen content increases, the iron oxide content of the slag increases. Slag composition is not, however, very important in acid steel making since neither dephosphorization nor desulfurization is carried out.

Basic steel-making slags are formed by the initial slag taking lime into solution, so that while the final slag is strongly basic in character, the initial slag is acid. This clearly presents a problem with regard to refractory attack. The best refractory for the basic slag is magnesite or dolomite, and reference to Fig. 9 shows that at steel making temperatures severe attack can be expected from a $FeO–MnO–SiO_2$ slag that is close to silica saturation. The effect of this is most pronounced in the case of the basic Bessemer process. As mentioned earlier, solution of lime in this process takes place comparatively slowly and slag attack can be severe. In order to minimize this, the silicon content of the hot metal must be kept to a minimum, and this reduces the flexibility of the process. Phosphorus and silicon are the chief fuels in the converter but, because silicon must be kept to a minimum, they are not interchangeable and this means that the phosphorus content of metal for the basic Bessemer process must be quite high, viz. $\geq 1.5\%$.

This limitation does not apply to the top-blown oxygen converter because the solution of lime takes place so quickly, even so a high Mn/Si ratio in the hot metal to give a less silicous initial slag is certainly to be preferred. The problem is less serious in the open-hearth furnace. Because of the large surface area the relative area subject to slag attack is small and the construction allows the hearth to be maintained by fettling.

G. $CaO–FeO–SiO_2–P_2O_5$

In an earlier section basic steel-making slags with a low phosphorus pentoxide content were considered in terms of the simple $CaO–FeO–SiO_2$ system. This approximation cannot be used for slags high in phosphorus pentoxide for which the quaternary $CaO–FeO–SiO_2–P_2O_5$ is the simplest system to have any relevance. Data on this system are few, but the more important features for the range of compositions that includes steel-making slags at a temperature of 1600°C have been determined in the last few years largely by Trömel *et al.* (1967) and Drewes and Olette (1967).

The most important features from the metallurgists point of view can perhaps best be described by Figs. 11 and 12. In the $CaO–FeO–P_2O_5$ system

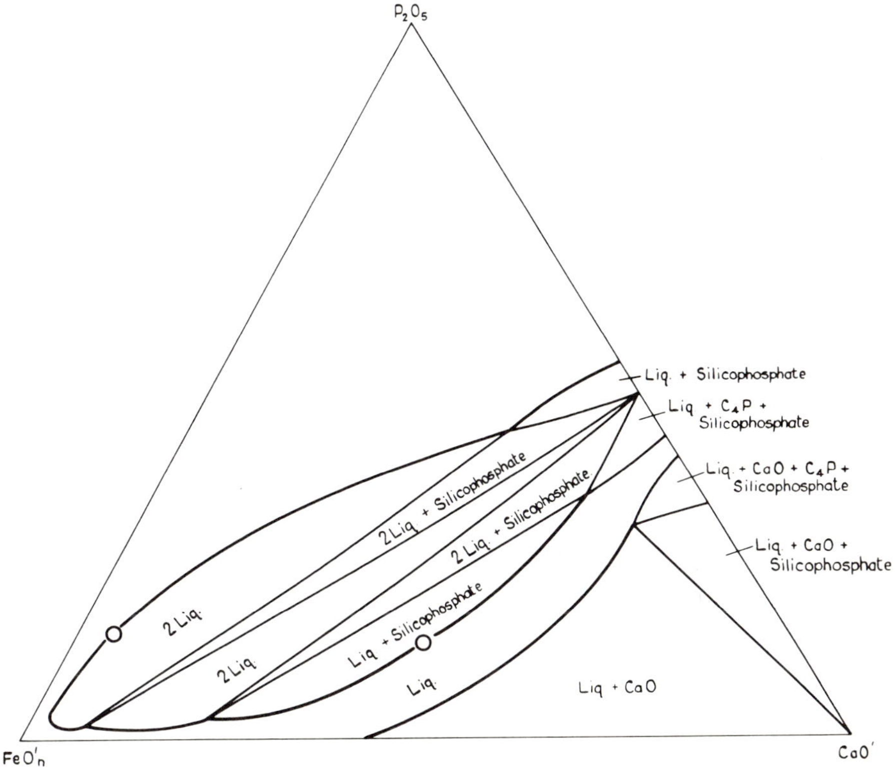

Fig. 11. $CaO–FeO–P_2O_5$ at 1600°C. From G. Trömel, W. Fix, and K. Koch (1967). *Arch. Eisenhüttenwes.* **38,** 177–184.

(Fig. 11), the field of the high melting compound $3CaO·P_2O_5$ extends well into the ternary toward the FeO apex. Adjoining this field on the P_2O_5 side is a two-liquid miscibility gap. On the other side is a narrow liquid zone extending from the eutectic between $3CaO·P_2O_5$ and $4CaO·P_2O_5$ across the field. On the CaO side this homogeneous liquid field is bounded by the $4CaO·P_2O_5$ and CaO saturation surfaces. It is separated from the main homogeneous liquid region by the as yet somewhat ill-defined $3CaO·P_2O_5$

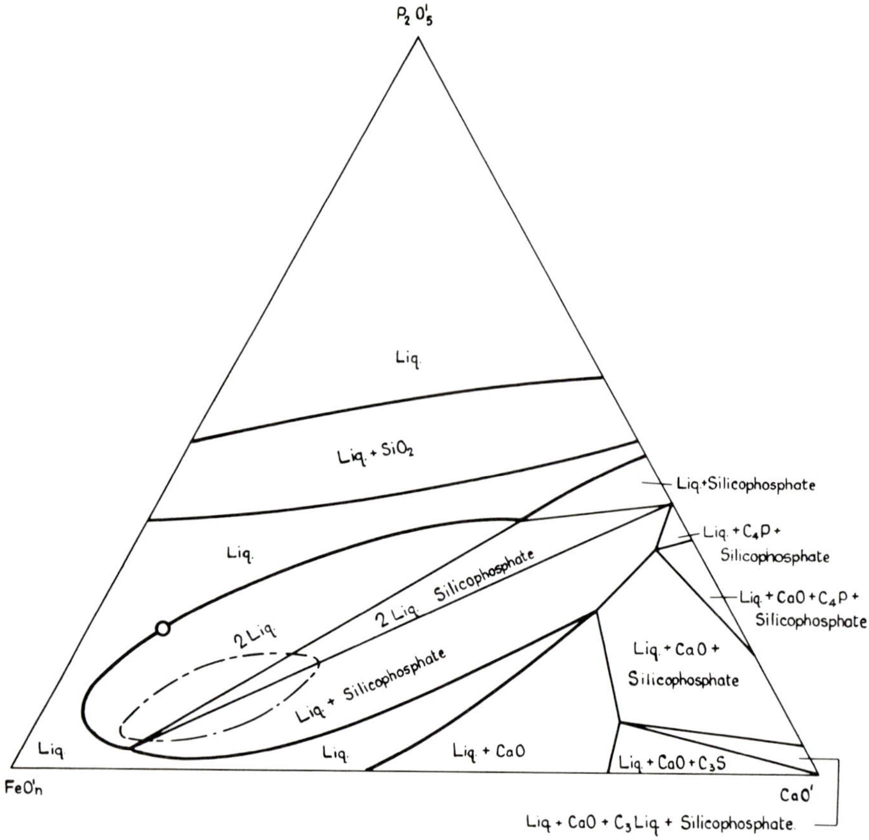

Fig. 12. $CaO–FeO–P_2O_5–SiO_2$ at 1600°C. Section at 5% SiO_2. —·—: Lens-shaped two-liquid field at 10% SiO_2. E. J. Drews, and M. Olette (1967). *Arch. Eisenhüttenwes.* **38,** 163–175

and two liquid fields. In this respect it shows a certain similarity with the $CaO–FeO–SiO_2$ field in which the dicalcium silicate "nose" intrudes between the main liquid field and the trough between the dicalcium silicate and lime fields.

The addition of silica to this system influences the portion of the diagram under discussion in two ways. Tricalcium phosphate forms a continuous series of solid solutions with the high-melting dicalcium silicate and the mutual solubility in the two-liquid field increases until at about 15% silica there is one homogeneous liquid. The effect of this is that with increasing silica concentration the two-liquid region detaches itself from the tricalcium phosphate, now silicophosphate solid solution field, first to form a lens-shaped region at between 5 and 10% SiO_2 and then to disappear at <15%

SiO_2. At the same time the silicophosphate ridge is enlarged and the liquid trough between this field and lime is reduced. Full details of these changes are not yet determined but the beginning of the process is shown in Fig. 12 after Trömel *et al.* (1967) for the $CaO'–FeO'–P_2O_5'$ system containing 5% SiO_2. The lens-shaped two-liquid field at 10% SiO_2 has been determined by Drewes and Olette (1967) and is superimposed on the Trömel diagram.

The chief metallurgical application of these data is to basic slags of high phosphorus pentoxide content, such as arise in the basic Bessemer and top-blown oxygen-plus-lime-powder, LDAC, OLP, processes. It can be shown that 5% P_2O_5, a concentration reached in many basic slags, can significantly emphasize the dicalcium silicate nose in the $CaO–FeO–SiO_2$ system and thus affect lime solution. But it is also known (Trömmel *et al.*, 1967) that manganese oxide has the opposing effect and, until the complex slag system is more thoroughly studied, too much emphasis cannot be laid on these minor effects in LDAC slags. As regards the high phosphorus pentoxide slags, chemical and mineralogical analyses have shown that saturation with the silicophosphate solid solution occurs in most cases and that the silicophosphate field acts as a barrier against the slag reaching lime saturation. It is difficult to see how this can be avoided in the case of a first slag, which of necessity contains a fair concentration of silica and there therefore seems little point in trying for higher lime concentrations, which can only result in undissolved lime particles. These not only represent a waste of lime but increase the tendency of the slag to foam. It should be possible to make the second slag lime saturated, particularly if the silica is kept to a low value, and this is, of course, simply a question of removing the first slag cleanly and efficiently. A low-silica slag can, however, encounter difficulties in another direction, namely through two-liquid formation, which also increases the foaming tendency.

This system is of great importance, particularly with regard to increasing demand for low phosphorus contents in strip steel. Low phosphorus is essentially a question of forming a slag with a high lime activity and high oxygen potential. The $CaO–FeO–SiO_2–P_2O_5$ system, incomplete as it is, at least allows the nature of the problem to be appreciated. Because of the effect of silica, it raises the question of whether sufficient emphasis is laid on the silicon content of the hot metal or rather the Si/Mn ratio, as it is this that controls the SiO_2/MnO(FeO) ratio in the slag. The effect of manganese oxide itself requires investigation; the indications are that it is helpful in surmounting the silicophosphate barrier, but quantitative data are lacking.

The systems discussed are by no means the only ones of interest to the extraction metallurgist, but they are those of the widest common interest. Probably on this account they are most widely studied and detailed. Even so, as has been pointed out, they are usually simplifications of the more complex industrial slags. Data are required on the effect of the lesser constituents in these systems although equally, more precise knowledge of conditions in many industrial processes is required to make the best use of thermal equilibrium data.

References

Bell, H. B. (1961). *J. Iron Steel Inst.* (*London*) **197**, 245.

Bell, H. B. (1963). *J. Iron Steel Inst.* (*London*) **201**, 116.

Drewes, E. J., and Olette, M. (1967). *Arch. Eisenhüttenwes.* **38**, 163.

Fincham, C. J. B., and Richardson, F. D. (1954). *J. Iron Steel Inst.* (*London*) **178**, 4.

Kalyanram, M. R., MacFarlane, T. G., and Bell, H. B. (1960). *J. Iron Steel Inst.* (*London*) **195**, 58.

Kay, D. A. R., and Taylor, J. (1960). *Trans. Faraday Soc.* **56**, 1372.

Muan, A. (1957). *J. Am. Ceram. Soc.* **40**, 429.

Osborn, E. F., De Vries, R. C., Gee, K. H., and Kraner, H. M. (1954). *Trans. AIME* **200**, 33.

Rein, R. H., and Chipman, J. (1965). *Trans. AIME* **233**, 415.

Sharma, R. A., and Richardson, F. D. (1961). *J. Iron Steel Inst.* (*London*) **198**, 386.

Sharma, R. A., and Richardson, F. D. (1962). *J. Iron Steel Inst.* (*London*) **200**, 373.

Taylor, J. (1962). *J. Iron Steel Inst.* (*London*) **200**, 701.

Taylor, J. (1964). *J. Iron Steel Inst.* (*London*) **202**, 420.

Trömmel, G., Fix, W., and Koch, K. (1967). *Arch. Eisenhüttenwes.* **38**, 177.

VII

Intermediate Phases in Metallic Phase Diagrams

T. B. MASSALSKI and HORACE POPS

MELLON INSTITUTE OF CARNEGIE-MELLON UNIVERSITY
PITTSBURGH, PENNSYLVANIA

I. INTRODUCTION

It is well known that alloying of different metallic elements results in the formation of solid solutions, alloy phases, and compounds. A complete range of solid solutions from one element to another can occur only if their crystal structures are basically the same, but it does not always occur even when this condition is fulfilled. More commonly, a number of intermediate structures are obtained in most alloy systems. Such structures, often called *intermediate phases*, may span wide ranges of composition (for example, β-brass) in a given phase diagram, or may be restricted to rather narrow ranges near simple stoichiometric ratios, such as AB, A_3B etc., and hence resemble chemical compounds. Whichever is the case, the majority of intermediate phases possess structures that are different from the structure of either of the component elements. In numerous metallic systems intermediate phases also form at compositions that would require unusual formulas to describe them, such as, for example, Cu_5Si, Fe_5Zn_{21}, or KHg_{10}, and they indicate a departure from the simple rules of chemistry that relate compound formation to chemical valence. In such cases more important than individual valence appears to be the total count of electrons in the unit cell of the structure, or the electron concentration.

The terminology of the whole field of intermediate phases is fairly arbitrary and one may speak of alloy phases, intermetallic compounds, metalloids, valence compounds, electron phases, or interstitial compounds, according to the bonding forces, the atomic sizes, or the general nature of the phase diagram involved. The knowledge of the phase diagram is clearly of importance because the information about the phases that are formed, their structure, stability, and composition constitutes the starting point both in the understanding of the technical properties and in the formulation of the theories of alloying.

The term *intermediate phase* may suggest that, in addition to an intermediate composition, the characteristics and properties of the material might also in some way be intermediate between those of the constituent elements. As is well known, the opposite behavior is very frequently the rule in metallic systems; there is nothing intermediate in the brittleness of the γ-brass formed by a union between the relatively soft and ductile metals Cu and Zn, in the negligible thermal expansion coefficient of the Invar alloy (Ni_3Fe), in the high superconducting transition temperature of the intermetallic Nb_3Sn, or in the crystal structure of the FeCr σ phase. These examples underline the fact that the rules that correlate properties of intermediate phases with the properties of the starting elements tend to be complex and that in general the interaction of several factors must be considered. Nevertheless the studies of numerous equilibrium and phase diagrams have shown that some predictions of the structure and properties of intermediate phases are possible and much progress has been achieved in this field in recent years.

The role of intermediate phases in materials technology is often crucial. Employed per se, or as major constituents, they can impart unique properties to specialized materials, such as crucibles, nozzles, tools, tunnel diodes, conductors, magnets, etc., and as minor constituents they play the critical function in many commercial alloys such as steels, duralumins, bearings, magnets, rotor blades, etc., to mention only a few. Phases limited to narrow composition ranges, the *intermetallic compounds*, are more likely to play a role as dispersed minor constituents whose presence may be beneficial to strengthening. Of interest then are the factors that affect their formation, distribution, and stability, and the way in which their appearance in the phase diagram affects the stability and ranges of the adjoining phases. On the other hand, intermediate phases with wide solid solubility are more likely to be involved as major constituents and their role in technology may be critically related not only to their variable composition and properties, but also to the various degrees of metastability and instability to which such phases are often susceptible. Phase diagrams, unlike equilibrium diagrams, often properly include information about unstable or metastable phases. Hence, the knowledge of martensitic structures, massive transformations, ordering reactions, and metastable states becomes of special significance.

The major objective of this chapter is to review the more typical groups of metallic intermediate phases and to point out the ways in which their structures, properties and stability relate to the bonding mechanisms and phase diagrams. Since many technological applications involve the presence of intermediate phases in various conditions of metastability, we shall include also some information about high temperature alloys and low-temperature nonequilibrium phases.

II. THERMODYNAMIC DESCRIPTION OF PHASE STABILITY

An equilibrium diagram represents the temperature–composition relations corresponding to the minimum free energy of all possible phases and compounds (Cottrell, 1948). By considering the relative positions of the free-energy curves at several different temperatures, one can determine the composition limits of the phases at each of these temperatures with the use of the common tangent principle and then assemble the results into a composition–temperature diagram.

The relationship between the equilibrium diagram and certain thermodynamic properties is illustrated in Fig. 1. The Gibbs free energy can be written as an extensive property function,

$$G = E + PV - TS = H - TS \tag{1}$$

where E, P, V, T, S, and H are the internal energy, pressure, volume, temperature, entropy, and enthalpy, respectively. Although the quantities in the

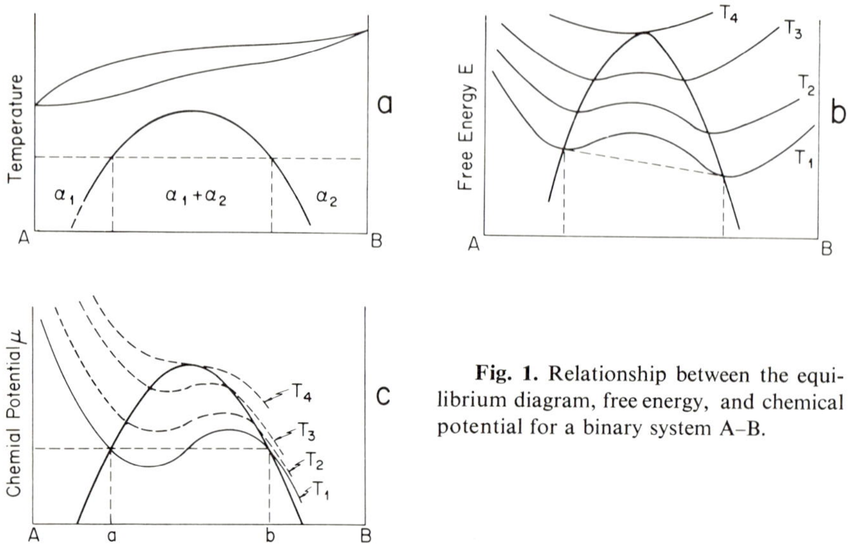

Fig. 1. Relationship between the equilibrium diagram, free energy, and chemical potential for a binary system A–B.

above equation can be measured experimentally with reasonable accuracy, their interpretation in terms of interactions between electrons and atoms is relatively little advanced and, even when properly conjectured, cannot be formulated into a quantitative theory.

From the point of view of thermodynamics a system in complete equilibrium may be said to be in a *mechanical* (uniform-pressure), *thermal* (uniform-temperature), and *chemical* equilibrium. Metallurgical phase diagrams are mostly constant-pressure diagrams, and the Gibbs free energy can be expressed as an extensive property function $G(P, T, N_1, N_2, \ldots)$ of the intensive parameters P and T and extensive composition parameters, the molar fractions $N_1, N_2, \ldots, N_i$ of the components in each phase. The uniformity of the chemical energy can be defined by the condition that the *chemical potential*, μ, of each component (or the *partial molal free energy*) is identical in all phases present. Thus, if N is the mole number of component I and the phases present are denoted 1, 2, 3, . . . , etc., the chemical potential of component I in phase 1 is given by

$$\mu_1{}^{\mathrm{I}} = \partial G / \partial N_1{}^{\mathrm{I}}|_{P,\,T,\ldots} \tag{2}$$

where

$$G = N_1\mu_1 + N_2\mu_2 + \ldots \tag{3}$$

at equilibrium,

$$\mu_1{}^{\mathrm{I}} = \mu_1{}^{\mathrm{II}} = \mu_{1\ldots}^{\mathrm{III}}, \qquad \mu_2{}^{\mathrm{I}} + \mu_2{}^{\mathrm{II}} = \mu_{2\ldots}^{\mathrm{III}} \tag{4}$$

The above requirement is the same as the condition that the composition of the coexisting phases be given by the common tangent to the free energy vs. composition curves. This is shown diagrammatically in Fig. 1 for a binary system A–B.

Some very successful free energy calculations have been attempted from first principles. References to such work and the general approach are discussed and illustrated in Volume I, Chapter II.

III. TYPES AND CLASSIFICATION OF INTERMEDIATE PHASES

Alloying produces solid solutions and alloy phases. If the size difference between the atoms involved is considerable, atoms of the one kind may be merely deposited in the holes (or the interstices) of the lattice formed by the other kind, resulting in *interstitial phases* or *compounds*. This occurs often when the nonmetallic elements, boron, oxygen, hydrogen, nitrogen, or carbon are alloyed with a metal.

Both interstitial and substitutional alloy phases can be *random*, with statistical distribution of atoms on lattice sites, or they may be partially or fully *ordered*. Most intermetallic compounds that occur at stoichiometric ratios of components are fully ordered and remain ordered up to the melting point. Phases with wider solubility may also include a stoichiometric composition and are then likely to be ordered only below a certain critical temperature when they undergo an order–disorder transition to form a *superlattice*. Superlattice formation changes the properties of intermediate phases because it affects numerous parameters such as stability, slip characteristics, transformation kinetics, lattice parameters, elastic constants and many others.

Within the vast number of intermediate phases now known, certain crystal types fall into groups that can be classified (to a varying degree of success) according to schemes that identify certain common features. One such classification may refer to the major binding forces involved: metallic, ionic, or covalent; another to the position in the periodic table of the starting elements; yet another to the major element that is involved. One can thus speak of the *alloys of the noble metals*, of the *transition elements*, the *rare earths*, etc., or of *Cu-based*, *Mg-based*, *Fe-based*, *Si-based*, etc., structures. More meaningful is a division of phases into similar structural types (isomorphous phases) or groups in which one major bonding factor appears to predominate. For example, one can speak of *normal valence compounds*, *electron phases*, *Laves phases*, *NiAs-type phases*, *coordination compounds*, *interstitial phases*, *close-packed structures*, and so on.

IV. BONDING MECHANISMS

Apart from van der Waals forces, which are weak and usually present only between inactive atoms or neutral molecules, crystals formed between metals, or between metals and nonmetals, may be associated with characteristic types of binding forces as follows:

(1) Typically *metallic* crystals consist of positive ions immersed in a "gas" of negative electrons. The attraction of the positive ions to the negative electrons holds the structure together and balances the repulsive forces of the ions for one another and of the electrons for other electrons. The electrons move freely through the lattice and provide good electrical conductivity. Their energies and properties are best described by wave functions that extend throughout the crystal. The electron band theory involves partially filled and overlapping energy bands and possible interactions between the Fermi surface representing occupied energy states and the Brillouin zones representing forbidden energies. In alloy phases the principal alloying parameter associated with the metallic bonding is the *electron concentration* (e/a), which denotes the number of the conductor (or valence) electrons per atom. Electron concentration can change on alloying, and hence, ranges of stability of intermediate phases often correspond to characteristic ranges of e/a. Among intermediate phases, the typically metallic electron phases possess the simple structures of metals such as fcc, bcc, or hcp, but electron concentration is also a useful parameter in a variety of broad alloy phases with complex structures such as the γ-brass phases of the noble metals and the σ, μ, χ, R, and P phases of the transition elements.

(2) Typically *ionic* crystals are bound together by the electrostatic attraction between positive and negative ions. This bonding occurs between strongly electronegative and strongly electropositive elements. In NaCl (rock salt), for example, the electron affinity of chlorine atoms causes a transfer of electrons from the electropositive sodium atoms to yield Na^+ and Cl^- ions. The strictly ionic bond is rarely found in the union between metals, but the structures* of typical salts (see Fig. 2) such as NaCl($B1$), CaF_2 (fluorite, $C1$), ZnS (zinc blende $B3$), CsCl ($B2$), and many other salts are frequently observed in stoichiometric valence compounds. The degree of ionic character in such structures depends of course on the elements and electrons involved (see below).

(3) *Covalent bonding* (or "homopolar") is produced by sharing of electrons between neighboring atoms. Diamond structure ($A4$) is a typical example

* The structure description such as $A1$, $C1$, $B2$, $C15$, $C14$, $C36$, etc. are "Strukturbericht" designations often used in nomenclature of alloy phases. For details see Pearson (1967).

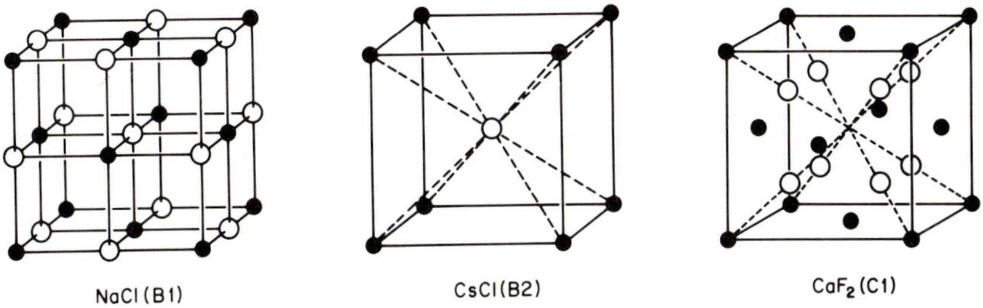

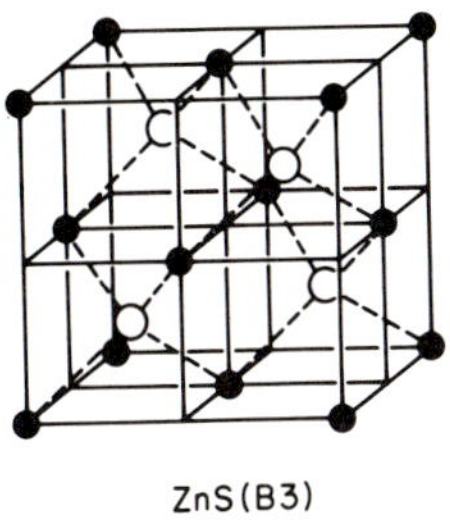

Fig. 2. Crystal structures of typical ionic compounds of type AX and AX_2.

(see Fig. 3) in which each carbon atoms shares its four valence electrons with the four nearest neighbors and thus completes an outer shell of eight electrons in each atom. Four-valent Si and Ge also have this structure. The crystals are characterized by poor conductivity and great hardness. It is possible to obtain this type of bonding among alloys by choosing equal amounts of two elements having valence other than four and being equally removed in the periodic table to the left and to the right of column IV. This is sometimes known as the Grimm–Sommerfeld rule (Grimm and Sommerfeld, 1926) and it emphasizes the existence of the III–V, II–VI, and I–VII compounds that obey normal valence rules. Some of these compounds are well-known semiconductor (Mooser and Pearson, 1960).

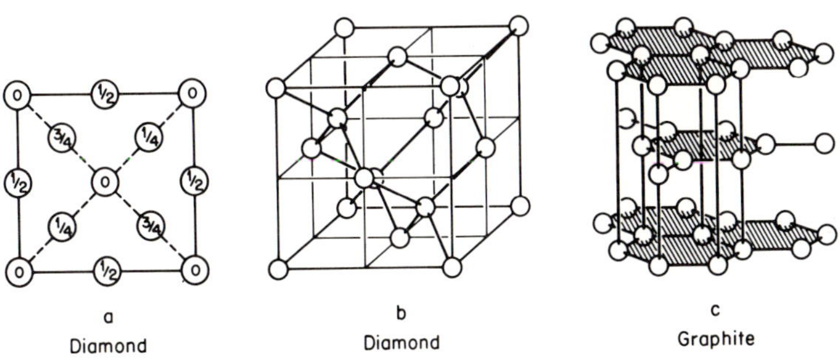

Fig. 3. The structure of diamond and graphite.

On the whole, the concept of *bond* conveys a notion of attractive forces between a *pair* of atoms. This is a chemical description and it is tempting to relate the "strength of bonding" to the number of electrons involved. In the case of metals this can be misleading because the cohesive energy (or bonding energy) may give little clue to the number of electrons involved. Although the cohesive energies of Ni and Ti are closely similar, the number of electrons involved is about 10 in Ni and 4 in Ti. The difference must be sought in the character of the wave functions of the electrons (Coles, 1967).

In discussing alloy phases it is sometimes useful to talk of mixed bonding. For example one can state that the compounds Mg_2Zn and Mg_2Pb possess different amounts of ionic and metallic character but it is not possible to express such a concept quantitatively, i.e., it is not possible to state that an $x\%$ of one type and a $y\%$ of another type exists. However, in compounds with layer structures it is possible to show that the bonding between the layers may be clearly different from bonding within layers (Drabble and Goodman, 1958). On the other hand similarities of crystal structure and interatomic spacing need not at all signify that the bonding character is the same. There is a profound difference between CsCl, for example, and CuZn β-brass.

V. METALLIC VALENCE

The usual concept of *valence* dates from the early studies of compound formation by chemists. The established rules require that molecules form by the union of elements at integral ratios, governed by the valence principles of electroneutrality for closed-shell configurations. In an ionic salt like CaF_2 calcium plays the role of a positive divalent element, and flourine that of a negative monovalent element. The compound is an insulator, and conduction can be obtained only by ionic mobility. On the other hand in the intermetallic compounds that are isomorphous with CaF_2, such as Mg_2Sn or Mg_2Pb, the presence of a certain amount of metallic conductivity suggests the presence of partially free electrons. Here the valence electrons responsible for bonding may differ in number from those responsible for electrical conductivity. Concepts of valence are complicated by the existence of extensive solid solubilities in many alloy phases and their occurrence at nonstoichiometric compositions. Chemical formulas like Fe_5Zn_{21} or KHg_{11} certainly do not resemble compounds that follow simple rules of chemistry. In KHg_{11} the valence of potassium may be reasonably taken as one but this would make the valence of mercury 1/11. Hence, the notion of valence needs a careful definition for the special type of bonding between metallic elements. A basically chemical approach is to consider that each atom in a metal shares its *bonding electrons* with the surrounding neighbors forming covalent links which resonate among the available atomic positions (Pauling, 1940, 1956). In this formulation the term "valence" may be a number that differs from the normally accepted valence familiar in the compound of chemistry of the element in question. For example, in considering the resonating bonds in copper Pauling ascribes a fractional valence of 5.56 and the *diminishing* valences of 4.56, 3.56, 2.56, and 1.56 to Zn, Ga, Ge, and As, respectively, which follow copper in the periodic table.

If electrons are treated collectively and are described by wave functions using band-theory concepts, usually only those electrons that contribute to *electrical conduction* and to the Fermi surface are considered (Mott and Jones, 1936). In this description the valences for Cu, Zn, Ga, Ge, and As are *increasing* from 1 to 5. Therefore, on alloying, electron concentration will be expected to increase according to one scheme and to decrease according to the other. Clearly there is need for a careful definition of valence in the description of alloying behavior.

VI. ATOMIC SIZES

In order to consider the influence of atomic sizes on crystal structure, it is first necessary to define the "atomic size." A number of schemes exist for describing the magnitude of atomic sizes in pure elements. The concept of atomic radius, defined as half the mean bond length, was introduced originally for ionic salts by Bragg and was extended to metals by Goldschmidt (1928). Since then, a number of excellent treatments and reviews have been published (King, 1965).

A number of elements crystallize in different allotropic forms and a study of atomic distances in such structures shows that the atomic radius diminishes with decreasing coordination number. The empirical correction for relating atomic radii to a standard close-packed structure of coordination number 12 is as follows:

	← decreasing radius				increasing radius →		
Coordination number (CN):	4	6	8	10	12	14	16
Required correction (%):	12	4	3	1.4	0	−1.2	−2.2

The above correlation indicates that the conservation of volume during structural changes is an important factor. Thus, for example, for the typical change fcc ⇄ bcc, if a and r denote the lattice parameter and atomic radius respectively, then assuming that volume per atom, Ω, remains unchanged, one has $(a_{\text{fcc}})^{3/4} = (a_{\text{bcc}})^{3/2}$, where $a_{\text{fcc}} = \sqrt{2}(r_{\text{fcc}})$ and $a_{\text{bcc}} = 2/\sqrt{3}(r_{\text{bcc}})$. Hence, $r_{\text{fcc}} \sim 1.03 r_{\text{bcc}}$, which gives the 3% correction for these two structures. It is important to realize that it is the atomic radius of the higher coordination structure, r_{fcc}, that becomes larger than the radius of the lower coordination structure, r_{bcc}, provided the volume per atom remains constant, or nearly constant (Rudman, 1965).

In principle, all atomic radii can be recalculated to correspond with a particular coordination, and this constitutes the basis of many tabulations of atomic radii. Schemes that employ coordination corrections sometimes break down for complex or noncubic structures such as those of Ga, Sb, Se, Zn, Cd, Be, α-Mn, Pu, etc., where the basic coordination number is difficult to define since there may be several bonds of slightly differing lengths.

A number of schemes exist to describe metallic radii of the elements. The *single-bond metallic radii* of Pauling (1940) are related to the resonating bonding valence mentioned above, while the *closest distance of approach* of atoms of Hume-Rothery and Raynor (1962) is derived from the interatomic distances in the structure. Occasionally, a good empirical measure of an effective size of an element in solid solution is obtained by extrapolation of lattice spacings of a series of alloys to a value corresponding to 100% solute (Axon and Hume-Rothery, 1948).

In another approach the volume per atom, Ω, or the radius derived from volume per atom, $r_\Omega = (\frac{3}{4}\Omega/\pi)^{1/3}$, also known as Seitz radius, has been used as a comparative measure of atomic sizes (King, 1965). The volume per atom is the volume of the unit cell divided by the number of atoms in the cell. It is one of the basic parameters in the equation of state, and hence, it is related to the energy of the metal or the alloy; in metallic structures it is relatively independent of coordination and crystal structure, not only in pure elements but also in alloys.

On forming a solid solution of element A with element B, two different kinds of atoms come in contact on a common lattice. This inclusion of new centers of disturbance will affect the existing electronic force fields between atoms, both short range and long range; the resulting effects will be of several kinds. On the atomic scale some atoms of the solvent and the solute will be shifted from the mean atomic positions on the lattice and thus suffer a permanent *static displacement*. The resulting average distance between any two neighboring atoms in a solid solution will depend on whether they are of the like kind, either both solvent or both solute, or of the opposite kind. We may thus talk of the average AA, BB, or AB *bond distances* which may, even for an identical pair of atoms, depend also on the direction in the lattice. The influence of the ratio of the radii of the components upon the choice of crystal structure has been recognized early in the theories of alloying. Hume-Rothery discussed the effect of this ratio as a possible hinderance to extensive solid solubility and introduced the concept of size factor in alloy systems. Size factors can be derived both from ratios of atomic radii or atomic volumes

(Massalski and King, 1961), and they can be related to the positive strain energy in the alloy lattice. Following analyses based on the theory of elasticity, the strain energy E_s, associated with a solid solution may be expressed by a general equation of the form

$$E_s(c) = A\mu\Omega \left(\frac{1}{\Omega}\frac{\partial\Omega}{\partial c}\right)^2 \cdot f(c)$$

where A is a numerical constant, μ the shear modulus, Ω the mean atomic volume, and c the concentration. Since in the case of the noble-metal primary solid solutions the variation of mean volume per atom with composition is very nearly linear,

$$\Omega = (1-c)\Omega_0 + c\Omega_a \qquad \text{and hence} \qquad \partial\Omega/\partial c = \Omega_a - \Omega_0$$

The fractional rate of change of atomic volume with composition is thus almost constant, and for dilute solutions one can write

$$\left(\frac{1}{\Omega}\frac{\partial\Omega}{\partial c}\right) \cong \left(\frac{\Omega_a - \Omega_0}{\Omega_0}\right) \cong (\Omega_{\text{s.f.}})$$

where $\Omega_{\text{s.f.}}$ denotes a size factor expressed in terms of volume. Hence

$$E_s(c) = \text{const}(\Omega_{\text{s.f.}})^2 f(c)$$

i.e., the strain energy per solute atom is governed by the volume size factor of the alloy system.

VII. ELECTROCHEMICAL FACTORS AND HEATS OF FORMATION

Metallic, ionic, covalent, and van der Waals types of bonds, as well as mixtures of these bonds, may exist in any single intermediate phase. As a rule an intermetallic compound will have a large exothermic (i.e., negative) heat of formation, if the difference in the electrochemical characteristics of the components is large. The heats of formation of many ionic compounds range between -40 and -8 kcal/g atom (Robinson and Bever, 1967), and of the covalent compounds between -8 and -3 kcal/g atom. With predominantly metallic bonding, the heats of formation are low and may be less than 1 kcal/g atom. Heats of formation clearly should bear relation to the electronegativities of the component elements provided that the latter can be defined satisfactorily. For example, Pauling (1940) has suggested the following equation for the heat of formation of an ionic compound AB

$$\Delta\, H(\text{kcal/g atom}) = 23.07 \times Z(AE_a - BE_b)^2$$

where Z is the number of valence links and E_a and E_b are the electronegativities of the component elements. A number of other electronegativity scales exist that have been used with reasonable success to predict the heats of formation of numerous predominantly ionic compounds. On the whole the agreement between the calculated heats of formation and the observed values shows scatter of about ± 2.5 kcal/g atom. Any equation based upon electronegativities becomes less applicable as ionic bonds are gradually replaced by covalent and metallic linkage. The heats of formation of a number of compounds in which the structure and bonding ranges from ionic to metallic are shown in Table I.

TABLE I

HEATS OF FORMATION AT 298°K OF SOME TYPICAL INTERMEDIATE PHASES AND COMPOUNDS[a]

Compound or phase	Structure	Predominant bonding	Heat of formation $-\Delta H$ (kcal/g atom)
MgSe	NaCl ($B1$)	Ionic	32.6 ± 2.0
MgTe	ZnS ($B4$)	Ionic	25.0 ± 2.5
ZnTe	ZnS ($B3$)	Ionic	14.4 ± 0.5
Mg_2Ge	CaF_2 ($C1$)	Partially ionic	9.2 ± 0.02
Mg_3Bi_2	Ca_2O_3 ($D5c$)	Partially ionic	7.4 ± 0.2
Mg_2Si	CaF_2 ($C1$)	Partially ionic	6.3 ± 0.3
InAs	ZnS ($B3$)	Covalent	7.4 ± 0.6
GaSb	ZnS ($B3$)	Covalent	5.0 ± 0.2
InSb	ZnS ($B3$)	Covalent	3.5 ± 0.1
NiTe	NiAs ($B8$)	Partially metallic	4.5 ± 1.5
CoSn	NiAs ($B8$)	Partially metallic	3.6 ± 0.3
Co_3Sn_2	NiAs ($B8$)	Partially metallic	2.7 ± 0.2
$CaMg_2$	$MgZn_2$($C14$)	Metallic	3.2 ± 0.1
Ag_5Zn_8(0.61 Zn)	γ-brass ($D8c$)	Metallic	1.1 ± 0.05
AgZn (0.50 Zn)	β-brass ($B2$)	Metallic	0.75 ± 0.05

[a] Data taken from Robinson and Bever (1967).

It is important to remember that the heat of formation will yield only an approximate indication of the stability of a given phase or compound. On the whole, large negative heats of formation are indicative of compounds with a rather restricted solid solubility and a high melting point, but, as the temperature is increased in any phase diagram, the importance of the entropy contribution must not be overlooked. Stability is related to the free energy and only at relatively low temperatures does the ΔH component play the predominant role. At high temperatures the tendency to form more complex molecules becomes quite pronounced, and it should not be surprising if some of the reactions become endothermic. Similarly, entropy considerations encourage substantial departures from stoichiometry in many intermetallic compounds.

In general, the more purely ionic the bonding, the more restricted is the homogeneity range. Intermetallic compounds with large numerical heats of formation are essentially "line compounds" in the phase diagram. Predominantly ionic compounds of the type AB usually crystalize with structures resembling ionic salts.

VIII. COMPOUNDS WITH METALLIC BONDING—ELECTRON PHASES

Alloy phases with predominantly metallic bonding are often formed between metallic elements that have approximately the same atomic sizes and possess similar electronegativities. Numerous typically metallic alloy phases with wide ranges of homogeneity occur in systems formed between the noble metals Cu, Ag, and Au and the B-subgroup elements of the periodic table as follows:

	A	B			
	Cu	Zn	Ga	Ge	As
	Ag	Cd	In	Sn	Sb
	Au	Hg	Tl	Pb	Te
Group valence	1	2	3	4	5

Intermediate phases in a system such as CuZn, which may be considered to be the prototype of the above group, are typically metallic and show a very clear dependence upon the electron concentration. Since, in the alloys of the noble metals, the inner *d*-band levels in the atoms may be considered to be occupied, the main effect of substituting a solute for a solvent atom is a change of the electron/atom ratio. Many electron phases that possess identical or nearly identical structures in different systems are observed to occur within the same ranges of electron concentration. They are of special interest in the theory of alloy phases because their stability and ranges of occurrence can be related to the interaction between the Fermi surface of the conduction electrons and the Brillouin zone. Typical examples of the electron phases are the β-, ζ-, and γ-brasses. Their properties and electronic features have been described in detail in the literature (Massalski and King, 1961). Ionic contributions to the bonding of electron phases may arise from the difference of the electronegativities of the component elements. For example, in the case of the β phases in the systems CuZn and AuCd, although in both systems an ordered CsCl-type β structure exists at low temperatures, the exothermic nature of the heat of formation in the CuZn β phase merely increases with the

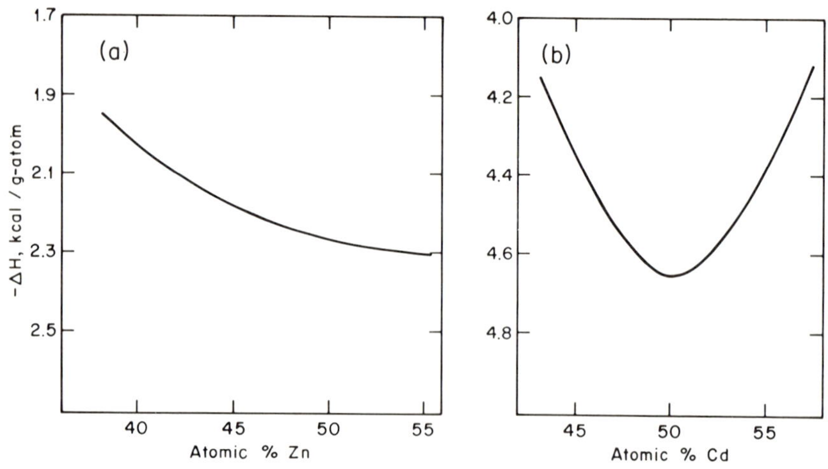

Fig. 4. Heats of formation of the β phase in the systems (a) CuZn at 1100°K and (b) AuCd at 700°K (after Robinson and Bever, 1967).

increasing zinc content [see Fig. 4(a)], but it does not go through a minimum like that found in the system AuCd where the electronegativity difference is more pronounced [see Fig. 4(b)].

IX. LAVES PHASES

Structures which are in many ways similar to the electron phases are the so-called Laves phases of which very numerous examples are found in alloy systems. Laves phases resemble electron phases in that the electrochemical nature of the components is usually rather similar and the electron-to-atom ratio is an important factor. However, in addition to these two features, Laves phases represent an example of a special atomic packing and are usually formed between atoms that permit a denser packing than that resulting from close packing of spheres of equal size. The ideal radius ratio is 1.225 (see Fig. 5). The exothermic heats of formation of Laves phases (see Table I) tend to be numerically smaller than -5 kcal/g atom, but a few exceed the value of -10 kcal/g atom. The three main structural types of the Laves phases are based on magnesium as follows: the cubic $MgCu_2$ ($C15$), the hexagonal $MgNi_2$ ($C36$), and the hexagonal MgZn ($C14$). Despite the fact that the structures of Laves phases do depend upon certain electronic features and hence resemble electron phases, most Laves phases are of fairly limited solid solubility. The restriction to the narrow homogeneity ranges is most certainly due to the rather strict packing requirements between atoms of different sizes.

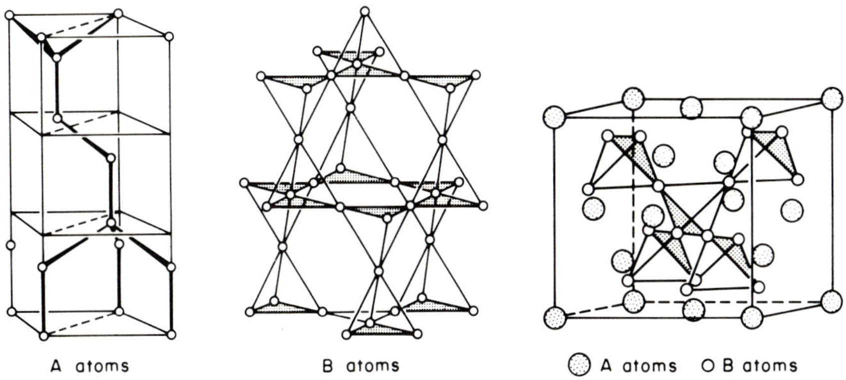

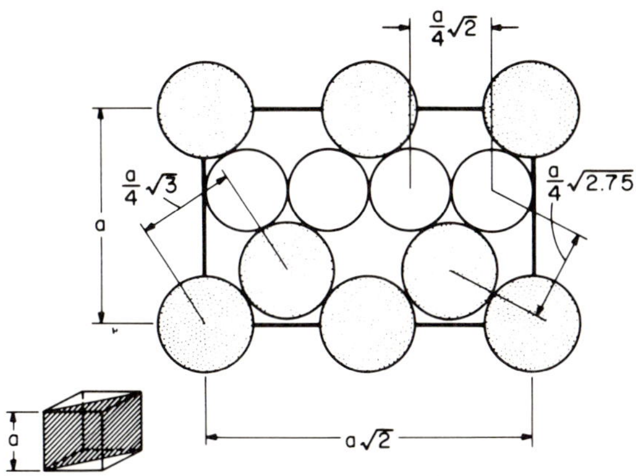

Fig. 5. Distribution of A and B atoms and stacking of layers in the $MgCu_2$ Laves phase.

X. PHASES WITH MIXED IONIC, COVALENT, AND METALLIC BONDING

A typical structure in which the electronegative atoms form one sublattice and the electropositive atoms form another is provided by the NiAs structure type. Here a close-packed hexagonal sublattice of metalloid atoms is coupled with metallic atoms situated in the octahedral interstices. Compounds having this structure are based on the general AB-type formula where A is usually a transition metal and B belongs to groups 3–6 of the periodic table. Many nickel arsenide structures are known to extend over wide ranges of solid solubility, and the structure can generally change from the A_2B type to the AB_2 type merely by changing the ratio of the atoms in the hexagonal sublattice and in the octahedral interstices. This change also changes the axial ratio of the structure. Exothermic heats of formation are generally associated with large axial ratios and are found in those compounds in which the electro-positive element is a transition element with a small number of *d* electrons, while the atoms on the close-packed hexagonal sublattice have a large electronegative character. In such compounds there is a considerable ionic contribution to the bonding which is probably also responsible for the rather restricted homogeneity ranges. Naturally, on moving from the formula A_2B to AB_2 the general content of the metallic element gradually decreases with the increasing degree of ionic character as further and further metal atoms are omitted from the octahedral sites. In the limiting case of the AB_2 structure the resulting compound may be quite ionic and resemble for example the cadmium iodide structure (*C*6) (Hume-Rothery and Raynor, 1962).

More typically covalent bonding occurs in crystals in which an atom has four nearest neighbors as in certain elements of group IVB where the ratio of valence electrons to atoms is 4:1 making each atom tend to form four covalent bonds. Intermetallic phases in this group take various modifications of the zinc blende (*B*3) structure and are often referred to as tetrahedral structures (Parthè, 1964). The heats of formation of the zinc-blende-type structures are generally small, indicating that there is little truly ionic bonding.

XI. PHASES WITH FIXED STOICHIOMETRY

Structures that basically correspond to various stoichiometric ratios of atoms such as A_4B, A_3B, AB are very numerous, and the review of these would be beyond the scope of the present chapter. Some excellent reviews of such structures have been published in various books and symposia proceedings (Beck, 1963 and Westbrook, 1967) along with attempts to summarize

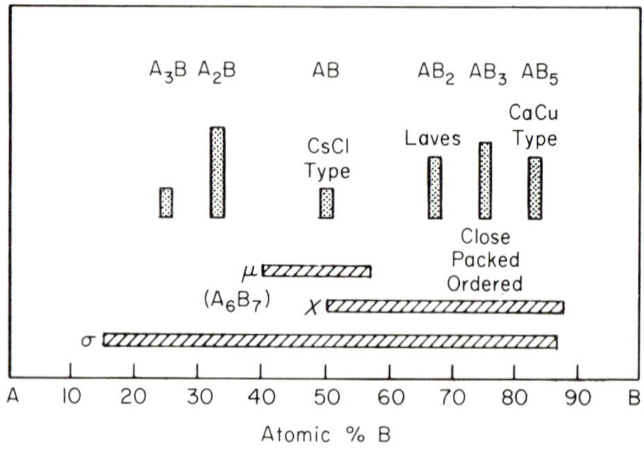

Fig. 6. Occurrence and distribution of some typical intermediate phases with fixed stoichiometry and wide solubility.

their properties and stability. Many of the fixed-stoichiometry compounds owe their stability to the electrochemical factors and may be regarded as an extension of the crystal chemistry of compounds into alloy structures. Numerous other structures show dependence on such factors as electron concentration and hence bear relation to alloy phases with wide solubility (see Fig. 6). Finally, a very large number of alloy structures clearly owe their stability to size relationships and the convenient way in which atoms of certain sizes can be assembled together into close-packed structures.

XII. COORDINATION STRUCTURES

When atoms of pure elements are assembled into crystal structures, the closest packing obtained corresponds to a coordination of twelve (the close-packed *A*1 and *A*3 structures). On alloying, atoms of substantially different sizes may come into play and the general tendency for the maximum filling of available space may be reflected by a further increase in the coordination number. Taking only two types of atoms, A and B, and disregarding the chemical difference between them, one finds that groups of such atoms can be arranged into one of four types of polyhedra of the general formula AB_n in which nB atoms surround an A atom. The value of the coordination number n can be 12, 14, 15, or 16, giving rise to the well-known *coordination polyhedra* discussed by Kasper (1956) well over 14 years ago. They are shown diagramatically in Fig. 7. Each polyhedron is formed by a network of B atoms around an A atom. The network of B atoms consists of triangular faces with five or six triangles meeting at each corner.

The geometry of each polyhedron involves a modified concept of coordination. Referring to atoms of equal sizes, the centers of any three atoms in the closest approach will form a triangle and the centers of any four atoms a tetrahedron. A coordination shell consisting of equilateral triangles only makes it possible to construct a tetrahedron, an octahedron or an icosehedron (coordination 4, 8, and 12, respectively). However, if it is required also that all atoms in a coordination shell make equilateral triangles with the central A atom, no solution is possible: no structure can utilize icosahedral coordination and provide twelvefold coordination to *all* atoms. Hence, in the coordination polyhedra observed in nature, a moderate variation from equality of interatomic distances is observed—a condition particularly well suited for mixing atoms of different sizes or ones that undergo small modification of size due to localized transfer of electrons.

If the domain of an atom is described as the space in which all points are nearer to that atom than to any other atom, we arrive at a concept of a *coordination shell.* Together with the central atom, the coordination shell of

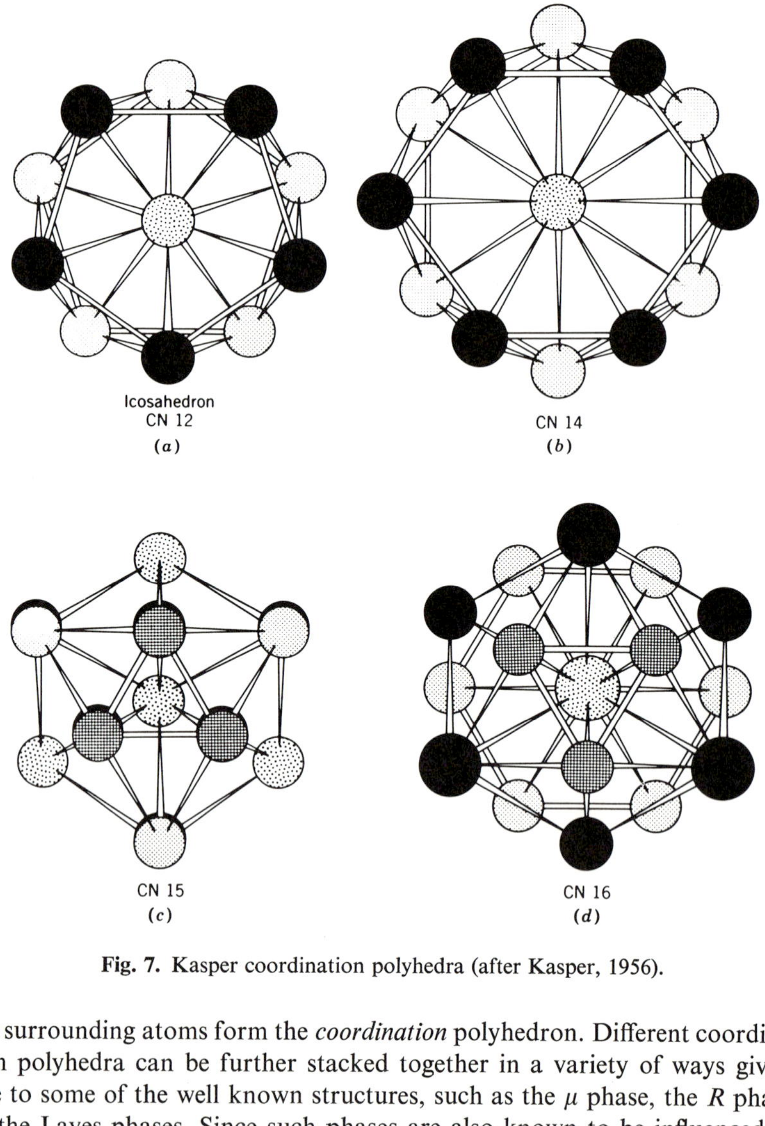

Fig. 7. Kasper coordination polyhedra (after Kasper, 1956).

the surrounding atoms form the *coordination* polyhedron. Different coordination polyhedra can be further stacked together in a variety of ways giving rise to some of the well known structures, such as the μ phase, the R phase, or the Laves phases. Since such phases are also known to be influenced by the electronic features, they provide a good example of the interplay that occurs among the various factors that influence the choice of the crystal structure in intermediate phases. It is sometimes convenient to illustrate and summarize the changing character of the influencing factors by means of a diagram such as that shown in Fig. 8. It emphasizes the fact that no particular structure is likely to be solely dependent on a singular factor but

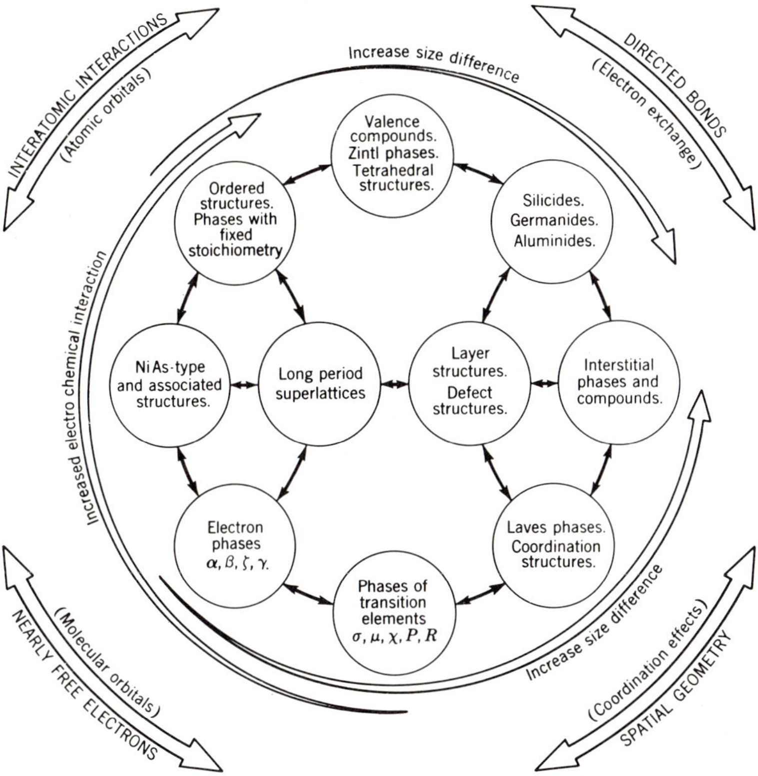

Fig. 8. Crystal families in alloy phases (after Barrett and Massalski, 1966).

rather that there is a continuous interplay of different factors according to the alloying conditions.

XIII. FACTORS GOVERNING THE CRYSTAL STRUCTURE

Although thermodynamically the structure of greatest stability is determined by the free energy principles, it is possible to single out certain principles that determine the choice of crystal structure under various conditions. Basically, the determining factors can be divided into two main groups, those depending upon *electronic features* such as electron concentration, electronegativity and various types of electron sharing and exchanges, and *geometrical features* that emphasize such parameters as the radius ratio, the accommodation of lattice strain, and the tendency to fill space in the most dense way (*space-filling principle*). Sometimes in preference to the most convenient space filling a tendency to form arrangements of high symmetry becomes apparent (*symmetry principle*), and this in turn may be overcome by yet another principle, the *connection principle*, which describes the tendency of the atoms to form as many three-dimensional connections (links) as possible (Laves, 1963).

XIV. TECHNOLOGICAL ASPECTS RELATED TO INTERMEDIATE PHASES

Technologically important commercial alloys are usually quite complex and more often than not are based on multicomponent systems. Many of the intermediate phases discussed previously may occur as minor constituents of commercial alloys. The possible equilibria between the various phases that may be involved are usually approximated from the knowledge of the binary, ternary or quaternary phase diagrams. The materials scientist faces a very difficult problem if he is to predict the constitution of such complex alloys, i.e., predict the number of phases involved and their proportion and composition. His task is compounded by the fact that technologically attractive alloys are often in a metastable state, even after long times at quite elevated temperatures, i.e., their usefulness is associated with the presence of metastable structures or with continuing, but not necessarily completed, precipitation hardening reactions.

Numerous alloys that are used at high temperatures reaching well over 1000°C are usually ferrous-type alloys, and they may contain various quantities of Cr, Ni, and Co in solid solution added for adequate oxidation resistance and in order to maintain the fcc. "austenitic" structure. Elements such as Ti and Al may also be added in small quantities to produce additional strength through subsequent aging heat treatments. The intermediate phases which have been observed in complex multicomponent commercial stainless steels and other iron-based alloys are often basically similar to those found in binary and ternary systems. Consequently, concepts that are applicable in interpretation of alloy phase stability of the less complex intermediate phases may be used as semiempirical guidelines for the design of improved high-temperature materials. The increased demands for new and improved materials in such fields as, for example, space vehicles, gas turbines, and steam power units, have resulted in the development of many complex alloys that are based predominantly upon the transition elements from groups VI–VIII. The knowledge of intermediate phases formed between these elements is, therefore, essential to the understanding of the more complex systems. Although the major improvements in high-temperature alloys susceptible to

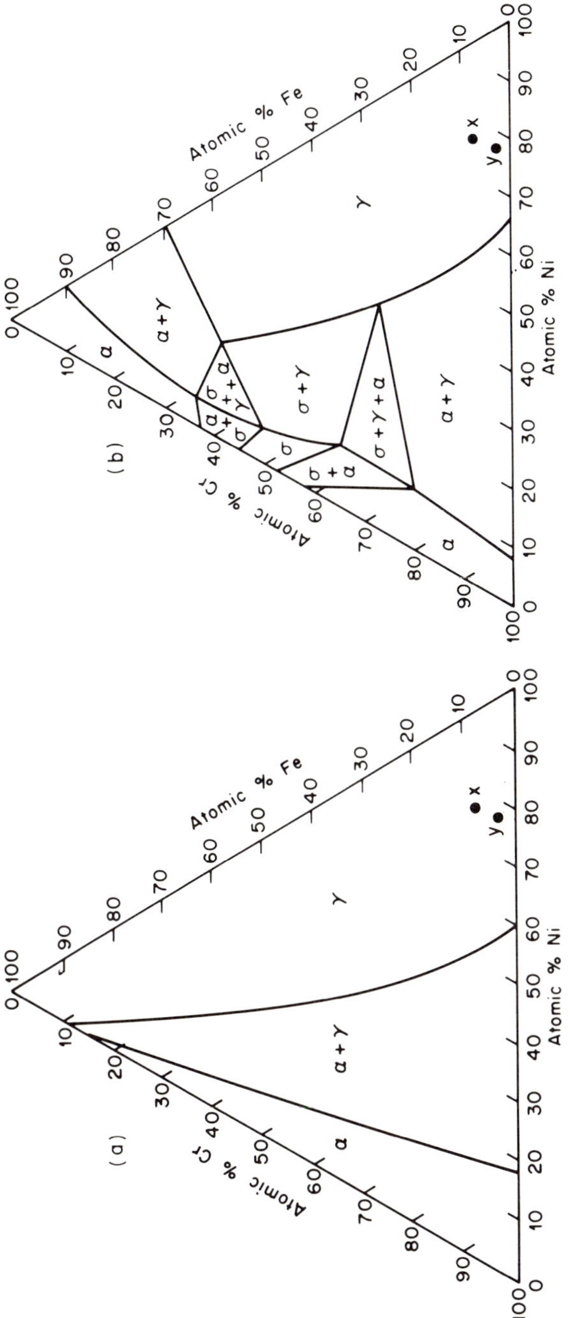

Fig. 9. Isothermal sections at (a) 1100°C and (b) 400°C for the Fe–Cr–Ni equilibrium diagram (after Pugh and Nisbet, 1950).

precipitation hardening have occurred through empirical experimentation, it is helpful to approach the understanding of these alloys through the knowledge of the phase relationships, crystal structures, and, resulting from them, stability or instability of the various components. The precipitation of the intermediate phases from solid solutions constitutes the most important strengthening mechanism, and sometimes also grain-refining mechanism, in commercial alloys. Clearly, not all precipitation reactions are beneficial. For example, a decrease in the high-temperature rupture strength will usually result if the brittle σ phase precipitates in the form of large particles from the austenitic matrix. The theory and mechanism of precipitation hardening has been discussed in many excellent review articles (ASM, 1959). In this chapter we merely consider a few typical relationships between phase diagrams and intermediate phases.

XV. ENHANCEMENT OF PROPERTIES WITH DISPERSED INTERMEDIATE PHASES

The important group of Ni-based alloys normally contains Fe and also Cr for oxidation resistance at elevated temperatures. The isothermal sections at 1100°C and at 400°C (Pugh and Nisbet, 1950) for the Ni–Cr–Fe ternary system is shown in Fig. 9. The basic matrix here is the fcc γ phase, which extends over a wide range of compositions. Compositions marked x and y, representative of commercial alloys such as Inconel and Nimonic 75, fall into single phase fields at both temperatures. Thus, the precipitation hardening mechanism does not apply directly but the limit of solid solubility for NiCr based alloys is decreased when Ti or Al are added. Figure 10 shows portions of isothermal sections at 1150°C and 750°C (Taylor and Floyd, 1953) for the Ni-rich Ni-Cr-Al alloys. The solubility of Al in the Ni-rich γ phase decreases from about 10 wt % at 1150°C to a minimum value of approximately 2.5% at 750°C. Hence, the high temperature "austenitic" γ phase containing Al may be initially maintained in a supersaturated condition by cooling to room temperature and subsequently the alloy may be strengthened by heat treatment that promotes the precipitation of the γ' phase, which has an ordered fcc structure. γ' may be represented by the stoichiometric formula Ni_3Al. Commerical alloys based on Ni may be strengthened by the γ' phase when it is dispersed uniformly throughout the matrix as small, spheroidal particles. Cellular growth, orginating from the grain boundaries of the γ matrix, may occur when the difference in lattice parameters between the γ and γ' phases exceeds approximately 1% (Hagel and Beattie, 1959). A cellular distribution of the γ' phase in the matrix is almost always unfavorable because it provides easy paths for crack propagation. If the precipitating γ' particles agglomerate,

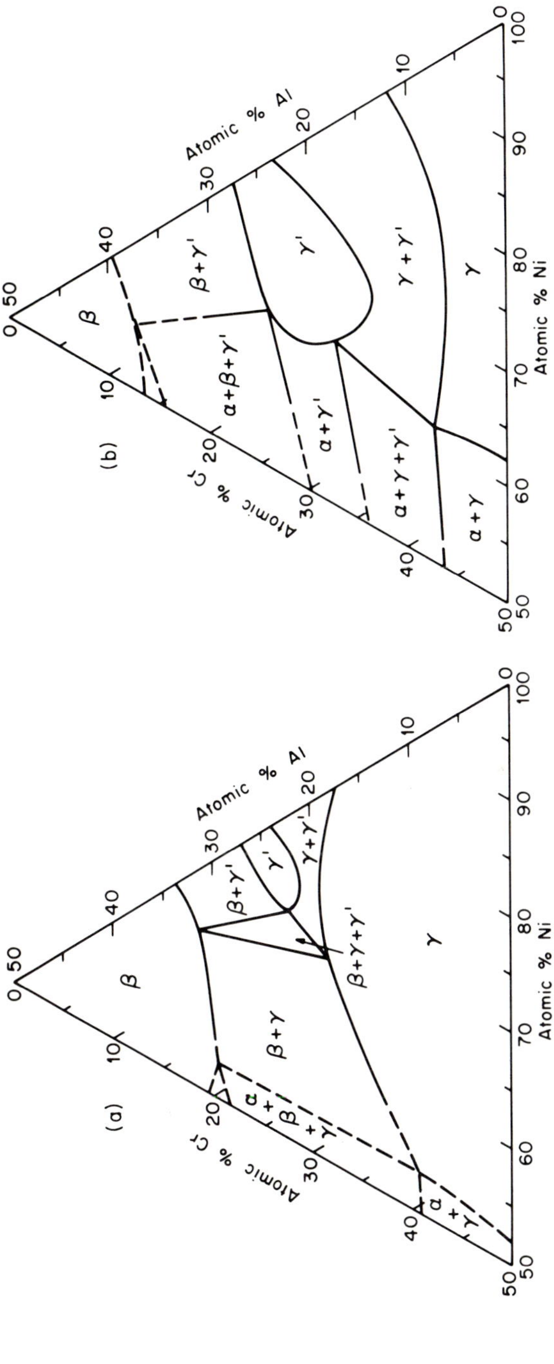

Fig. 10. Portions of isothermal sections at (a) 1150°C and (b) 750°C for Ni-rich, Ni–Cr–Al alloys (after Taylor and Floyd, 1953).

or if their shape changes, the rupture strength at high temperatures deteriorates (Murphy *et al.*, 1967).

When Ti is added to the Ni-based alloys, the precipitating phase becomes Ni_3Ti (η) which has the close-packed hexagonal structure. Since most high-temperature commercial alloys contain both Ti and Al, it is useful to consider ternary sections involving both of these elements. Such sections at 1150°C and 750°C (Taylor and Floyd, 1952) are shown in Fig. 11. In contrast to the NiCr-based systems shown in Figs. 9 and 10, the single phase γ matrix region is now limited to higher Ni contents. When the additions of Al and Ti exceed approximately 6% and 7.5%, respectively, the intermediate phases γ' and η may be precipitated. Since as much as 60% of the Al atoms in γ' can be replaced by Ti (Nordheim and Grant, 1954), the precipitating phase that contributes to the strengthening is an intermetallic phase of the form $Ni_3(Al_x,Ti_y)$ (Bieber and Raudebaugh, 1959). Precipitation of transition metal intermediate phases depletes the matrix of alloying elements, and in some alloys this allows the γ phase to undergo a martensitic transformation on cooling (Lena, 1959). Strengthening then occurs by a combination of martensitic and precipitation reactions. In this area belong also for example the stainless metastable stainless steels (Bressanelli and Moskowitz, 1966) and the *TRIP alloys* (Zackay *et al.*, 1967) in which the deformation process is aided by the occurrence of a *strain-induced martensitic transformation*, but the ultimate strength of the TRIP-type alloys depends upon prior carefully selected heat treatments that promote precipitation of finely dispersed intermediate phases and their interaction with martensite. The occurrence and importance of the strain-induced martensite transformations emphasizes the importance of the knowledge of the M_d as well as M_s temperatures. This information is rarely provided as yet in published phase diagrams, and the knowledge of the M_d temperatures in intermediate phases is particularly scarce.

The γ' and η phases are equilibrium phases in the respective equilibrium diagrams. Their formation follows, therefore, phase-rule principles. With metastable transformations, however, the phase rule does not apply. When attempts are made to obtain an optimum microstructure via a suitable heat treatment, the knowledge of both equilibrium and metastable conditions is important. As is well known, the lowering of the free energy of one phase can affect the range of stability of adjoining phases. This is illustrated in Fig. 12 for a hypothetical system involving two phases, α and γ. At some high temperature T_1, the stable state of an alloy of composition X is the γ phase. On cooling to temperature T_2, the free energy of the γ phase is still lower than that of the α phase of the same composition. For compositions in the range between a and b, however, the stable condition corresponds to a phase mixture of α_a and γ_b phases. The equilibrium formation of such a

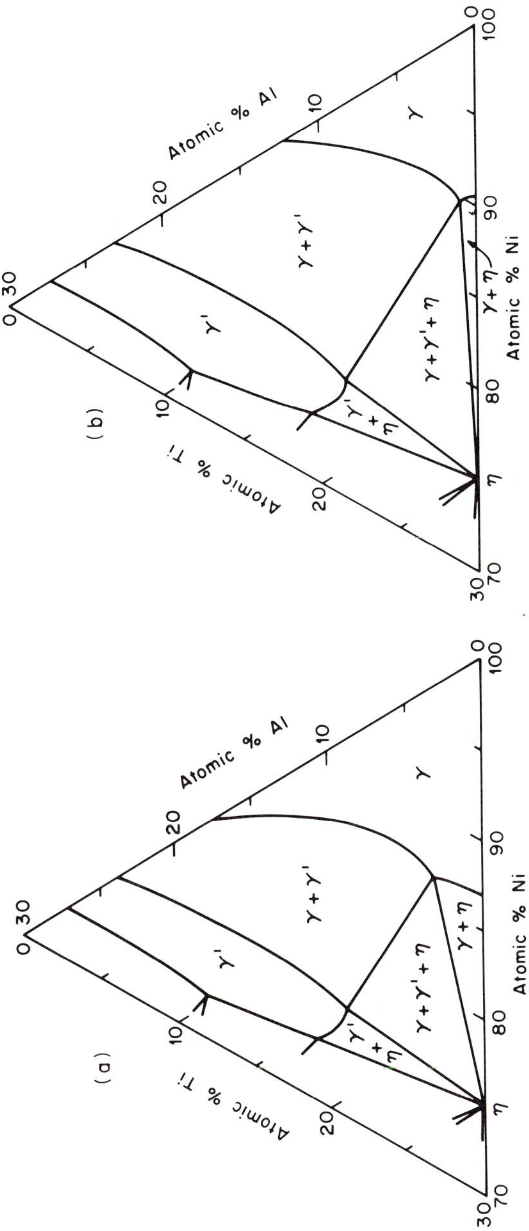

Fig. 11. Sections at (a) 1150°C and (b) 750°C of the Ni–Al–Ti system (after Taylor and Floyd, 1952).

mixture involves, of course, changes in composition, and therefore, requires diffusion. However, it may be suppressed if reaction kinetics become sluggish on rapid cooling. At a temperature T_3 the free-energy curves F_x^{α} and F_x^{γ} intersect at composition X and at this temperature (that may be designated T_0), the free energies of both phases are equal, but the *stable* condition is still a two phase $\alpha_a + \gamma_b$ alloy. However, at any temperature $T_4 < T_0$ the γ_x phase may transform by a martensitic diffusionless reaction into α_x of the same composition. The martensitic phase is of course *metastable*, and it may subsequently decompose into the stable mixture of α and γ phases of different compositions a and b provided that the long-range diffusion is still possible, given sufficient time. In some cases the transformation $\gamma_x \rightarrow \alpha_x$ occurs rapidly by a "massive" transformation (Massalski, 1958), as for example, in FeNi alloys (Gilbert and Owen, 1962). The massive reaction usually involves sufficient short-range diffusion to allow a structure change but not enough to allow any long-range atomic transport necessary for compositional changes. In certain nonferrous alloys such as CuGa, the massive transformation occurs from an equilibrium phase to a nonequilibrium phase, but the free-energy principles are similar to those in Fig. 12; the resulting product phase must be the most stable phase at the *given composition* in order for a massive transformation to occur.

Recent advances in the development and use of FeNi based *maraging steels* (Decker *et al.*, 1962) have spurred a renewed interest in the studies of the massive and martensitic transformations in many commercial alloys, such as those, for example, based on the binary FeNi system (Wayman, 1965). From the point of view of technological applications, the formation of metastable phases is of extreme importance since it may yield alloys that are both strong and tough.

XVI. SOME DETRIMENTAL EFFECTS

Many transition-metal intermediate phases that occur in dispersed form within the matrix of commercial alloys produce detrimental effects, being sources of weakness rather than strength. Some of these phases appear to owe their stability to electron concentration dependence, and hence, the likelihood of their formation can be studied systematically. In most of the complex phase diagrams of high-temperature alloys, the phase boundaries of the intermediate phases such as σ, μ, χ, etc. appear to follow approximately isoelectronic lines, i.e., it is possible to postulate that provided one defines the electron concentration in a suitable way its overall value appears to remain fairly constant along lines parallel to the phase fields (Barrett and Massalski, 1966, pp. 226, 354). Sigma phase is particularly interesting in the

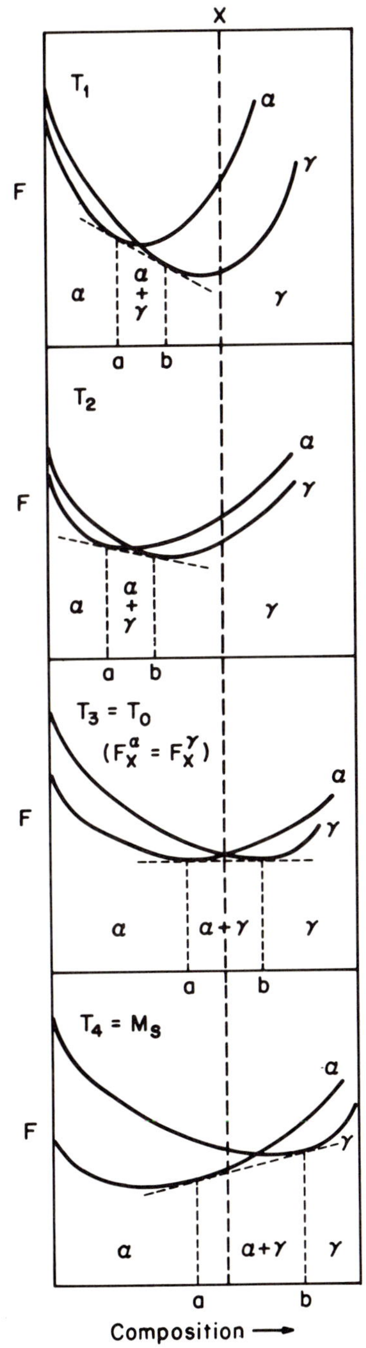

Fig. 12. Hypothetical free energy curves as a function of temperature for a binary system involving two phases, α and γ. See text for discussion.

science of alloy crystal structures because it is representative of many structures that occur over a wide range of homogeneity and involve the transition elements. It is also important technologically because it can have a very harmful effect upon the mechanical properties. Being extremely brittle, it cracks easily, causing a reduction in the rupture strength at elevated temperatures and an increased brittleness at low temperatures. The morphological distribution of the σ phase in the matrix is almost always unfavorable, i.e., long thin plates run generally along certain crystallographic directions in the matrix, or cellular aggregates grow in a lamellar manner from the matrix boundaries (Pickering, 1959). Both types of morphology provide easy crack propagation. Another reason for avoiding the σ phase is that its formation depletes the matrix of important alloying elements needed in solid solution to provide resistance to oxidation and high-temperature rupture strength. The structure of σ phase is usually ordered with a complex tetragonal unit cell containing 30 atoms. It may be represented by the general formula A_xB_y, where x and y can vary from 1 to 7. A is usually a transition element from groups III-A through VI-A (such as Cr or Mo), and B is from group VII-A or VIII-A, as, for example, Ni or Co. As a rule, as the number of electrons outside filled shells increases in the B elements, the composition of the observed σ phase shifts toward A. Many authors (Nevitt, 1963) regard σ phase as similar to electron phases based on the noble metals, because structurally related phases occur at similar values of electron concentration in different alloy systems. In Fig. 13 the elongated phase fields of the σ phase tend to follow lines of approximately constant number of unfilled states in the d bands of the transition elements involved. This idea defines a way of expressing a constant electron concentration.

Commercial high temperature alloys are also prone to embrittlement if the transition-metal phases μ, χ or some types of Laves phases precipitate from the matrix. μ phase has the structure of β-manganese and contains 13 atoms in a rhombohedral (or trigonal) unit cell, and hence, may be represented by a formula such as A_7B_6 (Das *et al.*, 1952), where A is usually Fe or Co, and the B atoms are the refractory metals Mo or W. μ phases occur over an extended range of compositions in the ternary systems Fe–Co–Mo and Fe–Ni–Mo (see Fig. 13). According to one view point (Duwez, 1956) they are somewhat related to the σ phases. σ and μ phases occur when the number of unfilled states in the d bands is low. Again, the phase boundaries tend to follow lines of approximately constant electron concentration. There is a tendency for the μ phase to precipitate intragranularly in complex austenitic alloys (Beattie and Hagel, 1961) in the form of thin plates that are aligned parallel to the $\{111\}$ planes of the matrix.

χ phase, with the structure of α-manganese, is another complex transition-metal phase that may form in some high-temperature alloys containing iron

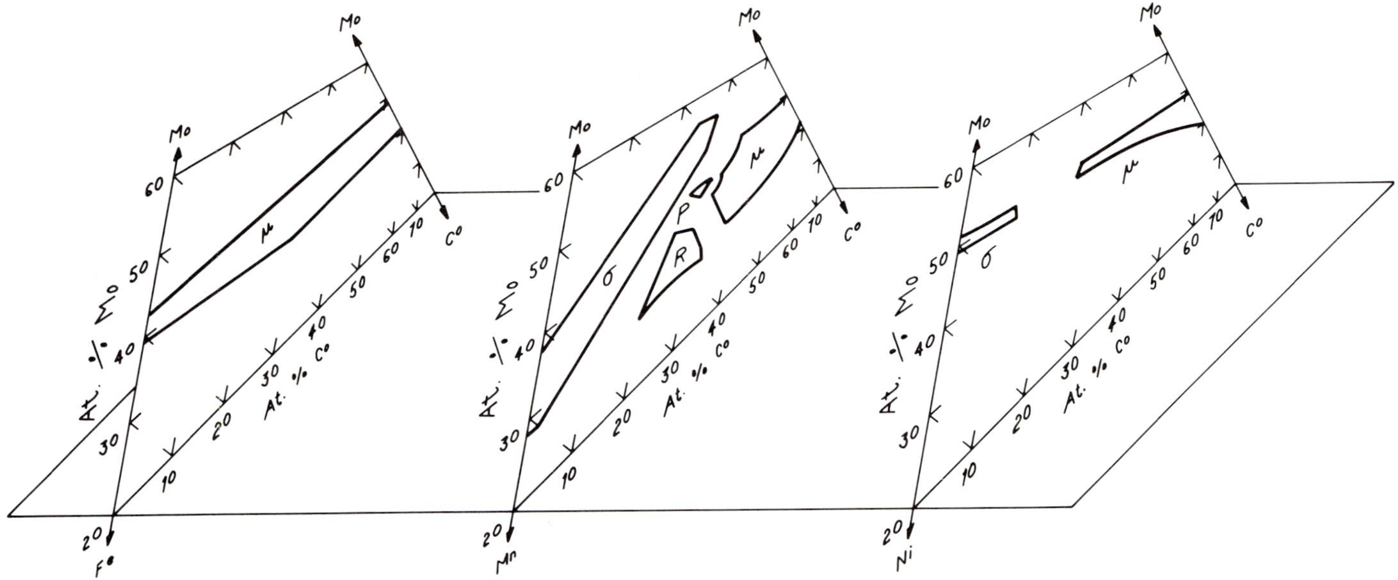

Fig. 13. Alignment of μ, σ, δ, P, and R phase fields along contours of electron vacancy concentration for some ternary systems.

(Hagel and Beattie, 1959). The particular formula analyzed chemically was $(Fe_{27}Ni_8)Cr_{13}(Mo_{5.5}Ti_{4.5})$. This composition is similar to the ternary compound $Fe_{36}Cr_{12}Mo_{10}$, known to have the α-manganese structure (Andrews and Brookes, 1951). Kasper (1945), using x-ray and neutron-diffraction studies, has shown that the atoms are ordered. χ phase can contain carbon in solid solution and apparently act as an intermediate stage for the formation of the interstitial carbides of the general formula M_6C (Goldschmidt, 1957). Serious embrittlement occurs in some ferrous alloys containing more than approximately 5% Mo (Pops, 1966b) due to precipitation of the χ phase. As mentioned earlier the σ, μ, χ, and Laves phases have certain common geometrical features that are responsible for structure stability. They are densely packed and have coordination numbers which are larger than those of the close-packed fcc and hcp elements (CN = 12), namely 14, 15, or 16.

A number of the other structures exist in complex alloys whose crystal structures are somewhat related to the σ phase. Reported in the literature are, for example, the *P*, *R*, *E*, etc. phases (Beck, 1963; Barrett and Massalski, 1966, pp. 226, 354). The *P* phase is known only in some ternary alloys which contain Mo, Ni, and Cr. The crystal structure has been reported to be orthorhombic with 56 atoms per unit cell (Brink and Shoemaker, 1955). Figure 13 again illustrates the general tendency of the *P* and *R* phases to form narrow elongated phase fields within a relatively small range of electron concentration.

A silicide, called the *G* phase, is prominent in several iron- and nickel-based high-temperature alloys (Lena, 1959). This phase has a complex bcc crystal structure with ordered vacant sites, and is isomorphous with the γ-brass structure (Hagel and Beattie, 1959). It forms primarily at the austenite grain boundaries, but does not contribute appreciably to strengthening. It may be described by a stoichiometric formula such as $Si_6Ti_8Ni_{13}$.

XVII. SEMIEMPIRICAL GUIDELINES RELATED TO PHASE PRECIPITATION AND PHASE STABILITY

The studies of the precipitation hardening process alone are usually insufficient for a significant improvement of the design or development of commercial alloys. They must be coupled with the understanding of the constitution of multicomponent systems and the related knowledge of the phase diagrams. In this section we discuss briefly some semiempirical concepts that serve as guidelines for predicting the structure and composition of precipitating phases.

Examination of intermediate phases that form in many different commercial and experimental high-temperature alloys indicates that they may be characterized according to the composition of the matrix. For example, the phases with high coordination numbers that tend to cause embrittlement—σ, μ, χ, and Laves—show strong preference for an Fe-based matrix. As the matrix composition is adjusted from Cr to Mn to Fe to Co to Ni the nature of the precipitating phase shows a related trend; the close-packed γ' and η phases have a high frequency of occurrence in nickel-based alloys. Cobalt appears to be a borderline element, and densely packed compounds as well as close-packed fcc and hcp structures are quite common with this element as a major matrix constituent. To illustrate these tendencies, when elements of groups IV-B (Ti, Zr, Hf), V-B (V, Nb, Ta), or VI-B (Mo, W) combine with Cr, Mn, and Fe, only those intermediate phases which have the $C14$, $C15$, $C36$, $A12$, $D8_b$ structures (i.e., the Laves phases, μ and σ) are observed, as for example, Cr_2Zr and Cr_2Nb. On the other hand, compounds having the structure types DO_{24} and $L1_2$ such as Pt_3Zr, Pt_3Ti, and Cu_3Ti appear more frequently in nickel-based alloys. The structures of the precipitating phases in stainless steels and superalloys are consistent with the findings of Laves and Wallbaum (1939), who determined some thirty years ago the structures of some binary compounds formed between transition metals from the central portion of the periodic table.

The L–W effect is most likely an expression of the electrochemical principle suggesting that elements that are not too far apart in the periodic table will tend to form the more metallic phases as opposed to the more ionic bond likely to occur between more widely separated elements. The suggestion of L and W has been used quite successfully for interpreting the structures of precipitating phases in Fe and Ni-based high-temperature alloys. Using the commercial Ni-based alloy Rene 41, it may be shown for example that the first phase to precipitate from the matrix is γ'. Since this phase is rich in nickel, the matrix will become enriched in cobalt. Therefore, precipitation of a phase which is somewhat more dense and related to the σ phase may be expected. The actual phase found in this alloy has the μ structure.

A different approach has been used recently to predict precipitating intermediate phases in high temperature alloys. It is based primarily upon the data of Beck and co-workers (1963), who reported that the important factor governing the formation of the σ phase and other related structures is the number of unfilled states in the $3d$ shell.* Prediction of precipitating phases can be made with the use of a computer program, and therefore, has been called "Phacomp" (phase computation) in the literature (Woodyatt *et al.*, 1966). Application of Phacomp was first reported in 1964 to control composition in an Ni-based alloy; precipitation of σ phase was successfully prevented (Boesch and Stanley, 1964). Similar results have since been obtained for

* The number of unfilled states is sometimes called the number of electron vacancies.

several hundred experimental and commercial high-temperature alloys (Woodyatt *et al.*, 1966; Sims, 1966). Since a value for the number of unfilled states in the 3*d* shell *Nv* is needed, the following numbers first assigned by Pauling (1940) are used in the computation:

Group No.:	VI	VII	VIII-A	VIII-B	VIII-C
Element:	Cr	Mn	Fe	Co	Ni
Nv:	4.66	3.66	2.66	1.71	0.66

One can calculate the average number of unfilled states, $\bar{N}v$, for a solid solution alloy by simply adding together the effects of each particular element using the expression $\bar{N}v = mNv$, where m is the atom fraction of the element. Single phase fields of σ, μ, R, and other electron phases of the transition metals occur in the phase diagrams within a very narrow range of $\bar{N}v$. In order to apply the Phacomp method to precipitating "electron phases" in commercial alloys, the composition of the alloy must be initially adjusted to allow for possible precipitation of borides, carbides, and γ'. The type and composition of these phases are obtained by a judicious guess, based upon experience, metallographic studies, etc. For example, the amount of Cr which is present in any phase such as a boride, $M_{23}C_6$ or γ' is subtracted from the atomic percent Cr in the alloy. Similar adjustments are made for A1, Ti, Mo, Co, Ni, etc., in order to obtain the amounts of residual elements in the matrix. The residual elements in the matrix are then scaled to 100% and the new matrix composition (expressed in atomic percent) is used to compute the average $\bar{N}v$ value from the following expression: $\bar{N}v = \sum_{i=1}^{n} m_i\,(Nv)_i$, where n is the number of elements in the matrix. Examination of a large number of compositions in austenitic high-temperature alloys (Sims, 1966) indicates that σ phase forms when the electron-vacancy number is greater than approximately 2.52. Laves phases and μ phases occur with lower $\bar{N}v$ values of about 2.30 Phacomp has been used to evaluate potentially new alloys without resorting to costly and long-time laboratory tests, to predict whether or not undesirable phases might form during aging, and to provide guidelines for long-time stability at service temperatures. Within its particular field of application the Phacomp method has certainly removed some of the guesswork in alloy development, but its use constitutes an enlightened and desirable empiricism rather than a basic development.

Although recent studies of transition metals suggest that in the crystal structures some of the energy states such as *s*, *p*, and *d* may exist in hybridized states (Hume-Rothery, 1965) and that the *Nv* values used by Pauling are not constant (Hume-Rothery, 1966), the concept of the importance of the *d* bands is clearly of interest. It should be noted that the principle of the *Nv* values merely reflects a difference of approximately one unfilled state in the *d* band of each atom between the adjoining columns in the periodic table. The actual

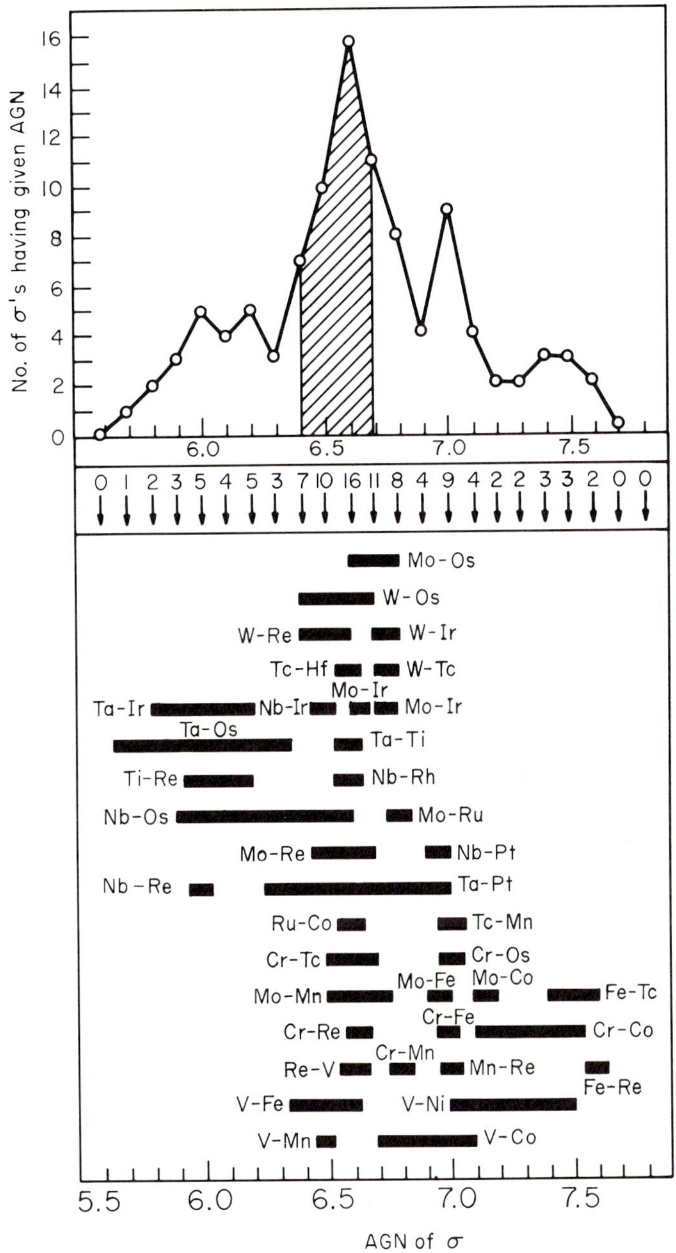

Fig. 14. Composition ranges of most of the σ phases in terms of average group numbers (AGN) (after W. Hume-Rothery, 1966).

values do not affect the proposed schemes. In fact different numbers for *Nv* than those suggested by Pauling may also be used. Since the valences of transition metals do not always appear to be constant, it may be desirable to consider the transition metal compounds in terms of their average group number (AGN). In this case, for example, the σ phase occurring in an Fe (group VIII)–Cr (group VI) alloy of equiatomic composition has an AGN of 7, i.e. $\frac{1}{2} \times (8 + 6)$. Figure 14 shows composition ranges of most of the known σ-phase in terms of average group numbers as well as the number of σ phases having a given AGN. It suggests that σ phases tend to form at an AGN of between 6.4 and 6.7.

Multicomponent phase diagrams between a number of transition metals have been constructed recently by Brewer (1965) in an attempt to predict the structures of phases that are most likely to occur in binary systems. By combining some known data with predictions of stability based on the total

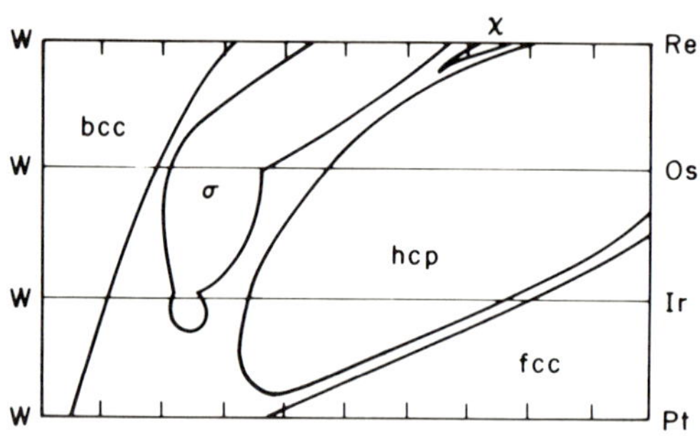

Fig. 15. Correlation of alloy structures for systems of W with Re, Os, Ir, and Pt (after Brewer, 1965).

number of $(s+p)$ electrons (Engel, 1954; 1967), Brewer has been successful in correlating and predicting many alloy structures, particularly for alloys of the second and third transition series metals of groups IV-A, V-A, and VI-A with those of groups VII-A, VIII-A, B, and C. The method of representation is illustrated in Fig. 15 for systems of W with Re, Os, Ir, and Pt. The three metallic structures bcc, hcp, and fcc as well as the σ and χ phases are shown. Horizontal lines represent the binary systems, and the ranges of stability corresponding to the full lines in Fig. 15 represent maximum solubilities of the given phases irrespective of the temperature at which they occur. Thus, the diagrams are partially empirical in that they incorporate points taken from known diagrams, but they allow useful extrapolation into multicomponent regions. The fundamental postulate, that the three simple metallic structures fcc, bcc, and hcp are simply related to the numbers of $(s+p)$ electrons, has met, however, with a number of objections (Hume-Rothery, 1968).

XVIII. SOME NONFERROUS ALLOYS

When technological applications do not require relatively high temperatures ($<1000°C$), intermediate phases based upon nonferrous components generally play an important role. The most significant metallic phases are found in commercial alloys of Al, Mg, Ti, and Cu. In this chapter we comment on only a few of the phases which may be found in the brasses, bronzes, and their analogs since the stability of these important commercial alloys may be described to some degree of success in terms of present theories of alloying.

Many of the copper-based alloys which are used commercially contain as the predominant phase the fcc α primary solid solution that may be followed in the phase diagram by electron phases such as β, ζ, or μ.* As discussed previously, the limits of solid solubility may be evaluated by the common tangent construction to the free energy curves of the α phase and of the adjoining intermediate phases. As an example, Fig. 16 shows several isothermal sections of the copper-rich region of the Cu–Zn–Si equilibrium diagram (Pops, 1964). The shape of the hcp ζ phase is unusual in that it is very narrow and follows a line of approximately constant silicon content while dissolving as much as 16.5 at. % Zn. It is apparent from this figure that the primary solubility is nearly a straight line which runs parallel to lines of constant electron concentration, but the boundaries are slightly convex, away from the copper-rich corner, when the adjoining intermediate phases are ζ or μ. Near the CuSi binary system the boundary of the α phase no longer runs parallel to the lines of constant electron concentration. As the temperature is lowered, the α phase enters into equilibrium with the μ phase, and the α phase boundary becomes more restricted. Because of this restriction, precipitation of brittle μ, χ, or ζ phases with complex crystal structures may occur at low temperatures. They may decrease the ductility of the alloy and cause serious embrittlement, since they provide easy paths for crack propagation. Most commercial silicon brasses such as, for example, Tombasil, contain less than about 5 at. % silicon in order to restrict the microstructure to the presence of only the α or β phases. As is well known, intermediate phases having the γ-brass-type structure are extremely brittle and generally precipitate from the β phase with the characteristic "rosette" shape. Occasionally, however, the γ phase forms as small spheroidal particles uniformly dispersed throughout the matrix. In this case the alloy exhibits extremely high wear resistance. This desirable combination of properties is used to advantage in certain of the aluminum brasses, for example Narite, which is a commercial

* This phase has a complex cubic structure, and it is different from the alloy phase based on the transition metals.

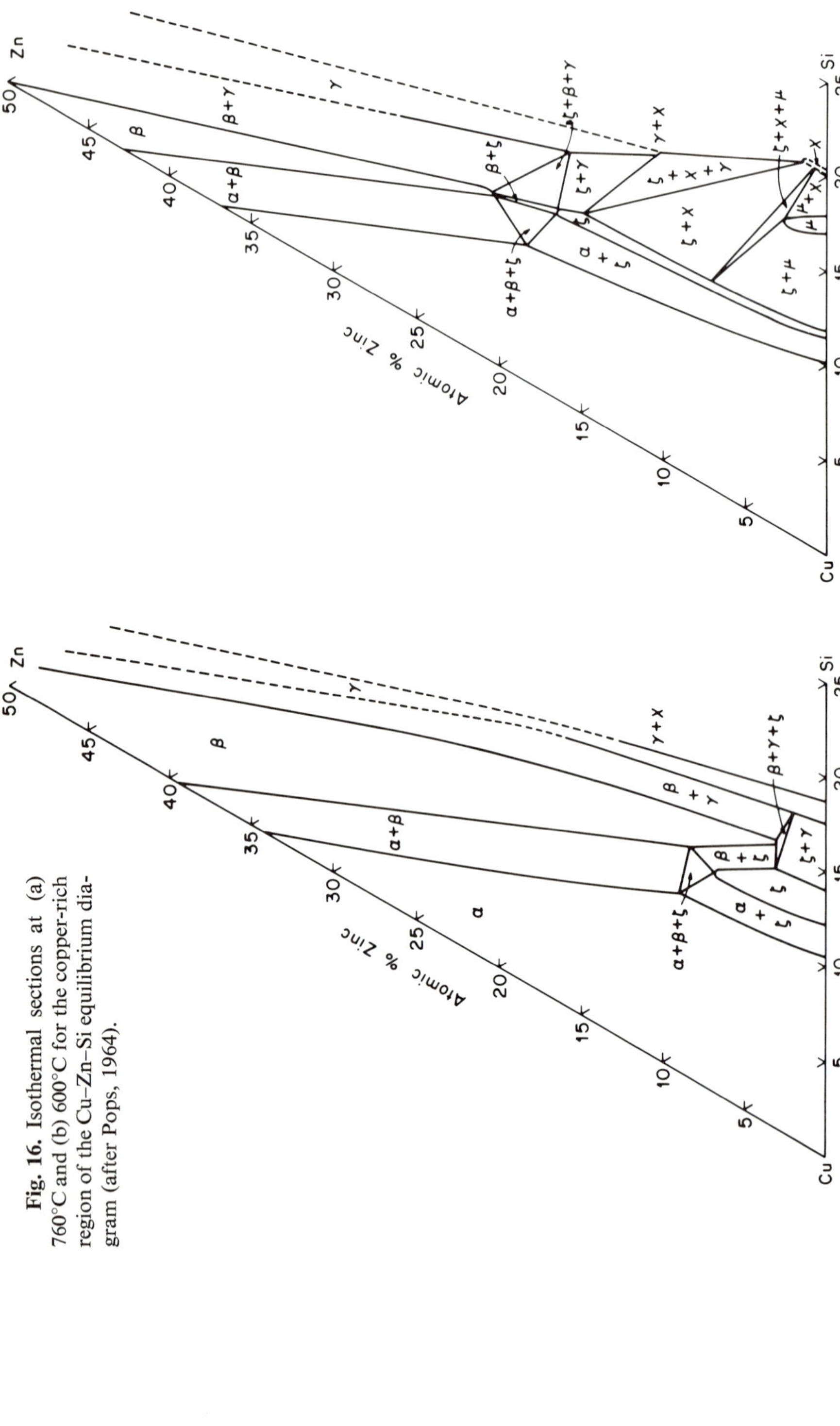

Fig. 16. Isothermal sections at (a) 760°C and (b) 600°C for the copper-rich region of the Cu–Zn–Si equilibrium diagram (after Pops, 1964).

alloy of Cu, Al, and Ni presently being used for press tools and deep-drawing dies. Since a brittle intermediate phase can be used successfully provided its morphology is controlled, it should also be possible to have high-temperature commercial "superalloys" that contain σ phase.

The bcc β' (ordered β) phase that occurs in most phase diagrams based upon Cu, Ag, and Au is unstable and usually decomposes by martensitic or massive transformations on rapid cooling or deformation. Zener (1947, 1967) has discussed the instability of the bcc lattice in terms of low resistance to shear along the (110) planes in the direction $[1\bar{1}0]$. On the whole a martensitic transformation will tend to occur when the elastic anisotropy is large. In contrast to the martensites that occur in steels containing carbon, substitutional martensites may be both strong and "tough" as for example in Fe–Ni–Mo–Co maraging steels.

In some instances the brass-type alloys show "superelastic" behavior, as in the case of AuCd, NiTi, and Cu–Al–Ni, which transform martensitically during loading, and tend to revert to the parent phase when the applied (elastic) load is released or when the alloy is heated. Alloys of this kind show elastic properties that exceed by an order of magnitude the usual elasticity observed in bronzes (Rachinger, 1960). Clearly, information concerning the variation of the transformation temperature with composition is needed, and could best be represented as M_s and M_d lines on the appropriate phase diagrams. This is illustrated in Fig. 17 for some typical binary β-phase fields based upon Cu, Ag, and Au. Metastability may be either increased or decreased (i.e., a raising or lowering of the M_s may result) when third elements are added to these binary systems (Pops, 1966a). This is shown in Fig. 18, which contains plots of M_s versus solute content for different ternary CuZn-based alloys. A marked variation occurs within each constant valence subgroup, and no systematic dependence upon valence is evident. It is important to remember that although stability of β phases are determined by electron concentration, electronic factors by themselves do not clearly influence *metastability*. Application of nonferrous martensites has not yet reached an advanced technological stage, although superelasticity offers a promising area for some of the more complex multicomponent β-phase alloys and opens a whole new area of specialized applications for intermediate phases and compounds. The nonferrous TiNi alloy is an extremely interesting alloy in this respect with many unique properties. It has high specific heat and relatively low density. It is resistant to corrosion and has high hardenability, but perhaps the most unusual is its ability to undergo a series of martensitic transformations which provide it with an unusual "memory effect."

The existence of a reversible martensitic transformation below 90°C in TiNi has been known for some time (Buehler and Wiley, 1962). As is typical of martensites the transformation is not isothermal and the largest amount

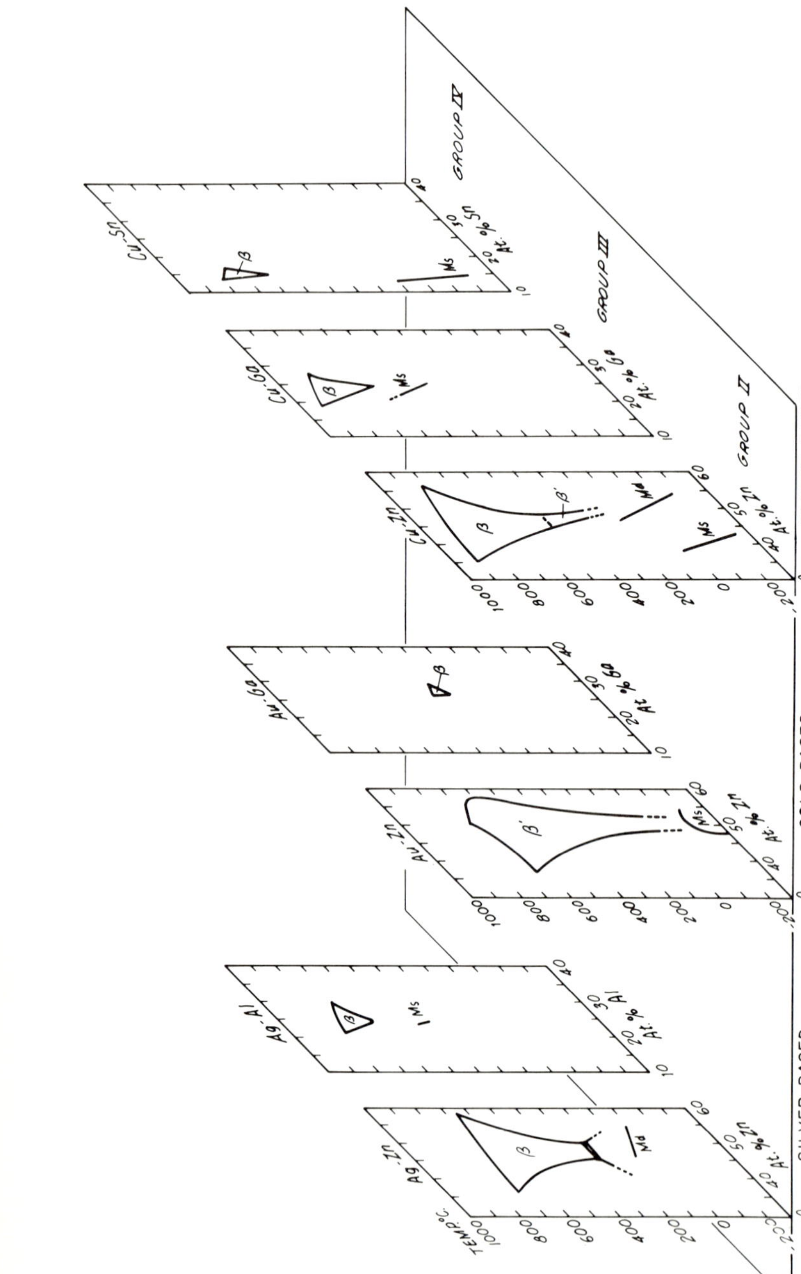

Fig. 17. Some typical binary β-phase fields based upon Cu, Ag, and Au, and their M_s and M_d lines.

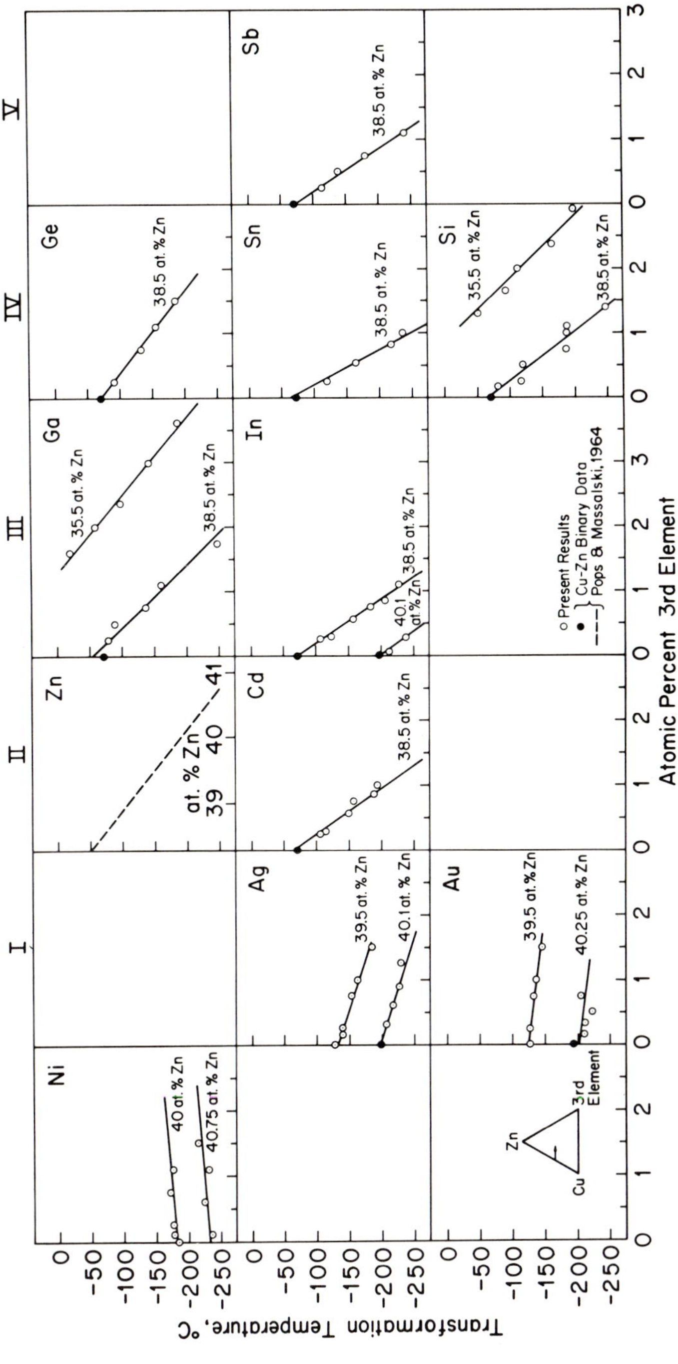

Fig. 18. Variation of M_s temperature with third element for ternary CuZn-based β-phase alloys (after Pops, 1966a).

of transformation occurs between 90° and about 30°C. This is indicated in Fig. 19, in which various forms of TiNi are indicated (Wang *et al.*, 1968). Even at room temperature the martensitic transformation is far from completion, but it may be further promoted by the application of external stress. If strain induced martensite is produced during stretching or bending, on subsequent heating above 90°C the reversal martensitic transformation takes place. The sudden change of the resultant shape into the original shape existing

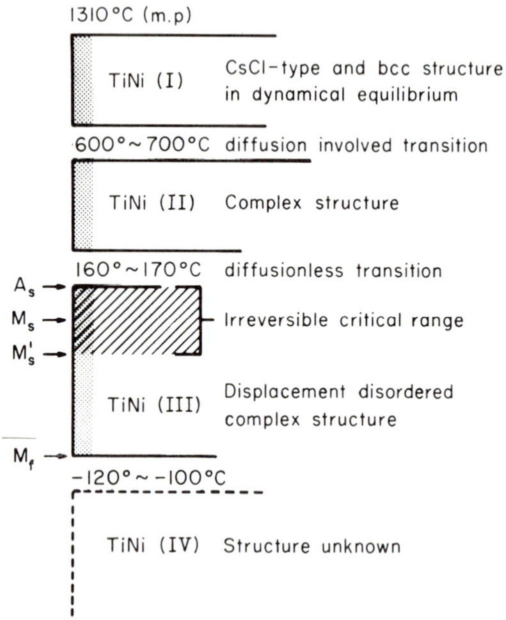

Fig. 19. Schematic diagram of the structural types and transformation temperatures for TiNi (after Wang *et al.*, 1968).

prior to the deformation accompanies the reversal transformation and has been called the "shape-memory effect." The diagramatic model for this effect in terms of the orientation of the martensitic products has been recently proposed by de Lange and Zijderveld (1968) (see Fig. 20). The rather remarkable properties of the TiNi alloy have stimulated interest in other compounds with similar structures such as TiFe–TiCo and TiCo–TiNi. This leads to the postulate that the transition observed is electronic in nature (the so-called Jahn–Teller effect associated with magnetostriction), and that similar transition might be found in some pseudobinary systems such as ZnRu–ZnRh, ZnRh–ZnPd, HfOs–HfIn, and HfIn–HfPt (Wang, 1965).

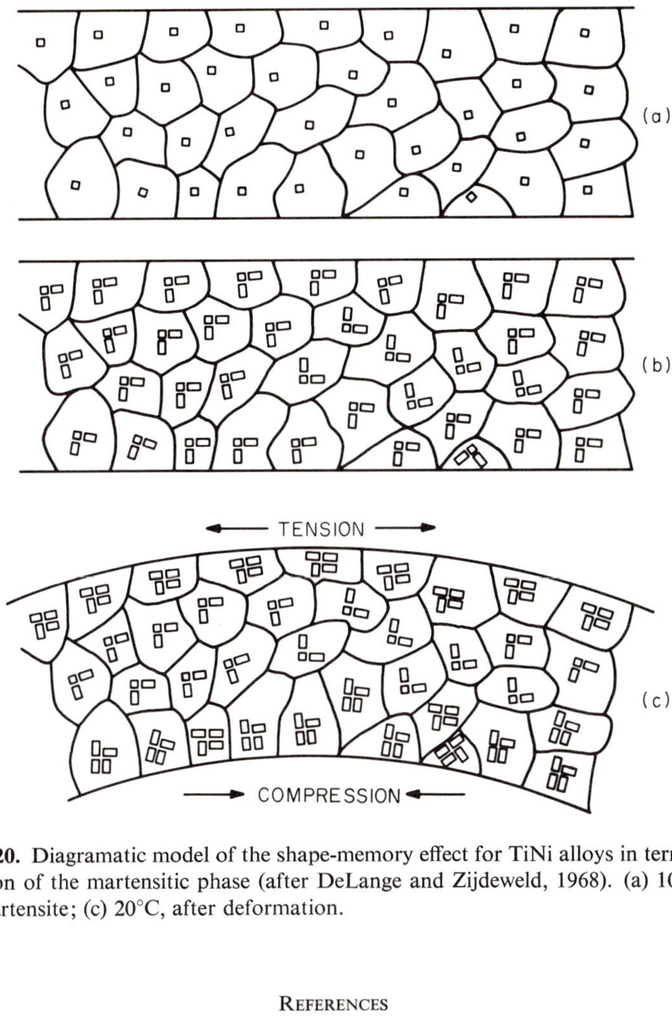

Fig. 20. Diagramatic model of the shape-memory effect for TiNi alloys in terms of the orientation of the martensitic phase (after DeLange and Zijdeweld, 1968). (a) 100°C; (b) 20°C, martensite; (c) 20°C, after deformation.

References

ANDREWS, K. W., and BROOKES, P. E. (1951). *Metal Treatment and Drop Forging* **18**, 301.
ASM (1959). "Precipitation from Solid Solution." Am. Soc. Metals, Cleveland, Ohio.
AXON, H. J., and HUME-ROTHERY, W. (1948). *Proc. Roy. Soc.* (*London*) **A193**, 1.
BARRETT, C. S., and MASSALSKI, T. B. (1966). "Structure of Metals," 3rd ed. McGraw-Hill, New York.
BEATTIE, H. J., and HAGEL, W. C. (1961). *Trans. AIME* **221**, 28.
BECK, P. A., ed. (1963). "Electronic Structure and Alloy Chemistry of the Transition Elements." Wiley (Interscience), New York.
BIEBER, C. G., and RAUDEBAUGH, R. J. (1959). *In* "Precipitation Processes in Steels," Spec. Rept. No. 64, p. 98. Iron and Steel Inst. (London).
BOESCH, W. J., and STANLEY, J. S. (1964). *Metal Progr.* **86**, 109.

BRESSANELLI, J. P., and MOSKOWITZ, A. (1966) *Trans. ASM.* **59**, 223.
BREWER, L. (1965). *In* "High Strength Materials" (V. F. Zackay, ed.). Wiley, New York.
BRINK, C., and SHOEMAKER, D. P. (1955). *Acta Cryst.* **8**, 734.
BUEHLER, W. J., and WILEY, R. C. (1962). *Trans. ASM* **55**, 269.
COLES, B. R. (1967). *In* "Intermetallic Compounds" (J. H. Westbrook, ed.), p. 79. Wiley, New York.
COTTRELL, A. H. (1948). "Theoretical Structural Metallurgy." Arnold, London.
DAS, D. K., RIDEOUT, S. P., and BECK, P. A. (1952). *Trans. AIME* **194**, 1071.
DECKER, R. F., EASH, J. T., and GOLDMAN, A. J. (1962). *Trans. AIME, Trans. ASM* **55**, 58.
DRABBLE, J. R., and GOODMAN, C. H. L. (1958). *J. Phys. Chem. Solids* **5**, 142.
DUWEZ, P., (1956). *In* "Theory of Alloy Phases." Am. Soc. Metals, Cleveland, Ohio.
ENGEL, N. (1954). *Powder Met. Bull.* **7**, 8.
ENGEL, N. (1967). *Acta Met.* **15**, 557 and 575.
GILBERT, A., and OWEN, W. S. (1962). *Acta Met.* **10**, 45.
GOLDSCHMIDT, H. J. (1957). *Metallurgica* **56**, 17.
GOLDSCHMIDT, V. M. (1928). *Z. Physik. Chem.* **133**, 397.
GRIMM, A. G., and SOMMERFELD, A. (1926). *Z. Physik.* **36**, 36.
HAGEL, W. C., and BEATTIE, H. J. (1959). *In* "Precipitation Processes in Steels," Spec. Rept. No. 64, p. 98. Iron and Steel Inst.
HUME-ROTHERY, W. (1965). *Acta Met.* **13**, 1039.
HUME-ROTHERY, W. (1966). "The Structures of Alloys of Iron." Pergamon, Oxford.
HUME-ROTHERY, W. (1968). *Progr. Mater. Sci.* **13**, 229.
HUME-ROTHERY, W., and RAYNOR, G. V. (1962). "The Structure of Metals and Alloys," 4th ed. Inst. of Metals, London.
KASPER, J. S. (1945). *Acta Met.* **2**, 456.
KASPER, J. S. (1956). *In* "Theory of Alloy Phases." Am. Soc. Metals, Cleveland, Ohio.
KING, H. W. (1965). *In* "Alloying Behavior in Concentrated Solid Solutions" (T. B. Massalski, ed.). Gordon and Breach, New York.
DELANGE, R. G., and ZIJDERVELD, J. A. (1968). *J. Appl. Phys.* **39**, 2195.
LAVES, F. (1963). *Advan. X-Ray Anal.* **6**, 43.
LAVES, F., and WALLBAUM, H. J. (1939). *Naturwissenschaften* **27**, 674.
LENA, A. J., (1959). *In* "Precipitation from Solid Solution." Am. Soc. Metals, Cleveland, Ohio.
MASSALSKI, T. B. (1958). *Acta Met.* **6**, 243.
MASSALSKI, T. B., and KING, H. W. (1961). *Progr. Mater. Sci.* **10**, 1.
MOOSER, E., and PEARSON, W. B. (1960). *Progr. Semiconductors* **5**, 103.
MOTT, N. F., and JONES, H. (1936). "The Theory of the Properties of Metals and Alloys." Univ. Press (Clarendon), London and New York.
MURPHY, H. J., SIMS, C. T., and HECKMAN, G. R. (1967). *Trans. AIME* **239**, 1961.
NEVITT, M. V. (1963). *In* "Electronic Structure and Alloy Chemistry of the Transition Elements" (P. A. Beck, ed.). Wiley (Interscience), New York.
NORDHEIM, R., and GRANT, N. J. (1954). *Trans. AIME* **200**, 211.
PARTHE, E. (1964). "Crystal Chemistry of Tetrahedral Structures." Gordon and Breach, New York.
PAULING, L. (1940). "The Nature of the Chemical Bond." Cornell Univ. Press, Ithaca, New York.
PAULING, L. (1956). *In* "Theory of Alloy Phases," p. 220. Am. Soc. Metals, Cleveland, Ohio.
PEARSON, W. B. (1967). "A Handbook of Lattice Spacings and Structures of Metals and Alloys." Pergamon, Oxford.

PICKERING, F. B. (1959). *In* "Precipitation Processes in Steels," Spec. Rept. No. 64, p. 118. Iron and Steel Inst. (London),
POPS, H. (1964). *Trans. AIME* **230**, 813.
POPS, H. (1966a). *Trans. AIME* **236**, 1532.
POPS, H. (1966b). *J. Iron Steel Inst.* **204**, 1117.
PUGH, J. W., and NISBET, J. D. (1950). *Trans. AIME* **188**, 268.
RACHINGER, W. A. (1960). *J. Aust. Inst. Metals* **5**, 114.
ROBINSON, P. M., and BEVER, M. B. (1967). *In* "Intermetallic Compounds" (J. H. Westbrook, ed.), p. 38. Wiley, New York.
RUDMAN, P. S. (1965). *Trans. AIME* **233**, 864 and 872.
SIMS, C. T., (1966). *J. Metals* **18**, 1119.
TAYLOR, A., and FLOYD, R. W. (1952). *J. Inst. Metals* **81**, 25.
TAYLOR, A., and FLOYD, R. W. (1953). *J. Inst. Metals* **81**, 451.
WANG, F. E. (1965). *Proc. Intern. Conf. Fracture 1st Tokyo, Japan* (T. Yokabori, T. Kawasaki, and J. L. Swedlow, eds.), Vol. 2, p. 899.
WANG, F. E., DE SAVAGE, B. F., and BUEHLER, W. J. (1968). *J. Appl. Phys.* **39**, 2166.
WAYMAN, C. M. (1965). *In* "Precipitation Processes in Steels," Spec. Rept. No. 64, p. 118. Iron and Steel Inst. (London).
WESTBROOK, J. H., ed. (1967). "Intermetallic Compounds." Wiley. New York.
WOODYATT, L. R., SIMS, G. T., and BEATTIE, H. J. (1966). *Trans. AIME* **236**, 519.
ZACKAY, V. F., PARKER, E. R., FAHR, D., and BUSCH, R. (1967). *Trans. ASM* **60**, 252.
ZENER, C. (1947). *Phys. Rev.* **71**, 846.
ZENER, C. (1967). *In* "Phase Stability in Metals and Alloys" (P. S. Rudman, J. Stringer, and R. I. Jaffe, eds.), p. 25. McGraw-Hill, New York.

VIII

The Use of Phase Diagrams in the Sintering of Ceramics and Metals

D. LYNN JOHNSON

DEPARTMENT OF MATERIALS SCIENCE
NORTHWESTERN UNIVERSITY, EVANSTON, ILLINOIS

and

IVAN B. CUTLER

DIVISION OF MATERIALS SCIENCE AND ENGINEERING
UNIVERSITY OF UTAH, SALT LAKE CITY, UTAH

I. INTRODUCTION

Sintering of powder compacts takes place at a temperature sufficiently high to achieve some type of atomic mobility. Under these conditions, there is at least an approach toward the equilibrium conditions described by phase equilibrium diagrams. The powder compact itself is an intimate, multiphase mixture of solid and vapor phases. The equilibrium state for a powder compact will exclude the vapor phase entirely. During sintering, the equilibrium state is ordinarily only partially achieved.

It is easily seen that phase equilibrium is very important in all sintering operations. Even though equilibrium conditions are not always achieved during sintering, the direction is continually toward elimination of excess surface energy. In order for the process to function rapidly at temperatures below the melting point, a powder of large surface area is utilized to obtain a large driving force toward equilibrium.

The sintering step in the processing of materials may be used with various objectives in mind. It might be the objective for the sintering process to increase the strength of the powder compact and, at the same time, retain porosity so that the final product might be used as a porous, self-lubricating bearing. Sintering is often the process used to consolidate a compact of small particles into a dense nonporous material. Ordinarily, during the consolidation process at high temperatures, shrinkage of the powder compact takes place. These dimensional changes add complexity to the sintering operation, especially when close dimensional tolerances are required. Commercial compositions of powders employed in forming useable shapes are often complex mixtures of sizes, shapes, and compositions designed to counteract shrinkage and promote rapid consolidation with a minimum of dimensional contraction. They may be multiphase systems with one or more components contributing to a liquid phase at the sintering temperature. When the sintered material is used for its stability at high temperature, the material may be a single phase with emphasis placed on its chemical purity to eliminate all traces of a liquid phase.

Much of the recent research directed toward understanding sintering has been designed to elucidate the path that atoms take during their mobile condition at the sintering temperature. Several excellent reviews of sintering are available (Thümmler and Thomma, 1967; Coble and Burke, 1963) and references are given here to assist the reader in gaining a broader background of information in this area.

This chapter will briefly review some of the important findings concerning sintering and relate them to the use of phase diagrams. Even though all compacts are in a nonequilibrium condition, classification is made in this chapter according to the final state as if equilibrium were achieved. Classification will be made as single and multiphase systems, recognizing that this is the direction that powder compacts go during the sintering operation, even though some trace of a vapor phase invariably remains in the sintered material as closed, isolated pores.

II. REVIEW OF SINTERING THEORY

A. Introduction

The purpose of this section is to present briefly the theory of sintering of pure crystalline and amorphous or glassy materials, while the following section introduces the modifications required for discussing sintering of multicomponent single phase materials. The discussion is confined to sintering without the application of pressure during the sintering process.

The driving force for sintering is the excess surface energy present in a powder compared with a single crystal of the same composition. As the temperature is increased, the atoms acquire some degree of mobility and a compact of particles undergoes changes of morphology that reduce the overall surface free energy. Three stages of sintering have been defined. In the initial stage, necks form at points of interparticle contact and usually, but not always, the particle centers approach each other. The relative amounts of neck growth and shrinkage that takes place depend upon the mass-transport mechanisms that are operative. As the necks impringe upon each other, an interconnected pore structure known as the intermediate stage of sintering develops. Eventually the pores become isolated from each other at the beginning of the final stage. Shrinkage is usually observed during all stages of the process, although under some conditions pore growth and compact swelling (overfiring) can occur during the final stage.

The various mechanisms of material transport and their influence on each stage of sintering will now be discussed.

B. Initial Stage

The most rapid changes in a compact occur during the initial stage of sintering since the driving force is high and transport distances are small. The possible mechanisms that might provide mass transport have been more or less understood for some time (Kuczynski, 1949; Kingery and Berg, 1955; Coble, 1958), although an unambiguous determination of the relative importance of various of these mechanisms has often been difficult if not impossible. This is not surprising when one considers the complexity of the problem.

Figure 1 shows two sintering crystalline particles in contact. The grain boundary that exists between the particles is both a vacancy sink and a diffusion path. Surface-tension forces produce compressive stresses under convex surfaces and tensile stresses beneath concave surfaces. These stresses are at least partially hydrostatic. The resulting stress differences cause chemical potential gradients that tend to drive atoms to the neck surface. The variety

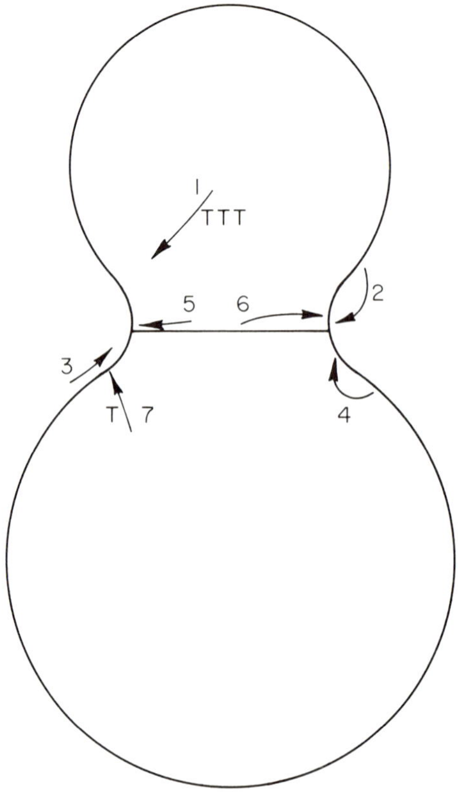

Fig. 1. Mechanisms of material transport during the final stage of sintering. 1. Plastic deformation (dislocation motion). 2. Vapor transport. 3. Surface diffusion. 4. Volume diffusion from particle surface to neck surface. 5. Grain-boundary diffusion. 6. Volume diffusion from grain boundary to neck surface. 7. Volume diffusion from dislocations or internal grain boundaries to neck surface.

of paths that the atoms might take gives rise to the variety of sintering mechanisms.

Dislocation motion (plastic deformation) can cause densification through bulk flow of material to the neck surface if the stresses are sufficiently high. It may be that this mechanism is important in the very earliest sintering of many crystalline materials, when the calculated stress in the neck can be well above the macroscopic flow stress. This is the least understood of the sintering mechanisms, partly because little is known about the plastic flow behavior of particles of the sizes commonly employed in sintering practice. There is no kinetic model to describe sintering by plastic deformation, and the evidence for its occurrence is at best qualitative (Lenel and Ansel, 1967; Morgan and

Hall, 1967; Moore, 1968). If plastic deformation does contribute to sintering, its effect will diminish rapidly with neck growth and shrinkage since these will cause a reduction in the stresses in the neck. Thus a transition from plastic flow to purely diffusive mechanisms has been predicted by Lenel and Ansell (1967) and observed by Moore (1968).

The chemical potential gradient from the particle surface to the neck surfaces produces three types of mass transport. These are vapor transport, surface diffusion, and volume diffusion from the particle surface to the neck surface (see Fig. 1). These three mechanisms can cause neck growth and rounding of pores, but cannot result in shrinkage of a powder compact since they do not remove material from between the centers of adjacent particles. Rather, these mechanisms actually retard densification because the excess neck growth results in a more rapid reduction in the driving force (curvature of the neck surface).

The tensile stress immediately beneath the neck surface is balanced by a compressive stress at the center of the grain boundary according to Johnson and Clarke (1964). This stress gradient gives rise to a chemical potential gradient which tends to drive atoms from the grain boundary to the neck surface by grain boundary and volume diffusion. Similar but lesser gradients can produce atom migration from internal defects, such as dislocations, to the neck surface (see Fig. 1).

The sintering of glassy or amorphous materials is similar to that described for crystalline materials with the exception of the impossibility for grain boundary diffusion or plastic deformation by dislocation motion to be active mechanisms. Volume diffusion is equivalent to viscous diffusion in amorphous materials. Frenkel (1945) derived an analytical expression for initial sintering by viscous diffusion which was the first attempt to describe sintering in terms of model equations.

It is quite probable that more than one of the mechanisms discussed above will be active during the initial sintering of most materials, and the over-all behavior of the compact will be the result of the interplay among these mechanisms. High relative surface diffusion or vapor transport could result in entry into the intermediate or even the final stage of sintering after relatively little shrinkage. On the other hand, a low surface diffusivity and vapor pressure would result in maximum shrinkage before the onset of the intermediate and final stages. The former case would tend to give some strengthening without shrinkage, and, therefore, good dimensional stability, while the latter would tend to result in high density. There are, of course, other phenomena that can occur during the intermediate and final stages that can prevent the attainment of theoretical density. Some of these are discussed below.

It is frequently desirable to know the mechanisms of material transport and the effect of variables such as temperature, atmosphere, and impurities, on each. The models which have heretofore been published do not allow such a determination because only a single mechanism is assumed to be operating (Kuczynski, 1949; Kingery and Berg, 1955; Coble, 1958; 1961; Nichols and Mullins, 1965; Rockland, 1966; 1967). For the few materials for which surface diffusion coefficients have been published, it appears that surface diffusion ought to predominate during the initial stage of sintering (Seigle, 1964; Wilson and Shewmon, 1966). However, powder compacts of these same materials show substantial shrinkage, indicating that volume and/or grain boundary diffusion, or possibly plastic deformation, are also active. A model proposed by Johnson (1969) permits identification of all of the significant material transport mechanisms and, in many cases, makes it possible to

TABLE I

SINTERING MECHANISMS FOR "PURE" MATERIALS

Material	Mechanisms
α-Fe	Volume, grain boundary, and surface diffusion (data of Shingu, 1967)
Cu	Volume, grain boundary, and surface diffusion (data of Kingery and Berg, 1955; Wilson and Shewmon, 1966)
Fe_2O_3	Grain boundary and surface diffusion (Smart, 1968)
LiF	Initially plastic deformation followed by volume diffusion after about 1 % shrinkage (Moore, 1968)

calculate the volume grain boundary, and surface diffusion coefficients. The procedure involves simultaneous measurement of shrinkage, shrinkage rate, and neck size on compacts of spherical particles. Table I lists some of the results of application of this method. Agreement between calculated and published diffusion coefficients was within a factor of 3, except in the case of the surface diffusion coefficient of copper, where the calculated value was an order of magnitude lower than the published value. Johnson and Clarke (1964) assumed that surface diffusion was insignificant in the sintering of silver in purified argon and obtained agreement with their volume plus grain boundary diffusion model. Their calculated diffusion coefficients are in very close agreement with the tracer values.

Many investigators have reported sintering mechanisms on a wide variety of metals and ceramics using the time dependence of shrinkage or neck growth as an indicator. Unfortunately, the significance of this time dependence cannot be interpreted unambiguously, and most conclusions about sintering mechanisms based upon this approach must be considered as tentative.

C. Intermediate Stage

As the necks formed during the initial stage impinge upon each other, the pore structure can be described as open pores interconnected by channels. All the mechanisms operable during the initial stage can continue, but the driving force is lower and transport distance are greater. Surface diffusion or vapor transport can move material from the surfaces of large pores to nearby small pores, resulting in loss of small pores and an increase in the average pore size. At the same time, mass transport from the grain boundaries will result in continued shrinkage and, therefore, lower total porosity.

The large interparticle necks and reduced porosity make grain boundary migration more likely. When the grain boundary between two particles is eliminated, shrinkage is stopped at that point in the compact (see Alexander and Balluffi, 1950). The adjacent pores will continue to shrink or grow, depending upon the geometry, until they become isolated by channel closure. Closure of the channels and isolation of pores from each other lead to the final stage.

D. Final Stage

The elimination of interconnectivity of pores eliminates surface and vapor transport as significant mechanisms; these can change the shape but not the size of a closed pore. Grain growth will seriously reduce densification rates and can prevent attainment of theoretical density (Coble, 1962; Jorgensen and Westbrook, 1964). Closed pores isolated from grain boundaries shrink very slowly because grain boundary diffusion is eliminated and the vacancy sinks (grain boundaries) for volume diffusion are far away from the pores. Of primary importance in the attainment of high density is retardation of grain growth, so that the pores can be in close proximity to the grain boundaries (Jorgensen and Westbrook, 1964).

An additional factor in the attainment of theoretical density is the permeability of the material to the ambient gases during firing. Thus, alumina can be fired to theoretical density in hydrogen, oxygen, and vacuum, but not in helium, argon, or nitrogen (Coble, 1962; Warman and Budworth, 1967a,b). Residual gases can also cause bloating as the gas diffuses from small pores, where surface-tension forces produce high pressure, to large pores where the pressure is less.

III. SINTERING OF SOLID SOLUTIONS

A. Introduction

The foregoing section dealt with sintering in general; it is now desirable to examine the effects of solutes on sintering of single-phase crystalline solids. Although considerable work has been done in this area, and many of the underlying principles are understood, there are few, if any, systems in which the effects of the additives can be definitely stated. Most experiments have involved mixing powders containing various additives, pressing, and firing for specified times and temperatures. This approach reveals the overall effects of the additives during the course of sintering (frequently into the final stage), but is not conclusive as to the effect of the additives on material transport mechanisms. Nevertheless, most investigators have felt impelled to describe their results in terms of the supposed material transport mechanism responsible for sintering. Thus, Reijnen (1969) found that TiO_2 additions resulted in lower density of Al_2O_3 samples fired at 1850°C in air, and concluded that TiO_2 retards sintering of Al_2O_3 by reducing the concentration of cation vacancies (assumed to be rate controlling in this case), whereas Jones *et al.* (1958) observed enhanced densification in the same system at lower temperatures. The latter authors explained that adding TiO_2 increases the concentration of cation vacancies and thereby promotes sintering. However, the densification of compacts of alumina is the result of grain boundary diffusion (Johnson and Cutler, 1963) and the data of Bagley (1964) seem to indicate that small amounts of TiO_2 cause greatly enhanced diffusion in a relatively wide region adjacent to the grain boundaries (see also Johnson and Berrin, 1967). It is also well known that TiO_2 enhances grain growth in Al_2O_3. Therefore, Reijnen's results can be explained tentatively. The high firing temperature caused very rapid densification and entrapment of air in closed pores. Rapid grain growth would prevent much further densification after pore closure, and any redistribution of the residual gases from small to large pores would cause bloating. The net effect could easily be a density lower than that observed for pure Al_2O_3.

The opposite effects are observed for additions of MgO to Al_2O_3. In this case, initial sintering is retarded (Johnson and Cutler, 1963) while grain growth is also retarded and densification can continue to the theoretical limit (Coble, 1962; Jorgensen and Westbrook, 1964).

Unfortunately, in neither of the preceding examples is the role of surface diffusion, or the effects of the additives on surface diffusion, understood or investigated at all. In fact, all conclusions concerning the sintering mechanisms of solid solutions, as well as pure crystalline materials, must be regarded as tentative and incomplete until experiments that can distinguish among the various mechanisms are completed. The discussion that follows must of necessity, therefore, deal with generalities in most instances.

B. Effects of Solutes on Sintering Mechanisms

In phenomena involving mass transport of ionic compounds it is necessary that all elements of the compound be transported at the same rate, and the net rate of material transport will be determined by the rate of the slower moving species. Solutes can thus affect sintering rates of ceramics by altering the diffusion coefficients of the slower ions. Readey (1966a,b; Readey and Jech, 1968) has derived volume-diffusion sintering flux equations for pure metals and ionic crystals, impure ionic crystals, and nonstoichiometric crystals. If the added impurity or deviation from stoichiometry increases the concentration of slower vacancies, the volume-diffusion flux will increase. However, at the same time the concentration of the faster vacancies will be reduced. Since the flux will be controlled by the slower moving ionic species, a maximum and subsequent decrease in flux is expected if the concentration of faster vacancies is reduced to sufficiently low levels by increasing the impurity content. On the other hand, an impurity that increases the concentration of vacancies of the faster species will reduce the concentration of slow vacancies and will cause a monotonic decrease in the flux.

As a possible example of some of the foregoing ideas, Peterson and Cutler (1968) observed that the shrinkage of compacts of CaO appeared to follow the volume-diffusion model, and that water vapor enhanced the sintering. Calcium vacancies would be created if water disolves as hydroxyl ions which substitute for oxygen ions. However, they were unable to determine the rate-controlling species or the actual diffusion coefficient because their powder was uncharacterized. Also, they did not investigate the contribution of surface diffusion. Beyond this, the authors are unaware of any well-defined examples of impurity effects on the volume diffusion flux during sintering.

The influence of solutes on grain boundary and surface diffusion has received little theoretical or experimental attention. This is unfortunate because these are probably the most important sintering mechanisms for the particle sizes usually employed in practice. The work of Keski and Cutler (1968) and Johnson and Cutler (1963) was mentioned in the Section A. Eastman and Cutler (1966) found that water vapor enhanced the apparent grain boundary diffusion coefficient in MgO, but here, again, the powder was not characterized and the surface diffusion contribution was not investigated.

The low temperature sintering of tungsten is apparently by grain boundary diffusion (Hayden and Brophy, 1963), although the possibility of a surface diffusion contribution has not been studied. Most of the group VIII elements enhance the sintering of tungsten, while iridium retards it (Hayden and Brophy, 1963, 1964; Brophy *et al.*, 1961). The manner in which the modification of sintering rate depends upon additive content can be correlated to the phase diagrams of the respective systems (Prill *et al.*, 1964).

In systems where the additive was insoluble in tungsten but the opposite solid solution exhibits a wide range of solubility of tungsten, the degree of enhancement (or retardation) depended upon additive concentration until a monolayer was formed on the tungsten particle. Additions beyond this amount had no effect. Where tungsten is insoluble in the additive (or adjacent intermetallic compound), the degree of enhancement went through a maximum and decreased with increasing concentration beyond a monolayer. Some type of grain boundary or interfacial diffusion is being affected, but many details of the mechanisms are still lacking.

Finally, the problem of solute distribution is important, particularly in the final stages of sintering. Solutes which are preferentially located on the grain boundaries can exert a drag on the motion of the grain boundaries (Cahn, 1962). Jorgensen and Westbrook (1964) have shown that there is a direct correlation between solute segregation at grain boundaries and ultimate densification of Al_2O_3. Jorgensen and Anderson (1967) have presented similar results for Y_2O_3 doped with ThO_2. With grain-boundary motion inhibited, the pores remain on or near the grain boundaries until they are eliminated.

C. Areas for Further Work

The discussion thus far has shown that relatively few quantitative data are available from which to assess the effects of solute additions on the sintering of single phase materials. The difficulty is twofold. First are the relatively unknown effects of additives on the mass transport processes, particularly surface and grain-boundary diffusion. Second are the effects of solutes on other aspects, such as surface tension, grain-boundary migration, and vapor transport. It is desirable to know the defect structures of doped and nonstoichiometric materials, particularly for the defects that control sintering. Solubility limits in systems showing very limited solid solubility need to be investigated. Solute site and spatial distribution and solute valency are problems that require further study. Understanding the criteria for retardation of grain boundary mobility would aid greatly in selecting dopants to enhance ultimate densification. Finally, the vapor-pressure and vapor-transport parameters over solid solutions would help us to understand the sintering of some materials. Until substantial progress can be made in most of these areas, it will continue to be necessary to "cut and try" in the development of new sintering techniques and sintered products.

IV. SINTERING IN MULTIPHASE SYSTEMS

A. Introduction

Many sintered materials are multiphase compositions. In addition to closed porosity that is nearly always present in sintered materials, other phases are often present. Sintered tungsten carbide may contain liquid cobalt at the sintering temperature. The majority of porcelains, sintered aluminas, and other refractory oxides contain two or three phases in addition to the pores that are never entirely eliminated during the sintering process. It is easily seen that sintering in multiphase systems is of considerable industrial importance.

Following the work of Pines (1956), several general principles involving interfacial energies assist in describing sintering of two-phase systems. If we have an assembly of particles of A and B that represent two separate phases that are mutually insoluble in each other, three different types of grain boundaries will be present in the powder compact. There will be contacts connecting A particles with A particles and B with B particles and contacts joining A and B particles. If the interfacial free energy at the AB interface is smaller than interfacial energy at the A–A interface and the B–B interface, then the free energy of the system will be decreased if the A–B interface grows to replace the A–A and B–B interfaces. This is expressed in the equation

$$\gamma_{AB} - (\gamma_A + \gamma_B) < 0 \tag{1}$$

Under conditions of Eq. (1), shrinkage will take place at the A–B interfaces.

If Eq. (1) is not valid and the interfacial free energy γ_{AB} is larger than the sum of γ_A and γ_B, shrinkage will not take place at the A–B interface. This does not mean that a compact of a mixture of A and B particles will not shrink under these conditions, but it does signify that there will be no shrinkage that will be promoted at the A–B interface. In this assemblage of particles, shrinkage will take place only at the A–A interfaces and at the B–B interfaces, and there will be a trend toward segregation of the A composition away from the B composition by sintering and grain growth as the sintering process proceeds.

If the interfacial free energy, γ_{AB}, is sufficiently small, not only will Eq. (1) be valid, but Eq. (2) may also describe the system.

$$\gamma_{AB} < |\gamma_A - \gamma_B| \tag{2}$$

If Eq. (2) is valid for a particular two-phase mixture, the interfacial free energy between the two phases and the free energy of one of the phases will be smaller than the free energy of the other phase. Under these conditions

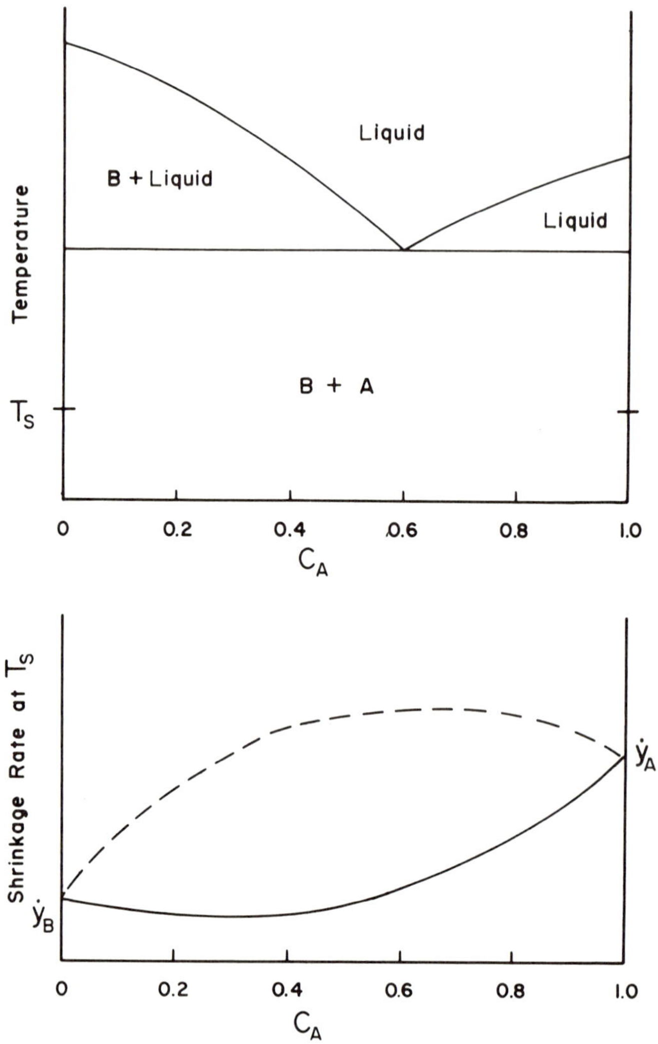

Fig. 2. Sintering of mixtures of two phases, A and B, that are mutually insoluble in each other but have an interfacial tension that satisfies Eq. (2). The upper convex curve (broken line) in the lower drawing signifies a shrinkage at the A–B interfaces that is greater than the average at the A–A particle contacts and the B–B contacts. The lower solid line is for the case where shrinkage at the A–B interfaces is smaller than the average.

the phase with the lower free energy may completely surround or wet the phase with the high free energy and lower the total free energy of the system through this process. The phase with the low free energy is often a liquid. There are many industrially important multiphase sintering systems that meet the conditions described in Eq. (2).

For two-phase mixtures that meet the conditions of Eq. (1), Pines and Sukhinin (1956) have described sintering in terms of shrinkage at particle contacts. Their analysis is reminiscent of the thermodynamic evaluation of mixtures of atoms given by Cottrell (1948). The fractional shrinkage y is shown to be a function of the volume fraction of A particles, according to the equation

$$y = y_A C_A^2 + y_B(1 - C_A)^2 + 2y_{AB}C_A(1 - C_A) \tag{3}$$

where y_A refers to the shrinkage at A–A contacts, y_B the shrinkage at B–B contacts, y_{AB} the shrinkage at A–B contacts, and C_A describes the volume fraction of A particles. In this analysis the particles are assumed to be of the same size and of spherical geometry so that similar packing can be obtained for each type of particle. Figure 2 illustrates either the total shrinkage for constant time and temperature or the rate of shrinkage encountered in a two-phase mixture of particles of A and B. Whether the shrinkage curve is concave upward or concave downward depends on whether the shrinkage at the A–B interface is greater or smaller than the average at the A–A and B–B interfaces. A linear amount or a linear rate of shrinkage would be obtained if shrinkage at the A–B contact were exactly equal to the average shrinkage at the other two types of contacts.

For two-component systems that are partially soluble in each other, the situation illustrated in Fig. 2 has to be modified somewhat. This is illustrated in Fig. 3. The modification arises due to the nonstoichiometry that may be introduced by a limited solid solution of A in B and B in A. For chemical compounds, solid solution may either enhance or decrease the concentration of vacancies of the slowest moving species. If the solid solution increases the vacancy concentration of the slowest moving species, the rate of shrinkage will very rapidly increase as the solute concentration increases. In the two-phase region of the equilibrium diagram, the same mixture law described in Eq. (3) will hold as illustrated previously.

It may be readily seen from Fig. 3 that there is an overlap between the principles described in Section III and the sintering involved with the mixtures of two different phases. In the sintering of compounds, it is very difficult to obtain intrinsic materials. Small amounts of impurities easily upset the stoichiometry; thus, sintering of two-phase materials described in Fig. 3 will more nearly be representative of the sintering of mixtures of compounds than Fig. 2.

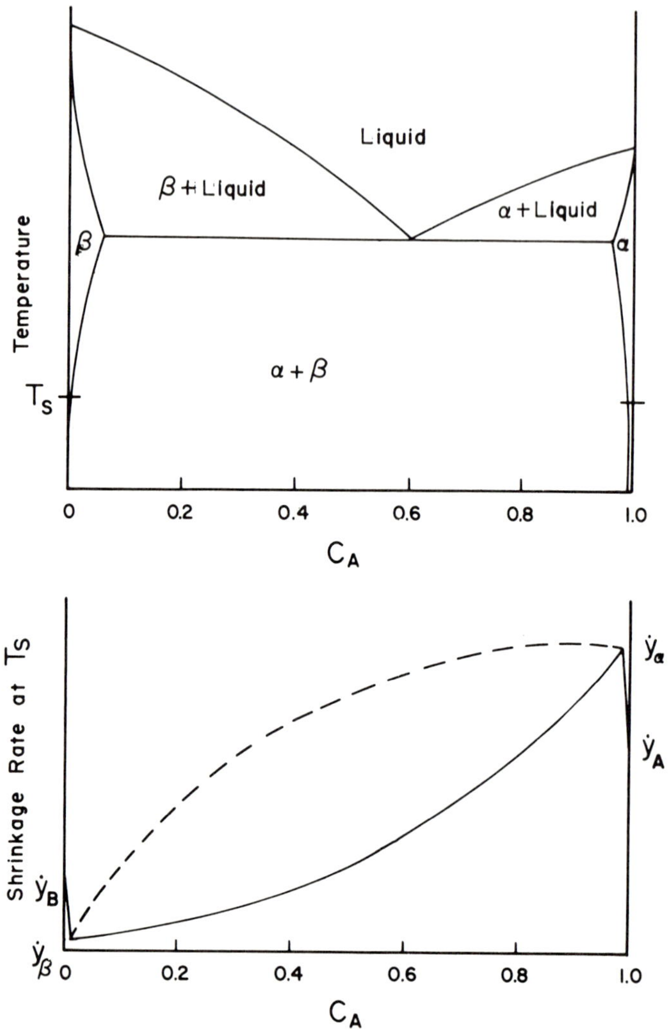

Fig. 3. Sintering of mixtures of two phases, A and B, that have limited solubility in each other. The shrinkage or shrinkage rate may be very different for the solid solutions, α and β, than for the pure materials, A and B, as outlined in Section III. The change in shrinkage rate between A and α and B and β is shown here as a straight line only for convenience. Interaction between mixed phases is similar to the previous drawing (Fig. 2).

B. Multiphase Systems Containing a Liquid Phase

There are many multiphase systems that are easily sintered because of the presence of a liquid phase at the sintering temperature. In order for the liquid phase to control the sintering operation, it must be present in sufficient quantity to fill the interstices surrounding the solid phase. Phase equilibria assists us in understanding this process because the quantity of liquid increases as temperature increases above the eutectic temperature. In addition, the viscosity of liquids is exponentially related to temperature, and diffusion in a liquid is much faster than it is in a solid. As a result, liquid phase sintering takes place rapidly because both the quantity of liquid increases as temperature increases and diffusion increases (or viscosity decreases) exponentially with increasing temperature.

Frenkel described sintering of a liquid with the very simple equation

$$y = \gamma_L t/2a\eta, \qquad T = \text{constant} \tag{4}$$

where η represents the viscosity of the liquid and γ_L is the surface free energy of the liquid. Because there are no fractional exponents modifying the viscosity during liquid phase sintering, the full activation energy of diffusion in a liquid controls the sintering process. This may be described by the approximate equation for constant rate of heating:

$$y \cong (\gamma_L RT^2/2acAQ)e^{-Q/RT}, \qquad T = ct \tag{5}$$

where the constant c represents the rate of heating and A is a pre-exponential factor for viscosity. Equations similar to (5) for crystalline systems contain an exponential factor that is either one-half or one-third of the activation energy for diffusion. For this reason, temperature control in systems depending on liquid-phase sintering must be much more accurate than for corresponding crystalline systems.

The phenomenological aspects of liquid phase sintering have been described by several investigators (Lenel, 1951; Kingery, 1959; and Gurland and Norton, 1952). On the basis that complete wetting of the solid by the liquid as described in Eq. (2) and in the presence of sufficient liquid to fill the interstices, shrinkage is generally observed to follow Eq. (4) as proposed by Frenkel. As the particles are drawn into contact with each other, the surface tension of the liquid exerts a pressure at the points of contact and material diffuses from the points of contact through the liquid into the interstices between the particles. This is analogous to grain boundary diffusion in a single phase, crystalline system. Kingery (1959) has proposed that this second stage of the sintering process follows an equation that is similar to the grain boundary diffusion equation

$$y = (Kt/a^4)^{\frac{1}{3}} \tag{6}$$

The constant K contains terms very similar to those for grain boundary diffusion, such as surface tension, diffusion in the liquid, average width of the liquid "grain boundary," solubility constants, etc.

If the liquid does not entirely wet and surround the solid particles, the densification may be considerably slower, requiring diffusion at the solid-particle grain boundaries.

The principles of liquid-phase sintering are best illustrated with a few examples. Several examples are well known and have been described in the literature, some more extensively than others. It will not be possible in this short chapter to describe in detail even the three examples that have been selected for illustration of the principles of liquid phase sintering. The reader should seek more information in the literature if greater detail is needed, or refer to the reviews if additional systems are to be examined.

1. Iron–Copper System

Several metallurgists have observed the rapid sintering of iron powder with additions of copper. Approximately 20% by weight of copper is needed in order to densify iron powder to near pore free condition below 1200°C. From the phase diagram shown in Fig. 4, one may observe that liquid is formed above the peritectic temperature of 1090°C. However, with copper powder being mixed with iron powder in a nonequilibrium condition, the copper powder will no doubt begin melting at 1083°C with diffusion of iron into the copper increasing its melting point. Along with the formation of liquid at the sintering temperature, the simultaneous process of the formation of a solid solution of iron and copper will take place as the iron powder absorbs copper by diffusion. It is readily observed from the phase diagram that with 20% by weight of copper approximately 12 to 15% liquid will be formed at a sintering temperature of between 1120° and 1200°C.

Rapid sintering takes place in the iron–copper system because the copper-rich liquid completely wets the iron–copper solid solution and fulfills the conditions described in Eq. (2). Kingery and Narasimhan (1959) have shown that there is a limited amount of densification that may be described by Eq. (4), followed by an additional amount of shrinkage that may be correlated by Eq. (6).

After sintering is complete, there will be further adjustments in the phase composition as the sintered material cools to room temperature. There will be exsolution of iron from the copper solid solution, as well a precipitation of copper from exsolution as the gamma phase cools and, in addition, precipitation from the eutectoid during the transformation of gamma- to alpha-iron.

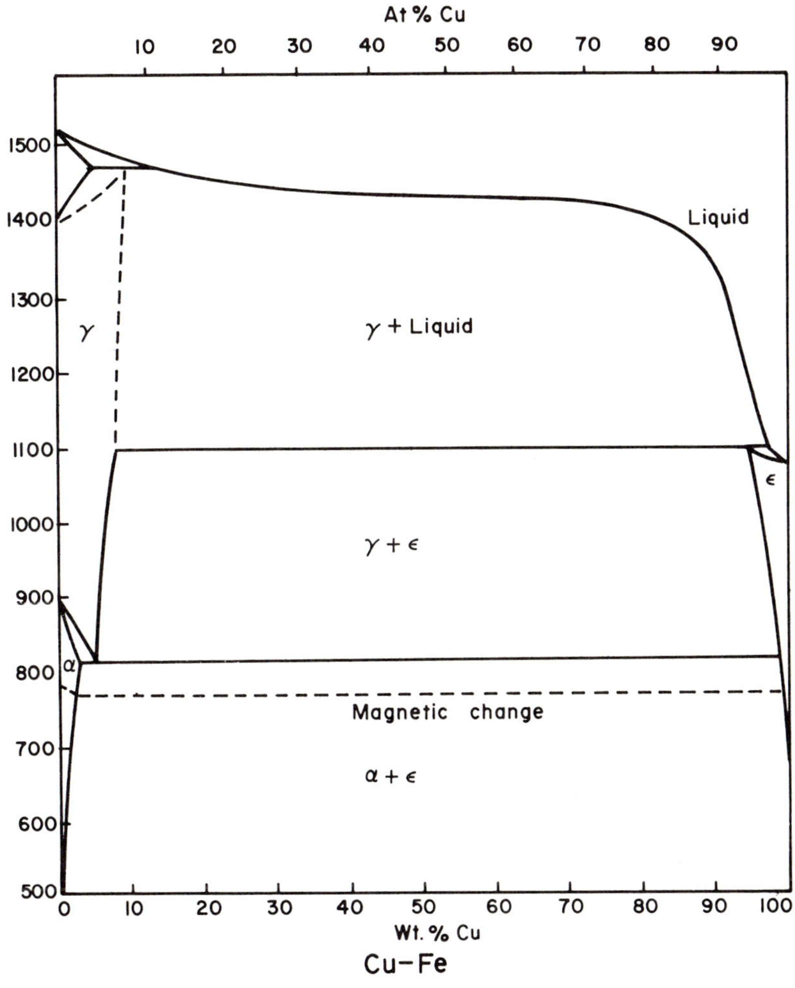

Fig. 4. Binary phase diagram for the Cu–Fe system. Sintering of iron powder containing copper additions occurs above the 1090°C peritectic. (After Smithells, 1962).

2. Tungsten Carbide–Cobalt System

Of great commercial importance are cemented carbides. The tungsten carbide–cobalt compositions are well known and illustrative of the cemented carbides. The maximum in many of the physical properties of the tungsten carbide–cobalt cemented carbides is achieved with a minimum amount of liquid. A composition of 94 wt % tungsten carbide with 6% cobalt is near this optimum composition.

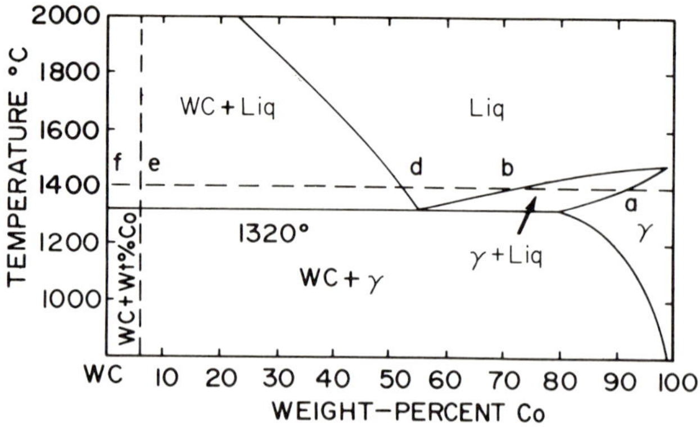

Fig. 5. Pseudobinary WC–Co diagram. (After Swarzkopf *et al.*, 1960).

The phase relationships may not be accurately described with a binary phase diagram because three components (W, C, and Co) are involved. For reasons of illustration, however, a pseudobinary shows the important features of the system and will be used here. This is illustrated as Fig. 5. A mixture of 94% tungsten carbide and 6% cobalt is shown in the diagram as a dotted vertical line intersecting the point e. The sintering temperature is illustrated as a horizontal line passing through point e as well as points a, b, d, and f. In order to understand the approach to equilibrium conditions, it is important to understand somewhat concerning the preparation and heating of the materials.

Tungsten carbide, being a brittle solid, is ball milled with cobalt powder in preparation for sintering. Because cobalt is ductile, the milling rapidly flakes the fine cobalt powder and intimately mixes it with the tungsten carbide particles. Milling is continued until the particles are in the subsieve range with many of the fine tungsten carbide particles only microns in diameter.

A compact of this intimate mixture of cobalt and tungsten carbide should form a liquid at the ternary eutectic between tungsten, carbon, and cobalt at about 1280°C. The liquid, however, actually forms at a temperature above this at the interface between the cobalt and tungsten carbide particles. At the sintering temperature, not only does the cobalt react to form a liquid, but tungsten carbide is also dissolved in the cobalt, whose melting point is above the 1400°C sintering temperature. Eventually, the liquid dissolves all of the cobalt and the cobalt–tungsten carbide solid solution, as well as many of the fine tungsten carbide particles, forming a liquid of composition illustrated at d. According to the phase diagram, the 6% cobalt composition forms about 12% by weight of liquid; however, because of the large density differences between tungsten carbide and the liquid, there will be closer to 20 vol % of liquid formed at the sintering temperature. The surface free energy of the liquid pulls the particles together and the powder compact shrinks until it is near theoretical density.

Measureable growth of the tungsten carbide particles occurs as the sintered material is cooled below the eutectic temperature. Rather than precipitate additional tungsten carbide particles, tungsten carbide precipitates from the eutectic mixture onto the surfaces of tungsten carbide already available. Additional tungsten carbide comes from the exsolution process occurring in the cobalt solid solution (gamma) as the temperature decreases below the eutectic temperature.

3. Porcelain

There are many ceramic compositions that illustrate the principles of liquid-phase sintering. The traditional European hard-fired porcelain is, perhaps, the best known and one from which many principles pertaining to liquid-phase sintering may be illustrated.

The composition of European porcelain today has not varied significantly over that utilized by the ancients. It represents a compromise between many desirable properties of the unfired and fired ware. The majority of porcelain compositions are in the neighborhood of one-half clay (kaolinite), one-quarter quartz or quartzite, and one-quarter potassium feldspar. The kaolinite gives the composition plastic properties desirable for the forming operation. The feldspar and silica, although finely ground, are coarse in comparison to crystals of kaolinite. The feldspar and quartz particles are ordinarily 10 to 100 times the size of the kaolinite crystals. This not only assists the reduction of drying shrinkage after forming, but it also reduces the firing shrinkage as well. First of all, the composition is nearly as dense as can be formulated by selecting proper proportions of fine and coarse fractions. Secondly, both feldspar and quartz increase in dimensions upon heat treatment. The melting of feldspar results in a permanent expansion and the transformation of quartz to any of the other polymorphic forms of silica including glass results in a decrease in density and an increase in volume. These volume increases counteract the firing shrinkage of the kaolinite and the sintering shrinkage of the composite powder compact.

An examination of the phase diagram, K_2O–Al_2O_3–SiO_2 (see Fig. 6) will indicate that a porcelain composition falls in the region of the diagram bounded by mullite, potassium feldspar, and silica. Upon firing this composition under equilibrium conditions, one would expect to obtain, upon cooling back to room temperature, a polyphase material consisting of about 45% quartz, 30% mullite, and 25% potassium feldspar. The composition is in the primary field of crystallization of mullite. This means that mullite will be a

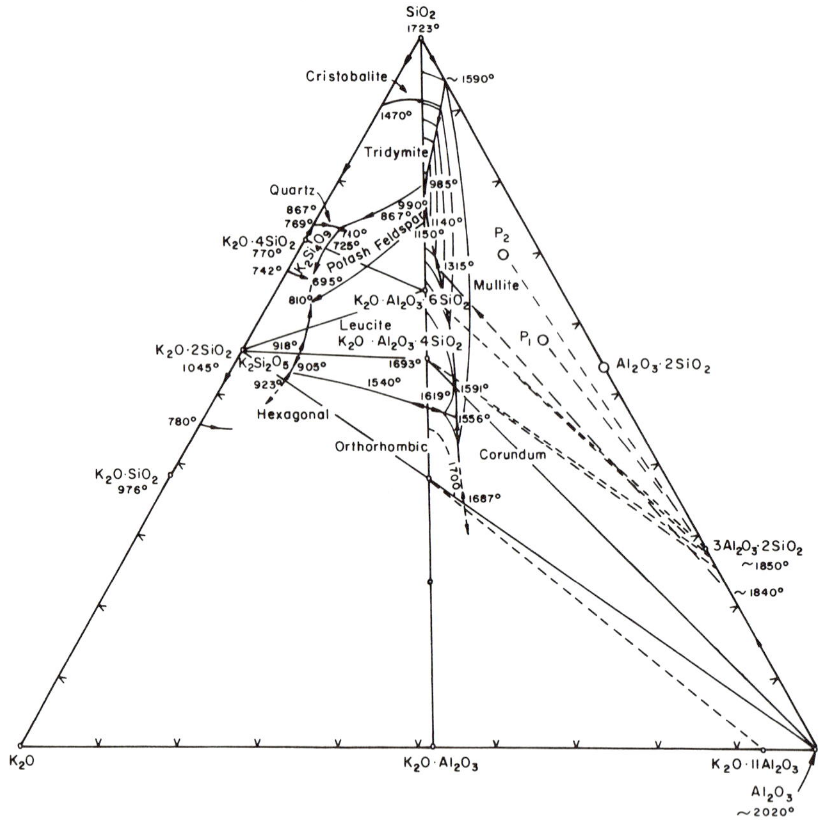

Fig. 6. System K_2O–Al_2O_3–SiO_2. (After Levin *et al.*, 1964.)

stable crystalline phase even at the high temperatures employed in firing porcelain. Upon cooling from the firing temperature assumed to be between 1300° and 1500°C, tridymite should crystallize as a secondary crystalline phase and, then as the ternary eutectic is obtained, potassium feldspar should crystallize out along with additional tridymite. Upon further cooling, tridymite should transform to quartz.

Experience has shown that equilibrium is never approached in the sintering of porcelain compositions. It is true that mullite forms at temperatures above 1000°C in nearly equilibrium quantities, but tridymite and potassium feldspar have never been observed to crystallize upon the cooling of a porcelain composition. The quartz that is sometimes found in a fired porcelain body can be traced to unreacted quartz crystals only partially dissolved during the firing operation. If the quantity of liquid is insufficient to dissolve all of the quartz, the quartz may transform in part or in entirety into cristobalite which does not transform to either quartz or tridymite during the cooling of the sintered porcelain.

The firing of porcelain is an excellent example of a multiphase system governed largely by kinetic processes rather than equilibrium conditions. The ceramic reactions occurring during the heating, firing, and cooling of a porcelain composition are varied in the speed or rate at which they occur. Decomposition of kaolinite crystals takes place rather rapidly at low temperatures. This follows the reaction

$$\underset{\text{kaolinite}}{Al_2(Si_2O_5)(OH_4)} \rightarrow \underset{\text{meta kaolin (amorphous)}}{Al_2O_3 \cdot 2SiO_2} + 2H_2O \tag{7}$$

Even more rapid is the transformation between alpha and beta quartz. Oxidation of the organic material contained with the clay is a somewhat slower kinetic process. Upon escape of the water from the clay structure, the material left behind is essentially amorphous near 1000°C. The amorphous decomposed clay mineral crystallizes first in a spinel structure and subsequently in the form of mullite crystals excluding silica from the structure. This is illustrated with the equation

$$\underset{\text{meta kaolin}}{3(Al_2O_3 \cdot 2SiO_2)} \rightarrow \underset{\text{mullite}}{3Al_2O_3 \cdot 2SiO_2} + 4SiO_2 \tag{8}$$

At about the same temperature, the feldspar in contact with the reactive silica available from the clay mineral decomposition forms a viscous liquid that promotes sintering and shrinkage of the powder compact. Between 1000° and about 1200°, depending on the rate of heating, the powder compact acts as though it were composed of clay and feldspar with the quartz remaining virtually inactive as far as the reactions are concerned. This is noted on the phase equilibrium diagram, Fig. 6, as point 1. As sintering draws the material together and as it shrinks, the viscous liquid comes in contact with quartz grains and at a temperature of about 1200°C and above, gradually dissolves the quartz. As quartz solution takes place, the composition in effect moves from point 1 to point 2. Melting of the feldspar and solution of the feldspar with silica along with formation of mullite crystals are slower processes requiring higher temperatures and longer times. Equilibrium is often closely approached at the firing or sintering temperature.

The simplified equations describing sintering cannot be readily applied to sintering of porcelain because the percentage of liquid phase, the composition of the liquid, and viscosity of the liquid all depend upon temperature. One of the slowest of the kinetic reactions taking place during firing is the solution of quartz crystals. Although quartz is thermodynamically unstable at the sintering temperature of porcelain, transformation to cristobalite or tridymite is slower than solution in the liquid or glassy phase of the porcelain body.

Rate of solution is dependent upon surface area and temperature. As quartz crystals grow smaller, their surface area decreases and rate of solution also decreases. Also, as the quartz crystals dissolve, they enrich the surrounding liquid in SiO_2 raising its viscosity and decreasing the rate of solution of more SiO_2. For these reasons, it is not uncommon to find quartz crystals that have survived the firing operation and are still present in the fired porcelain compositions. During cooling of the porcelain compositions, boundary stresses occur at the quartz–glass and mullite–glass interface. If the quartz crystals are sufficiently large, the boundary stress may be large enough to fracture the quartz crystals and separate them from the glass that envelopes them, particularly after the quartz has gone through the beta-to-alpha transformation upon cooling to room temperature.

The porcelain composition thus at high temperatures will be either liquid plus mullite or liquid, mullite, and cristobalite. Quartz will be present only if the composition has not been given sufficient time or temperature to reach equilibrium conditions. Upon cooling, even at reasonably slow rates of cooling, composition will essentially remain fixed as if it were still at high temperatures. No transformation from tridymite to quartz or cristobalite to tridymite or cristobalite to quartz occurs. No crystallization of the glass to leucite or feldspar has been observed. The final state at room temperature remains essentially the same as was obtained at the sintering temperature.

The amount of mullite in the porcelain can be determined largely on the basis of equilibrium conditions and can be computed from either point 1 or point 2, depending on whether or not the quartz has had time to dissolve into the liquid. As a result, compositions of the type described here have approximately 20 or 25% mullite. The balance is ordinarily liquid, although there may be some quartz or cristobalite, depending on how faithfully the compositions can be described in terms of the ideal composition we have described here.

C. Multiphase Systems Containing Solids Only

Goodison and White (1962) have summarized and classified sintering of multiphase systems containing crystalline material. They have classified them into

(1) mixtures of phases that interdiffuse to form solid solutions,
(2) mixtures of phases that react to form a compound or compounds,
(3) mixtures of phases that neither react nor interdiffuse.

Mixtures of phases that interdiffuse to form solid solutions can be treated in terms of the principles already outlined in this chapter. Solid solutions may change the vacancy or interstitial concentration so as to either enhance or retard the rate of sintering. The process, however, of forming the solid solution, i.e., the interdiffusion of the phases, constitutes still another problem. For metallic materials the rate of diffusion of one atom is not necessarily the same as the rate of diffusion of a different species of atom. The Kirkendall effect will influence the homogenization of powders. Kuczynski and Stablein (1961) have shown that homogenization of powder mixtures of two phases of metals results in the development of porosity as the faster diffusing species leaves pores behind as they diffuse away.

The problem of forming pores during homogenization of mixtures of powders is not always present in the case of compounds. In some oxides, for example, where the cation diffusion is much more rapid than the anion diffusion, homogenization takes place by counterdiffusion of cations of various species through a comparatively rigid oxygen lattice. In these instances, pores do not form and sintering is not as seriously interrupted by the homogenization, counterdiffusion process.

In all instances of sintering powders that form solid solutions, the problem really becomes one of kinetics rather than of phase equilibria. The homogenization process takes place much more rapidly as the diffusion distances decrease through using smaller particle sizes. When grain-boundary diffusion is the rate-controlling sintering step, homogenization appears to be very rapid and can take place by merely depositing the second phase uniformly on the surface of the primary sintering phase. Diffusion takes place evidently as rapidly as sintering takes place. These are the findings of Bagley in the case of small amounts of TiO_2 in solid solution with Al_2O_3.

Mixtures of powders that form a compound or compounds upon reaction during the sintering operation have been discussed by Kriek *et al.* (1959) and summarized again by Goodison and White (1962). For all systems that were studied, compound formation seemed to be a competitive process with diffusional shrinkage processes that are normal to sintering and interfered with densification. Some increased sintering occurred with small additions of Al_2O_3 and TiO_2 to MgO. In these cases, however, the small additions may have been within the solubility limit and may have promoted sintering through some vacancy formation mechanism. Large additions of Al_2O_3 and TiO_2 to MgO comparable to $MgAl_2O_4$ and $MgTiO_3$ formation interfered with sintering.

When the reaction between powders is highly exothermic, it may be possible to attain rapid densification through reaction, largely through the generation of very high temperatures that accelerate diffusion. These types of processes are very difficult to study and understand because of the difficulty of temperature control.

Of greatest interest in sintering is the use of two-phase mixtures that are entirely or nearly entirely insoluble in each other. As described earlier in this chapter, grain growth is important to the final densification. If grains grow so as to include pores, the opportunity for reaching near theoretical density diminishes. Just as pores normally inhibit grain growth, so is it possible for the second phase to be included into the sintering composition and arrest or at least partially arrest grain growth. The nature of grain growth inhibition with second phase constituents has been recently treated by Haroun and Budworth (1968). The ideal situation arises when powder of very small particle size is sintered in the presence of a few percent of a colloidal-size second phase that remains well dispersed and does not itself suffer from grain growth or coarsening. The process of coarsening of a dispersed phase is related to the solubility of the dispersed phase at the sintering temperature. Ordinarily, 2 to 10% of the dispersed phase is necessary to arrest grain growth in the sintered material.

An insoluble second phase is expected to interfere with the initial stages of sintering of powder compacts. Only in the very final stages of sintering will the dispersed phase assist in the densification as it inhibits grain growth and allows for disappearance of porosity by the retardation of grain boundary migration.

Several examples of dispersed phase materials are available. SAP (sintered aluminum powder) and TDNi (thoria-dispersed nickel) are cases where a dispersed stable oxide phase retards creep largely by interference in the

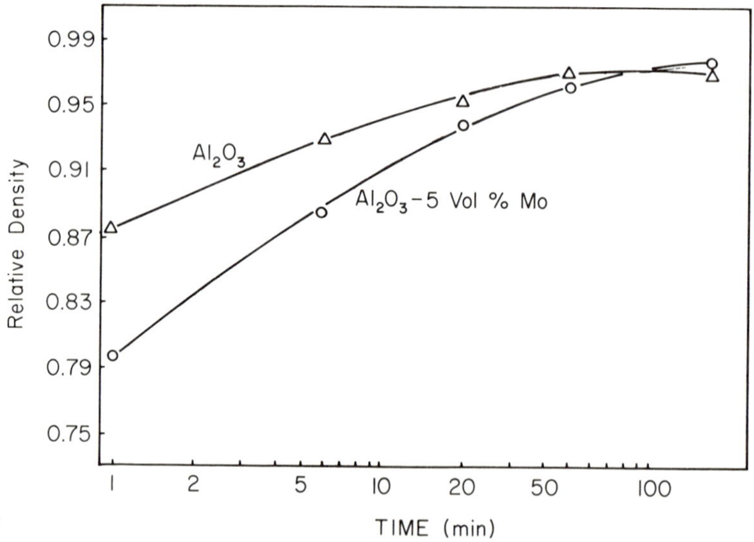

Fig. 7. Densification curves for Al_2O_3 and Al_2O_3–Mo materials at 2800°F in vacuum. (From McHugh *et al.*, 1966a).

movement of dislocation. Larger high-temperature strengths are noted in the presence of the dispersed phase. McHugh *et al.* (1966) have described dispersion strengthened sintered alumina using finely divided molybdenum as the dispersed phase. Once again, sintering was only accelerated in the very final stages. Their data illustrate this principle, as shown in Fig. 7. Grain growth was retarded through the use of the dispersed phase, and the higher strengths that were derived from the molybdenum dispersed alumina were accounted for on the basis of the smaller grain size. Hill *et al.* (1966) have achieved a similar result with beryllia containing silicon carbide as a dispersed phase with a small addition of magnesia for acceleration of sintering of the beryllia. Many more examples might be cited, and this description is not meant to be complete but merely illustrative of the principle involved.

V. SUMMARY

The sintering of metal and ceramic powders is driven by thermodynamic forces directing atomic movement toward equilibrium conditions. The rate of sintering is associated with the rate of atomic movement. The extent of completion of the sintering process must be judged in terms of phase equilibria. Thus, phase diagrams are intimately useful to those engaged in sintering studies. Although the principles of phase equilibria may not guarantee successful sintering processes, the equilibrium conditions do describe the end point for the processes and one is always well advised to keep the end point in view.

Acknowledgments

This chapter represents a special research interest that the authors gained through financial support of their respective institutions through ARPA Grant SD 67, through the Materials Research Center, Northwestern University and AEC Contract No. AT(11–1)–1122 with the Materials Science Division, University of Utah.

References

Alexander, B. H., and Balluffi, R. (1950). *Trans. AIME* **188**, 1219.

Bagley, R. D. (1964). Ph.D. thesis, Univ. of Utah, Salt Lake City, Utah.

Brophy, J. H., Shepard, L. A., and Wulff, J. (1961). *In* "Internal Powder Metallurgy Conference" (W. Leszynski, ed.), p. 113. Wiley (Interscience), New York.

Cahn, J. W. (1962). *Acta Met.* **10**, 784.

Coble, R. L. (1958). *J. Am. Ceram. Soc.* **41**, 55.

Coble, R. L. (1961). *J. Appl. Phys.* **32**, 787.

COBLE, R. L. (1962). *J. Am. Ceram. Soc.* **45**, 123.

COBLE, R. L., and BURKE, J. E. (1963). *Prog. Ceram. Sci.* **3**, 197–253.

COTTRELL, A. H. (1948). "Theoretical Structural Metallurgy," p. 154. Longmans, Green, London.

EASTMAN, P. F., and CUTLER, I. B. (1966). *J. Am. Ceram. Soc.* **49**, 526.

FRENKEL, J. (1945). *J. Phys.* (*USSR*) **9**, 385.

GOODISON, J. and WHITE, J. (1962). "Agglomeration" (W. A. Knepper, ed.), p. 251. Wiley (Interscience), New York.

GURLAND, J., and NORTON, J. T. (1952). *Trans AIME* **194**, 1051.

HAROUN, N. A., and BUDWORTH, D. W. (1968). *J. Mater. Sci.* **3**, 326.

HAYDEN, H. W., and BROPHY, J. H. (1963). *J. Electrochem. Soc.* **110**, 805.

HAYDEN, H. W., and BROPHY, J. H. (1964). *J. Less-Common Metals* **6**, 214.

HILL, N. A., O'NEILL, J. S., and LIVEY, D. T. (1966). Atomic Energy Res. Establishment Rep. 5056, Harwell, England.

JOHNSON, D. L. (1969). *J. Appl. Phys.* **40**, 192.

JOHNSON, D. L., and BERRIN, L. (1967). *In* "Sintering and Related Phenomena" (G. C. Kuczynski, N. A. Hooton, and C. F. Gibbon, eds.), pp. 445–469. Gordon and Breach, New York.

JOHNSON, D. L., and CLARKE, T. M. (1964). *Acta Met.* **12**, 1173.

JOHNSON, D. L., and CUTLER, I. B. (1963) *J. Am. Ceram. Soc.* **46**, 545.

JONES, J. T., MAITRA, P. K., and CUTLER, I. B. (1958). *J. Am. Ceram. Soc.* **41**, 353.

JORGENSEN, P. J., and ANDERSON, R. C. (1967). *J. Am. Ceram. Soc.* **50**, 553.

JORGENSEN, P. J., and WESTBROOK, J. H. (1964). *J. Am. Ceram. Soc.* **47**, 332.

KESKI, J. R., and CUTLER, I. B. (1968). *J. Am. Ceram. Soc.* **51**, 440.

KINGERY, W. D. (1959). *J. Appl. Phys.* **30**, 301.

KINGERY, W. D., and BERG, M. J. (1955). *J. Appl. Phys.* **26**, 1205.

KINGERY, W. D., and NARASIMHAN, M. D. (1959). *J. Appl. Phys.* **30**, 307.

KRIEK, H. J., FORD, W. F., and WHITE, J. (1959). *Trans. Brit. Ceram. Soc.* **58**, 1.

KUCZYNSKI, G. C. (1949). *Trans. AIME* **185**, 169.

KUCZYNSKI, G. C., and STABLEIN, P. F. (1961). "Reactivity of Solids," p. 91. Elsevier, Amsterdam.

LENEL, F. V. (1951). "The Physics of Powder Metallurgy" (W. E. Kingston, ed.), p. 238. McGraw-Hill, New York.

LENEL, F. V., and ANSELL, G. S. (1967). "Sintering and Related Phenomena" (G. C. Kuczynski, N. A. Hooton, and C. F. Gibbons, eds.), pp. 351–367. Gordon and Breach, New York.

LEVIN, E. M., ROBBINS, C. R., and MCMURDIE, H. F. (1964). "Phase Diagrams for Ceramists," p. 156. Am. Ceram. Soci., Columbus, Ohio.

MCHUGH, C. O., WHALEN, T. J., and HUMENIK, M., JR. (1966). *J. Am. Ceram. Soc.* **49**, 486.

MOORE, D. J. (1968). M. S. thesis, Northwestern Univ. Evanston, Illinois.

MORGAN, C. S., and HALL, L. L. (1967). *J. Am. Ceram. Soc.* **50**, 382.

NICHOLS, F. A., and MULLINS, W. W. (1965). *J. Appl. Phys.* **36**, 1826.

PETERSON, R. O., and CUTLER, I. B. (1968). *J. Am. Ceram. Soc.* **51**, 21.

PINES, B. YA. (1956). *Zh. Tekhn. Fiz.* **26**, 2086.

PINES, B. YA., and N. I. SUKHININ (1956). *Zh. Tekhn. Fiz.* **26**, 2100.

PRILL, A. L., HAYDEN, H. W., and BROPHY, J. H. (1964). *Trans. AIME* **230**, 769.

READEY, D. W. (1966a). *J. Appl. Phys.* **37**, 2309.

READEY, D. W. (1966b). *J. Am. Ceram. Soc.* **49**, 366.

READEY, D. W., and JECH, R. E. (1968). *J. Am. Ceram. Soc.* **51**, 201.

REIJNEN, P. (1969). "Reactivity of Solids" (J. W. Mitchell, R. C. DeVries, R. W. Roberts, and P. Cannon, eds.), p.99. Willey (Interscience), New York.
ROCKLAND, J. G. W. (1966). *Acta Met.* **14**, 1273.
ROCKLAND, J. G. R. (1967). *Acta Met.* **15**, 277.
SCHWARZKOPF, P., and KIEFFER, R. (with Leszynski W. and Benesovsky F.) (1960). "Cemented Carbides," p. 15. Macmillan, New York.
SEIGLE, L. S. (1964). *Prog. Powder Met.* **20**, 221.
SHINGU, P. H. (1967). Ph.D. thesis, Northwestern Univ., Evanston, Illinois.
SMART, J. S. (1968). Ph.D. thesis, Northwestern Univ., Evanston, Illinois.
SMITHELLS, C. J. (1962). "Metals Reference Book," 3rd ed., Vol. 1, p. 365. Butterworths, Washington, D.C.
THUMMLER, F., and THOMMA, W. (1967). *Met. Rev.* **12**, (No. 115), 69.
WARMAN, M. O., and BUDWORTH, D. W. (1967a). *Trans. Brit. Ceram. Soc.* **66**, 253.
WARMAN, M. O., and BUDWORTH, D. W. (1967b). *Trans. Brit. Ceram. Soc.* **66**, 265.
WILSON, T. L., and SHEWMON, P. G. (1966). *Trans. AIME* **236**, 48.

IX

Phase Diagrams and the Heat Treatment of Metals

GEORGE KRAUSS and JOSEPH F. LIBSCH

DEPARTMENT OF METALLURGY AND MATERIALS SCIENCE
LEHIGH UNIVERSITY, BETHLEHEM, PENNSYLVANIA

I. INTRODUCTION

The purpose of this volume is to discuss the importance of phase diagrams to the technologies associated with the development, production, and processing of glasses, ceramics, and metals. This chapter will discuss specifically just one aspect of this broad charge: the relationship of heat treatment of metals and alloys to their phase diagrams. We may define as heat treatment any thermal treatment that has as its purpose the production of a useful effect on the properties of a metal or alloy. The property changes to be effected by heat treatment of a given alloy might be widely divergent and are generally accompanied by changes in structure. For example, maximum ductility may be the desired goal during the fabrication of an alloy, but maximum strength may be desired for its final application. Heat treatment and an understanding of the structural changes that occur during heat treatment make both of these widely divergent goals possible.

Phase diagrams serve as a very important guide to the structural changes, and therefore the property changes, that are produced by a given heat treatment. The compositions, crystal structures and amounts of phases, and the temperature ranges over which the phases exist are all given by the phase diagram of an alloy system for equilibrium conditions. Heat treatment, however, may produce structures that only approach equilibrium or may even deliberately suppress the formation of equilibrium structures. It is the purpose of this chapter to consider various thermal treatments and the accompanying structural changes and to set these considerations into the framework provided by phase diagrams.

In order to relate heat treatment to phase diagrams, two major classes of thermal treatments have been chosen: those based upon a solid-state transformation and those that do not involve a phase transformation. The former heat treatments in one way or another correspond to heating and cooling through several fields of a phase diagram, and consequently the formation of new phases and the manner in which they form significantly influence the properties produced by such a treatment. The second type of heat treatment is performed entirely within a single- or two-phase field, and the property changes are therefore due to structural changes occurring in the phase or phases that are present at the start and persist throughout the duration of the treatment. A special case of heat treatment according to this classification is one that involves composition changes, usually on the surface of certain types of alloys. These latter treatments often involve phase transformation and will be discussed separately.

Extensive collections of metal phase diagrams have been prepared by the American Society for Metals (1948), Hansen and Anderko (1958), and Elliot (1965), and a recent compilation of detailed heat-treatment procedures for all major types of commercial alloys may be found in the American Society for Metals (1964) reference. Textbook discussions of structure and property relations as related to phase diagrams and heat treatment may be found in Brick *et al.* (1965) and Guy (1959).

II. HEAT TREATMENT WITHIN PHASE FIELDS

A. Homogenization

Homogenization is a heat treatment necessitated by compositional variations or segregation produced by nonequilibrium solidification of many single-phase alloys. Figure 1a presents a portion of an equilibrium diagram in which a single solid solution forms on solidification. The equilibrium diagram shows that an alloy of nominal composition C_0 should be completely

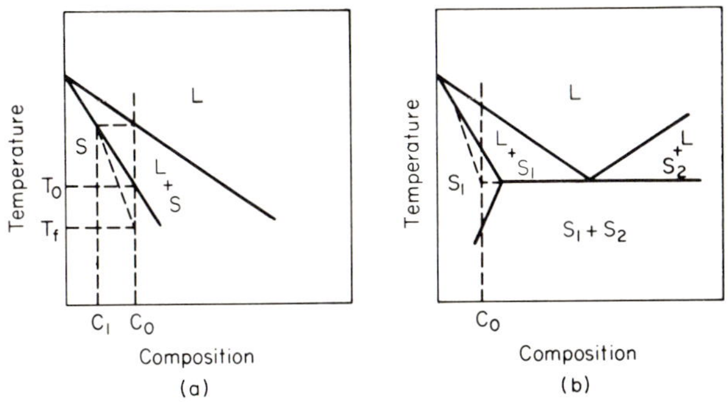

Fig. 1. (a) Displacement of solidus due to nonequilibrium solidification of alloy of composition C_0. (b) Nonequilibrium solidification that results in formation of two phases in alloy of composition C_0.

solidified at T_0 and should consist entirely of a single phase, uniform in composition throughout its microstructure. In many casting processes, however, the cooling during solidification of an alloy is too rapid and does not permit the solid-state diffusion necessary to homogenize the successive layers of solid that form on the first solid nuclei of composition C_1. As a result the average composition of the solid may be considered to follow a curve displaced from the solidus of the equilibrium diagram. The dashed line in Fig. 1 shows schematically this displaced solidus and shows that solidification is not complete until the temperature T_f is reached. The inhomogenities in composition are reflected in the interdendritically segregated or cored microstructures that develop under such cooling conditions. Figure 2 illustrates an extreme example of segregation where the dendrites are assumed to have the composition C_1 of the first solid that forms and the interdendritic areas the

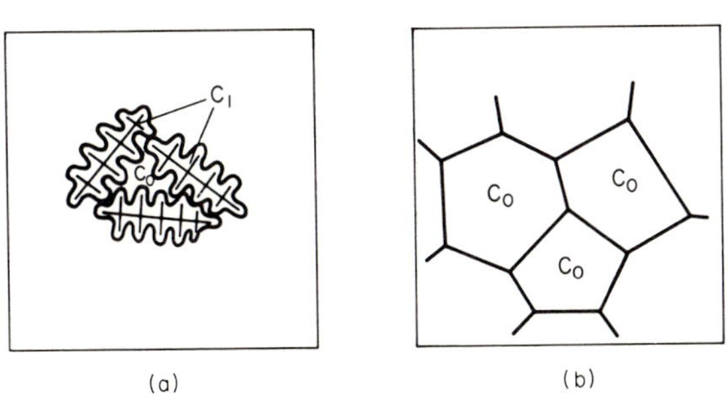

Fig. 2. Schematic representation of (a) as-solidified dendritic structure with composition variations between C_1 and C_0, and (b) homogenized grain structure of uniform composition C_0.

composition C_0 of the last solid that forms. The compositions will actually vary continuously from C_1 to C_0 in cast alloys.

In alloy systems such as shown in Fig. 1b, where solidification may occur by a eutectic or peritectic reaction over a certain range of compositions, alloys outside of this composition range that are expected to be single phase may nevertheless form a second phase in the interdendritic branches because of nonequilibrium solidification. In the case of a eutectic system as shown in Fig. 1b, solidification below the solidus enriches the liquid to the eutectic composition and final solidification takes place by the formation of two phases.

Homogenization is the heat treatment performed to eliminate the coring and segregation for alloys where such inhomogenity may be deleterious during subsequent deformation operations. The treatments are performed by heating for long times at high temperatures in order to ensure the development of the uniform, equilibrium composition throughout the microstructure by solid-state diffusion. Care must be taken to avoid heating above the

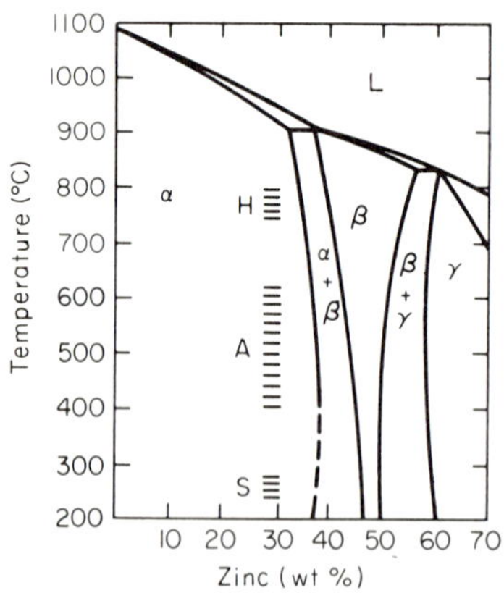

Fig. 3. Copper–zinc phase diagram. H, A, and S indicate temperature ranges for homogenizing, annealing, and stress relief for 70% Cu–30% Zn alloys.

nonequilibrium solidus point in order to prevent melting of the last solid to freeze. Such remelting at localized areas within the structure and oxidation of these areas would produce excessive brittleness and is referred to as burning. Figure 3 shows where homogenizing treatments would be performed in single phase Cu–Zn alloys containing approximately 30 wt% zinc. The lower-temperature heat treatments indicated in this diagram will be discussed in the following sections.

B. Annealing of Cold-Worked Single-Phase Alloys

An important approach to the production of sheet and products such as wire and tubing is the alternate cold working and annealing of as-cast structures of single-phase alloys. Homogenization treatments may be performed at the start of processing, but in many alloys the homogenization that occurs during the working and annealing treatments to be described provides adequate workability for the final stages of production. Annealing treatments of wrought single-phase alloys are necessitated by the work hardening that occurs during their deformation. Figure 4 shows the cold-rolled microstructure of essentially single-phase iron sheet whose microstructure prior to working consisted of equiaxed grains of ferrite with a high degree of internal perfection very similar to that indicated schematically in Fig. 2b. The grains

Fig. 4. Microstructure of Fe–0.003 C alloy that has been cold rolled 60%. Nital etch. 100 ×. (Courtesy D. A. Witmer, Bethlehem Steel Corporation.)

are elongated in the direction of rolling; deformation markings and a generally distorted structure are apparent within the grains. These microstructural features of a cold-worked metal are manifestations of the high density of imperfections introduced in a work-hardened metal to accommodate plastic flow. The most important of these imperfections are the line defects referred to as dislocations that have moved, interacted, and multiplied during working. An alloy in the cold-worked state may be considerably stronger and much less ductile in the cold-worked state than in the annealed state. For this reason, work hardening is frequently exploited as a strengthening mechanism for many single-phase materials not hardenable by any other technique. However, when improved ductility is required for either further processing or a given application, a cold-worked metal must be annealed.

Phase diagrams are useful in developing annealing treatments only insofar as they specify the temperature limits over which a single phase of given composition can exist. The structural changes that occur during annealing are dependent on atom mobility, and many combinations of heat-treatment times and temperatures within the single-phase field can achieve the same result. The annealing of a cold-worked metal consists of three distinct stages: recovery, recrystallization, and grain growth. Recovery processes involve the rearrangement and partial elimination of the imperfections, but do not change the large-angle grain-boundary configurations generated by cold work. Recrystallization consists of the nucleation and growth of strain-free grains in the cold-worked matrix and culminates with complete replacement of the distorted grains by an equiaxed grain structure of a high degree of internal perfection. The driving force for both recovery and recrystallization is the strain energy associated with the nonequilibrium atom displacements at the imperfections produced by plastic deformation. Grain growth involves the coarsening of the as-recrystallized grains by atom transfer across large-angle boundaries and is driven by the high grain boundary or interfacial energy associated with a fine equiaxed grain structure with many curved boundaries. If grain growth is permitted to proceed extensively, a coarse-grain structure with planar grain boundaries will replace a fine-grained as-recrystallized structure.

Commercial alloys that are subjected to cold work and annealing include the very-low-carbon steels or irons. Figure 5 shows the relevant portion of the iron–carbon diagram for this group of alloys. The microstructure of such an alloy in the cold-worked condition has been presented in Fig. 4. Figure 6 shows the microstructure of the same alloy after approximately 80% of the cold-worked structure has recrystallized. The recrystallized grains are equiaxed and much finer than the deformed, elongated grains, some portions of which are still present in the microstructure. The variations in etching response underscore the structural changes that occur during recrystallization. The deformed grains appear dark as a result of the surface roughening that accompanies the interaction of the etchant with the imperfections of the worked structure, while the relatively strain- or imperfection-free recrystallized grains are smooth etching and are marked primarily by the large-angle grain boundaries surrounding them.

The copper–zinc system is another commercially important alloy system in which single-phase alloys are processed by cold working and annealing. Figure 3 shows the temperature ranges within the alpha solid solution field for annealing 70% copper–30% zinc alloys or 70–30 brass. The lowest-temperature heat treatments are performed for the purpose of stress relief,

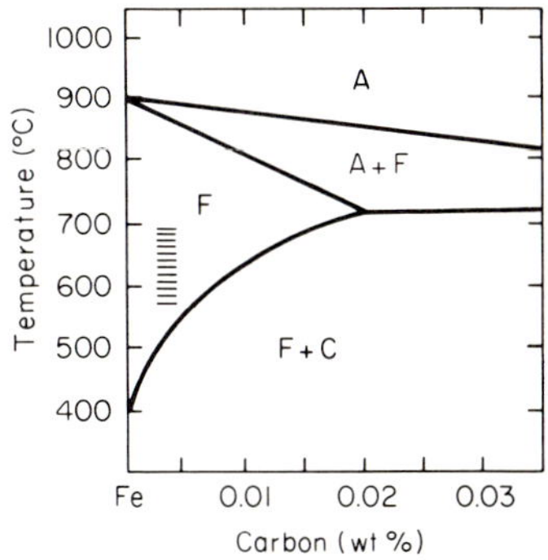

Fig. 5. Low-carbon portion of iron-carbon diagram with annealing temperature range for an iron–0.003 C alloy indicated. A, F, and C denote, respectively, the austenite, ferrite, and carbide (Fe_3C) phases.

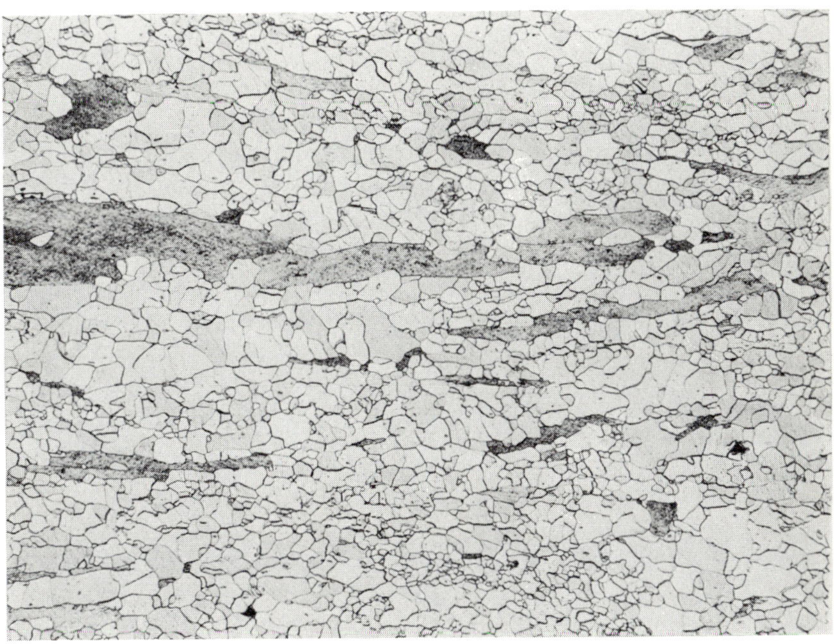

Fig. 6. Microstructure of Fe–0.003 C alloy that has been annealed at 1000°F for 2 h. Approximately 80% of the cold-worked structure has recrystallized. Nital etch. 100×. (Courtesy D. A. Witmer, Bethlehem Steel Corporation.)

which occurs by the recovery processes described above. In the absence of recrystallization, strength does not decrease significantly although the relief of the residual stresses associated with the as-worked state produces marked improvement in resistance to certain corrosive environments. The higher-temperature annealing treatments involve all three stages of annealing. The exact heat-treatment schedule depends on many factors such as the degree of cold work, deformed grain size, composition and impurity content, and the final properties and structure desired for further processing or a given application. Major decreases in strength and improvements in ductility are caused by recrystallization, but relatively small decreases in strength are also produced by grain growth. Too coarse an annealed grain size may, however, be undesirable because of surface unevenness that might develop on subsequent deformation due to the heterogeneous deformation of a coarse-grained structure.

In summary, phase diagrams are most useful for defining the acceptable temperature and composition limits for the annealing of single-phase alloys although many other factors must be taken into consideration. In the Cu–Zn system, there is a wide latitude in the heat treatments that can be performed to accomplish annealing, but in a system such as the Fe–C system the temperature limits are more closely defined by the equilibrium diagram. Annealing treatments of cold-worked metals may certainly be performed in two phase fields, but the kinetics of the annealing processes as well as the resultant structures may be significantly affected by the presence of a second phase.

III. HEAT TREATMENT INVOLVING PHASE TRANSFORMATION

A. Formation of One New Phase

1. Precipitation Hardening

The formation of fine precipitated particles of a second phase from a supersaturated solid solution by means of suitable heat treatment is an important method of developing strength in a very large number of alloys (American Society for Metals, 1959). Phase diagrams of the alloy systems are extremely important in defining the most suitable alloy compositions for precipitation-hardening response and in establishing the temperature limits for the heat treatments that will produce maximum strengthening. A general condition for precipitation hardening is large solubility of one component of a system within a phase at high temperatures and a substantially reduced solubility of this component at lower temperatures. Examples of portions of

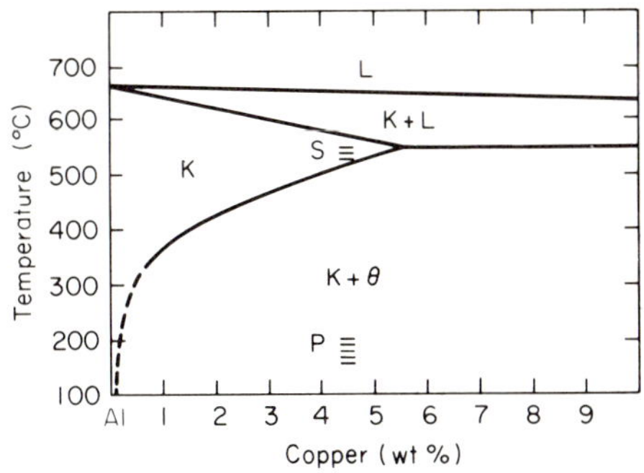

Fig. 7. Aluminum-rich portion of aluminum–copper phase diagram. S and P denote temperature ranges for solution and precipitation treatments for alloys containing about 4.5% Cu.

phase diagrams that satisfy this requirement of decreasing solid solubility with decreasing temperature are the aluminum–copper diagram shown in Fig. 7 and the copper–beryllium diagram shown in Fig. 8.

In order to exploit the potential for precipitation hardening of alloys based on these systems, two-step heat treatments must be performed. A treatment referred to as a solution treatment is necessary to produce a supersaturated solid solution from which a dispersion of fine precipitate particles can be produced by a second treatment, referred to as a precipitation or aging treatment. The temperature ranges for both of these treatments for aluminum–copper and copper–beryllium are indicated in Figs. 7 and 8, respectively. It can be seen that the solution treatment consists of heating into a well-defined temperature range of a single-phase field. Heating at higher temperatures leads to the danger of grain-boundary melting and attendant brittleness,

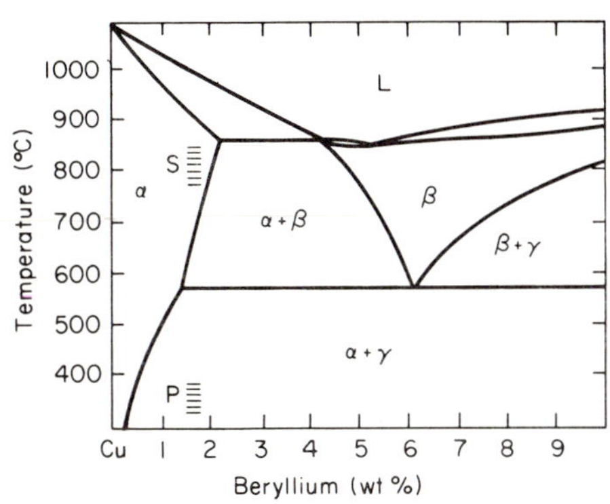

Fig. 8. Copper-rich portion of copper–beryllium phase diagram. S and P denote temperature ranges for solution and precipitation for alloys containing about 1.7% Be.

while heating at lower temperatures will result in the loss of some solute by the retention of a second phase. In the latter situation the morphology of the second phase will be too coarse for effective strengthening and, since solution is incomplete, some potential for strengthening during a subsequent aging treatment has been lost. Another important aspect of the solution treatment is the formation of a supersaturated solid solution at room temperature by rapid cooling from the solution temperature. Water quenching is frequently used to prevent the diffusion-controlled decomposition of the supersaturated solid solution into the equilibrium phases designated by the phase diagram.

The strengthening that develops during the precipitation treatment is due to the diffusion-controlled formation of fine precipitate particles in the supersaturated matrix. Frequently, the precipitation process begins with the formation of clusters of solute atoms or zones with maximum dimension on the order of a 100 Å. Precipitation proceeds with the growth of the nuclei, often through a series of transition structures that may be coherent with the parent solid solution (Nicholson *et al.*, 1958–1959). Maximum hardness develops when a dispersion of particles of critical size, shape, spacing, and degree of coherency is attained. Heat treatment beyond this point will result in a decrease of strength or overaging as a result of the formation of coarser distributions of particles of equilibrium structure and composition according to the phase diagram. In view of the fact that the precipitation hardening mechanism is primarily dependent upon the kinetics of nucleation and growth of nonequilibrium structures, phase diagrams provide only very approximate guides to the development of precipitation treatments, and serve primarily to establish the broad temperature range of two-phase equilibrium. The temperature for the aging treatment that produces maximum strength varies widely from one alloy to another. In the aluminum–copper system, natural or room-temperature aging is sufficient to produce strengthening, but in copper–beryllium alloys temperatures around 600°F are commonly used.

2. Martensitic Transformation

In many alloy systems it is possible to form a nonequilibrium phase by a diffusionless, shear-type mechanism referred to as a martensitic transformation. The product phases or martensites produced by such transformation vary widely in structure and properties, but in iron–carbon-based alloys or steels, the martensitic transformation produces a structure of very high strength, and as a result a highly developed technology of heat treatment has been developed to take advantage of this property. The martensitic transformation in iron–carbon alloys occurs when the stable high-temperature phase austenite, a solid solution of carbon in face-centered cubic iron, is cooled at

Fig. 9. Low-carbon martensite in Fe–0.2 C alloy after austenitizing at (a) 1900°F and (b) 2000°F. Coarsening of prior austenitic grain size with increasing austenitizing temperature is illustrated. Sodium bisulfite etch. 100 ×. (Courtesy of A. R. Marder and A. Benscoter, Bethlehem Steel Corporation.)

a rate rapid enough, usually by quenching in water, to prevent the diffusion-controlled decomposition of austenite to the stable low-temperature phase ferrite, a solid solution of carbon in body-centered cubic iron, and cementite or iron carbide (Fe_3C), as expected from the iron–carbon phase diagram. The martensite that forms is of the same composition as the parent austenite, and the carbon atoms trapped in interstitial sites between iron atoms cause the lattice to be body-centered tetragonal, the tetragonality increasing with increasing supersaturation of carbon.

Despite the fact that equilibrium phases are suppressed by martensitic transformation, phase diagrams may be useful in two respects in developing or evaluating heat treatments that are directed toward producing martensite. One such aid is the specification of the temperature limits of the stable high-temperature phase, austenite in the case of Fe–C alloys, in which the product phase forms. Not only is the minimum temperature at which complete transformation to the parent phase established, but also the effects of superheating above this minimum temperature on the grain size of the parent phase can be assessed. Figure 9 shows martensite formed in an Fe–0.2% C alloy austenitized for 1 h at two different temperatures in the single-phase austenite field. The martensite units in such low-carbon alloys form as small plates aligned parallel to each other in massive blocks or groups of plates in one of several different orientations (Marder and Krauss, 1967). Each austenite grain is therefore subdivided on martensitic tranformation into a number of blocks of martensite that are related to the orientation of the parent grain, and the straight-sided equiaxed grains of the austenite may be identified in a transformed microstructure by the change in orientation of the blocks between different austenite grains. Figures 9a and 9b show the significant coarsening of the austenite grain size that occurs when the austenitizing temperature is increased from 1900° to 2000°F.

Another important use of phase diagrams in the development of heat treatments for producing martensitic structures is the superposition of the important martensitic transformation characteristics on the temperature-composition axes of the equilibrium phase diagram. Figure 10 shows the martensite start and finish temperatures, M_s and M_f, respectively, as a function of composition for Fe–C alloys. Also included as a portion of Fig. 10 is a schematic diagram of the range of strengths that can be developed as a function of carbon content and structures ranging between completely martensitic and spheroidized Fe_3C in a ferritic matrix. The interrelationships between martensitic transformation characteristics, structure, and strength are illustrated by a consideration of the martensite strength curve. The effect of carbon on strength is most significant at low carbon contents. Above approximately 0.6% C the maximum strength attainable in a completely martensitic structure remains essentially constant with increasing carbon

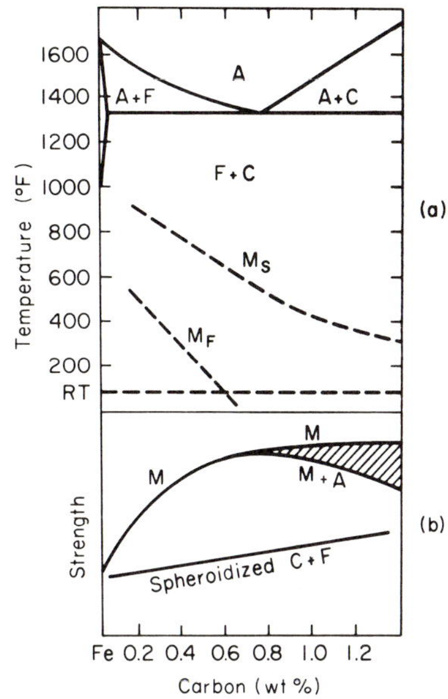

Fig. 10. Iron-rich portion of iron–carbon phase diagram and schematic diagram of strength as a function of carbon content for alloys in martensitic and spheroidized conditions.

content. However, strengths much lower than this maximum may be obtained on cooling medium- and high-carbon alloys to room temperature because of incomplete transformation of the austenite to martensite as indicated in the upper portion of Fig. 10 by M_f temperatures that are below room temperature for these alloys. The untransformed austenite is referred to as retained austenite. Figure 11 shows the acicular martensite plates that form in high-carbon alloys and the interspersed, white-appearing patches of retained austenite in an Fe–1.4% C alloy. The cross-hatched region in the bottom portion of Fig. 10 illustrates schematically the range of reduced strengths that are produced by various amounts of the ductile austenite. In order to eliminate retained austenite and increase strength in high-carbon alloys, subzero cooling is sometimes used to drive the martensitic transformation closer to completion.

Although the maximum attainable strength of any Fe–C alloy or steel is associated with a completely martensitic structure, in this condition the alloys are too brittle for direct application. As a result the martensite must be given an additional heat treatment referred to as tempering. Tempering of steels is performed at temperatures in the two-phase ferrite plus Fe_3C region of the Fe–C diagram and involves the decomposition of the as-quenched martensite into two-phase mixtures. Three progressive stages of structural change have been identified as occurring during a tempering treatment (Werner *et al.*, 1957): (1) the formation of a transition carbide, hcp

Fig. 11. Acicular martensite plates and retained austenite in an Fe–1.4 C alloy. Nital etch. 250 ×. (Courtesy A. R. Marder and A. Benscoter, Bethlehem Steel Corporation.)

epsilon phase, in the martensite as very fine platelets with largest dimension on the order of several hundred angstroms, (2) decomposition of retained austenite into a ferrite–Fe_3C aggregate, and (3) complete decomposition of tempered martensite into spherical particles of Fe_3C in a matrix of equiaxed ferrite grains. As tempering proceeds the carbon content and tetragonality of the matrix martensite decrease with attendant decreases in strength and increases in toughness except at a critical temperature range at the beginning of the third stage of tempering where an embrittlement is associated with the early stages of replacement of epsilon carbide by cementite. Figure 12 shows the microstructure of a Fe–0.6% C alloy that has been water quenched and tempered at 1300°F for 24 h in order completely to temper the martensite into spheroidal carbide and ferrite. This type of microstructure corresponds to the structure of steels in their most ductile, lowest-strength condition as indicated in the bottom portion of Fig. 10. All strengths intermediate between the maximum associated with untempered martensitic structures and the minimum associated with completely spheroidized structures are therefore attainable by appropriate control of the tempering process.

Titanium alloys with elements that stabilize the high-temperature-stable

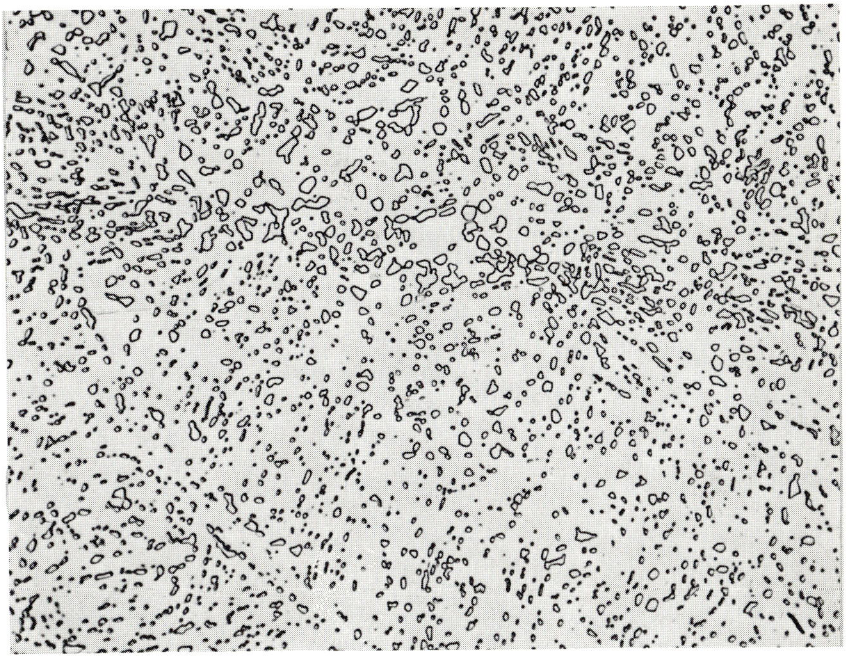

Fig. 12. Spheroidized carbides within a ferrite matrix in an Fe–0.6 C–1 Mn alloy formed by tempering martensite at 1300°F for 24 h. Picral etch. 1000 ×. (Courtesy A. R. Marder and A. Benscoter, Bethlehem Steel Corporation.)

bcc beta phase offer another example of alloy systems in which martensitic transformation may be important in developing and understanding heat-treatment procedures. Figure 13 shows an idealized phase diagram for a beta-stabilized system as would be characteristic of systems where elements such as Cb, Mo, V, or Ta are alloyed with titanium. The M_s and M_f temperatures for the martensitic transformation of beta to a nonequilibrium martensitic phase designated as alpha prime are superimposed on the diagram, and the lower portion of Fig. 13, after Seagle (1965), relates strength as a function of composition and structure to the diagram and the martensitic transformation characteristics. It can be seen that the M_s and M_f temperatures are quite sensitive to composition as in the Fe–C system, and that it is possible to stabilize completely the beta phase at room temperature by increasing the content of a beta-stabilizing element to a sufficiently high level. An important difference between Fe–C and titanium martensites, however, is the lower strength of the titanium martensites in comparison to other structures that can be developed in beta-stabilized alloys. The portion of the dashed parabolic curve that terminates at the composition that corresponds to an alloy

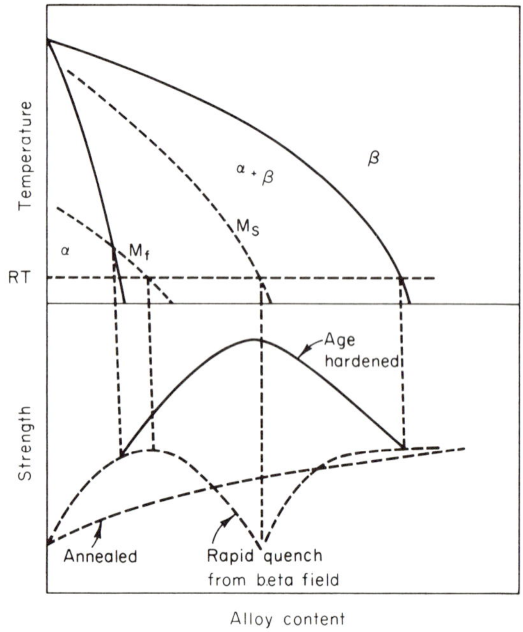

Fig. 13. Idealized phase diagram for a titanium-beta-stabilizing element system and schematic representation of strengths corresponding to various types of heat treatment applied to such a system. (After Seagle, 1965.)

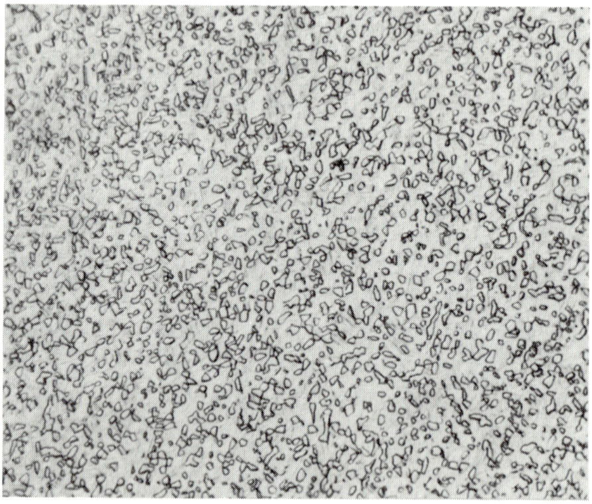

Fig. 14. Microstructure of a Ti–4 Al–3 Mo–1 V alloy showing alpha particles in a beta matrix. 2% HF–4% HNO_3 etch. 500 ×. (Courtesy L. J. Barto and H. B. Bomberger, Reactive Metals, Inc.)

with a room-temperature M_s shows the range of strength which can be developed by alpha prime or alpha-prime–beta structures. Strengths above these, as indicated in the lower portion of Fig. 13, are developed by heating metastable beta in the alpha–beta two-phase region in order to produce age hardening by the formation of equilibrium alpha-phase particles in the beta matrix. Figure 14 shows a microstructure consisting of alpha in the beta phase in a Ti–4 Al–3 Mo–1 V alloy where the composition of the beta phase exceeds that at which a room temperature M_s would exist. Heat treatment at higher temperatures would dissolve the alpha particles and lower the beta composition to levels that would enable the martensitic alpha prime to form on cooling (Bartlo, 1967).

3. Order–Disorder

Many solid-solution phases (Guttman, 1956; Barrett and Massalski, 1966) in which the component atoms are randomly arranged or disordered at high temperatures undergo with decreasing temperature a phase transformation that places each kind of atom on a specific set of sites or sublattice. Ordered solid solutions or superlattices form in alloys where the components are in ratios of 1:1 or 3:1 although ordering may exist or coexist with disordered structures in alloys in a range of compositions about the ideal. Figure 15 shows the phase diagram for the platinum-cobalt system in which at high temperatures random face-centered solid solutions exist for all compositions and an ordering transformation occurs at lower temperatures in alloys about the 50 at. % composition. The ordered structure is face-centered tetragonal with alternate (001) planes made up of only cobalt or platinum

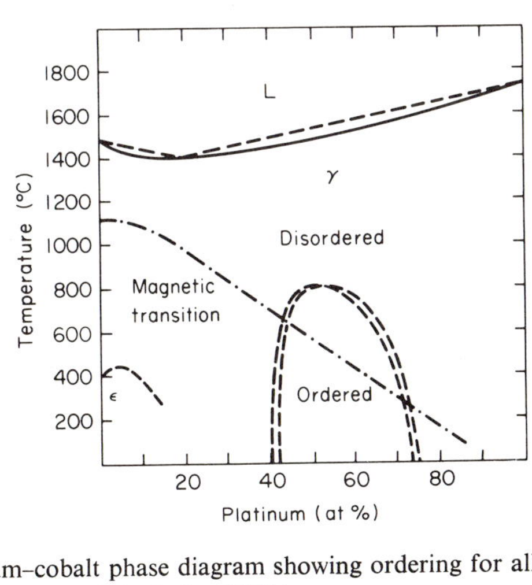

Fig. 15. Platinum–cobalt phase diagram showing ordering for alloys about the 50 at. % composition. (Hansen and Anderko, 1958.)

atoms. The development of ordered structures is analogous to precipitation from a supersaturated solid solution, and similar two-step heat treatments of quenching to retain the disordered structure and heating within the temperature region of order to develop the ordered structure are performed to effect desirable property changes. In the case of Pt–Co alloys, the magnetic properties are of interest and these change significantly together with hardness as the ordering transformation proceeds with increasing heating time at temperatures within the temperature range of order as indicated by the phase diagram (Newkirk *et al.*, 1950). Figure 16 shows the changes in coercive force

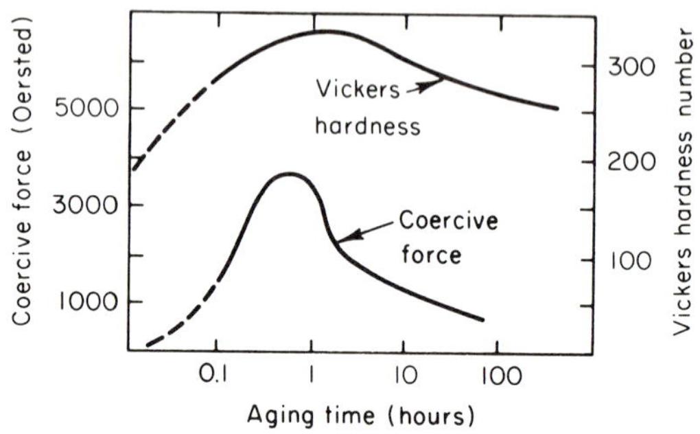

Fig. 16. Hardness and coercive force as a function of aging time at 700°C after water quench from 1000°C for a 52 at. % Pt–48 at. % Co alloy. (Newkirk *et al.*, 1950.)

and hardness that occur during the heating of a quenched 48 at. % Co–52 at. % Pt alloy at 700°C. The initial increases in the coercive force and hardness are attributed to the coherency strains developed in the early stages of the formation of the ordered structure in the disordered matrix.

B. Formation of Two New Phases

1. Eutectoid Transformation

A eutectoid transformation is one in which a single solid phase transforms to two product solid phases on cooling. The compositions of the two product phases usually differ significantly in composition from each other and the parent phase, and hence a eutectoid transformation, is one that is diffusion controlled. The Fe–C system offers an excellent example of how phase diagrams are used in the heat treatment of commercial steels in which a eutectoid transformation occurs. Throughout this discussion Fe–C alloys and steels will be used interchangeably, and it should be kept in mind by the reader that the alloying elements present in steels shift the boundaries of the Fe–C diagram somewhat. Figure 17a is a portion of the Fe–C diagram

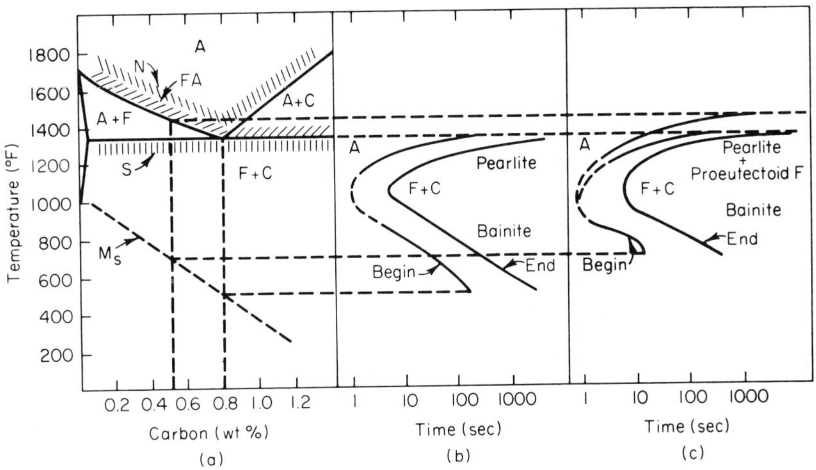

Fig. 17. (a) Iron-rich portion of iron–carbon diagram with temperature ranges for normalizing (N), full-annealing (FA), and spheroidizing (S) treatments. (b) Isothermal transformation diagram for 1080 steel. (c) Isothermal transformation diagram for 1050 steel.

on which the temperature ranges for heat treatments involving eutectoid transformation have been indicated. Figures 17b and c will be discussed in Section III,B,2. The phase diagram specifies the important boundaries between the phase fields of interest: the temperature of the isothermal eutectoid reaction where austenite transforms to ferrite and Fe_3C is identified as the A_1, and the boundaries between the austenite/austenite plus ferrite and austenite/austenite plus Fe_3C fields are identified as the A_3 and A_{cm}, respectively. Heating above these temperatures into the single-phase austenite field and subsequent cooling through the lower-temperature fields are the basis of what are referred to as full annealing and normalizing treatments for steels.

Full annealing treatments most closely approach equilibrium because very slow furnace cooling is used to bring the steel to room temperature from the austenitizing temperature range specified for full annealing as shown in Fig. 17a. Figure 18 shows the microstructure of an Fe–C alloy close to the eutectoid composition, 0.75% C, which has been given a full annealing treatment. The microstructure consists predominately of alternate platelets or lamellae of ferrite and Fe_3C, the characteristic microstructure produced by the eutectoid transformation in Fe–C alloys, and is conventionally referred to as pearlite. Alloys that are not of eutectoid composition will form either ferrite or cementite, referred to as proeutectoid constituents of the microstructure to differentiate them from the same phases in the pearlite, on cooling through either the A_3 or A_{cm} temperatures to the A_1 where the balance of

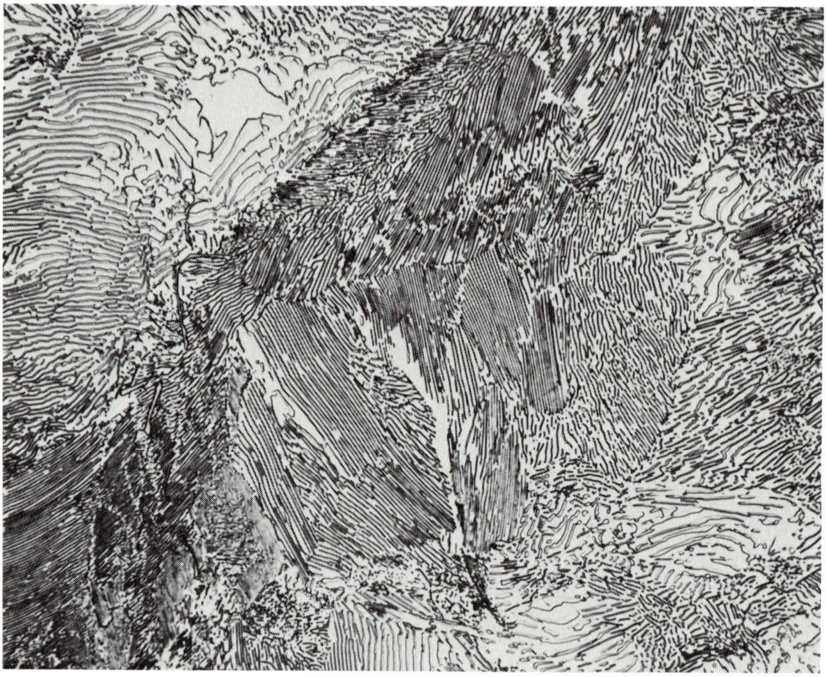

Fig. 18. Pearlitic microstructure of full annealed iron–0.75 wt % C alloy. Picral etch. 500 ×. (Courtesy A. R. Marder and A. Benscoter, Bethlehem Steel Corporation.)

the austenite will form pearlite. The proeutectoid phases nucleate and grow at the boundaries of the austenite and often form networks delineating the prior austenitic grain structure. Figure 19 shows such a proeutectoid ferrite network in a Fe–0.4% C alloy normalized at 1600°F. A normalizing treatment involves air cooling from the heating temperature and the faster cooling in comparison to furnace cooling allows less time for proeutectoid phase formation and produces a finer pearlite and therefore a slightly stronger structure than that produced by a full anneal. The strengths characteristic of annealed and normalized steels are dependent upon the relative amounts of proeutectoid phase and pearlite, and would lie somewhat above the strengths of the spheroidized structures as indicated in Fig. 10. Examination of Fig. 17a shows that the heat-treatment temperatures for normalizing and annealing diverge substantially for alloys containing more than 0.80 wt% C. The purpose of heating for full annealing in the two-phase austenite–Fe_3C field is to prevent the formation of the brittle carbide network that would otherwise be produced on cooling through this two-phase field. Agglomeration also plays a role in long-time spheroidizing heat treatments performed just below the A_1 as indicated in Fig. 17a. In these treatments the Fe_3C of the

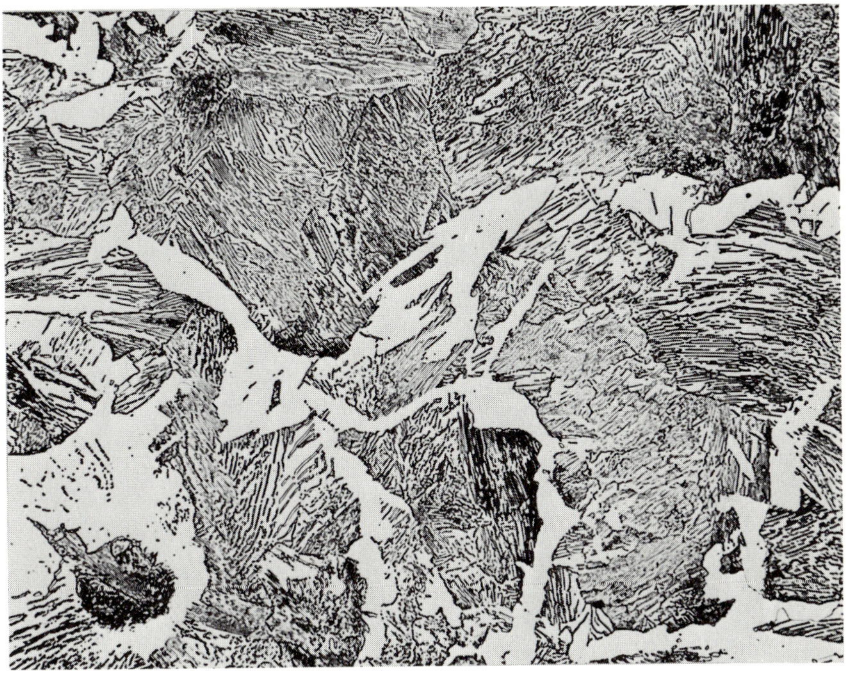

Fig. 19. Proeutectoid ferrite and pearlite of a normalized iron–0.4 wt % C alloy. Nital etch. 500×. (Courtesy A. R. Marder and A. Benscoter, Bethlehem Steel Corporation.)

pearlite as well as any proeutectoid carbide is spheroidized, and a microstructure similar to that shown in Fig. 12 is developed.

2. Isothermal Transformation

Two extremes in the decomposition of the high-temperature-phase austenite in Fe–C alloys have been discussed in previous sections. Section III,A,2 has discussed the diffusionless transformation of austenite to the nonequilibrium-phase martensite, and Section III,B,1 has discussed the near-equilibrium diffusion-controlled eutectoid transformation of austenite to a two-phase ferrite–carbide mixture as it occurs during a full annealing treatment. A systematic approach to the many intermediate structures that may be formed by decomposition of austenite during heat-treatment conditions between these extremes has been provided by the development of isothermal transformation diagrams for a large number of steels (United States Steel Corporation, 1951, 1953). Figures 17b and 17c are isothermal transformation diagrams for 1080 steel, a plain carbon steel of approximately eutectoid composition, and for a medium-carbon 1050 steel. The dashed horizontal lines between the parts of Fig. 17 relate the boundaries of the phase fields of the Fe–C phase diagram and the M_s temperatures of the two alloys to their respective isothermal transformation diagrams. The curves of Figs. 17b and 17c represent the beginning and end of isothermal decomposition of austenite between these temperatures. Not only are there large differences in the transformation kinetics as a function of temperature, but there are also major structural differences in the morphology of the ferrite–carbide aggregates. Above the nose (the temperature of most rapid transformation), of the isothermal transformation diagrams, the transformation produces pearlitic structures that become finer with more closely spaced lamellae as the nose of the curve is approached. Below the nose the lamellar morphology of pearlite is not produced, and the carbide phase forms in discrete particles rather than plates in a ferritic matrix. The latter structures are referred to as bainites, and these also become finer with decreasing transformation temperature until for transformation just above the M_s, the bainites are very similar in structure to slightly tempered martensites (Climax Molybdenum Company, 1967).

The great differences in transformation behavior with temperature reflect the very strong effect of temperature on atom mobility during diffusion-controlled austenite decomposition. The M_s temperature is the temperature at which the thermodynamic driving force for transformation is sufficient to cause transformation without diffusion. Whether or not a given cooling rate will bring a given alloy to its M_s without diffusion-controlled decomposition at higher temperatures is determined by many factors such as alloy composition, section size, and austenitizing temperature. The interrelationship of all of these factors and the ability of a given steel to form martensite is the subject of a well-developed branch of heat treatment known as hardenability. A discussion of hardenability is outside the purpose of this chapter except as it relates to phase diagrams as discussed above, and the interested reader is referred to more detailed discussions of this subject (Grossman, 1953; Bain and Paxton, 1961; Climax Molybdenum Company, 1967).

IV. HEAT TREATMENT INVOLVING COMPOSITION CHANGES

Many commercial processes are designed to change the composition of the metal surface to enhance the surface properties, i.e., hardness, corrosion resistance, etc. Such processes involve a means to bring the desired element to the surface of the part, often by surface-activated decomposition of a carrier gas or dipping in a liquid bath, and subsequent diffusion of the element into the surface to provide a suitable "case depth." When the purpose is to harden the surface as in nitriding (N) or carburizing (C), the process is called surface

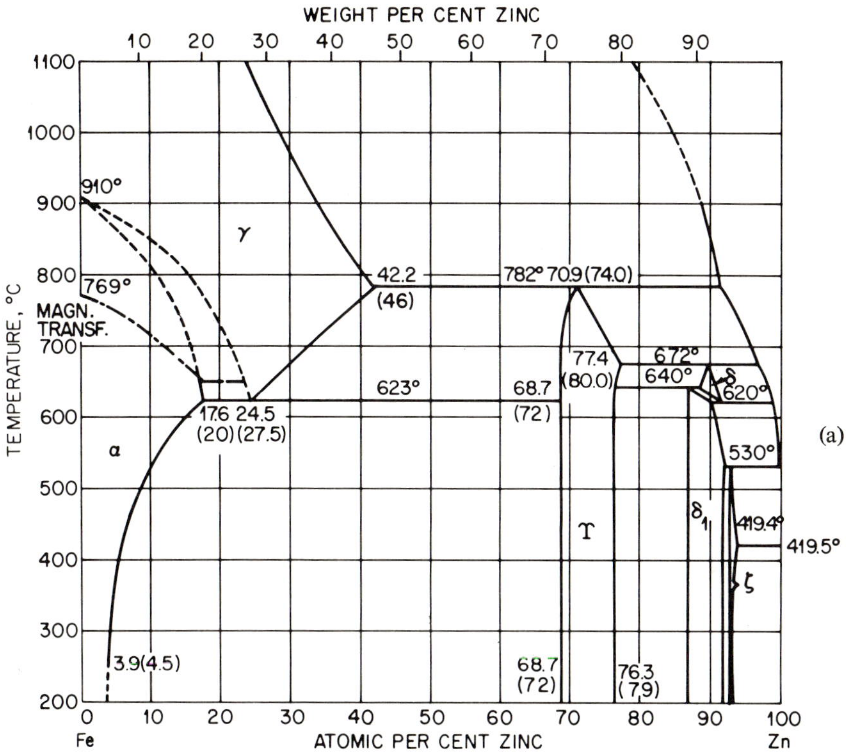

(a)

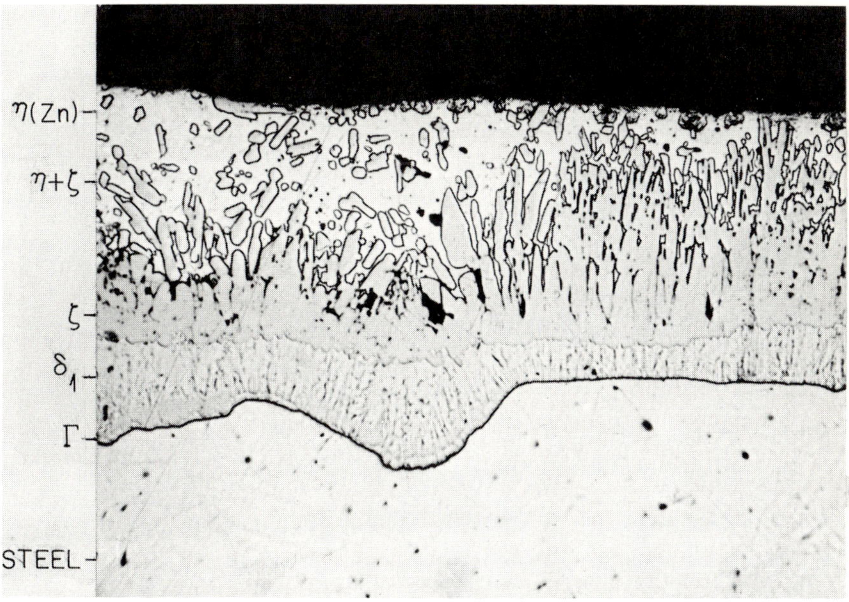

(b)

Fig. 20. (a) Iron–zinc phase diagram (Hansen and Anderko, 1958). (b) Galvanized coating on steel. The various layers corresponding to the phases of the Fe–Zn system are identified. 500 ×. (Courtesy A. Borzillo and J. Kilpatrick, Bethlehem Steel Corporation.)

hardening. In other instances, the diffusing element provides unusual corrosion resistance, as in the case of galvanizing or sherardizing (Zn), chromizing (Cr), calorizing or aluminizing (Al), or siliconizing (Si), etc., and the process is termed a "diffusion coating." In any event, the composition of the surface is changed in a significant way and the structure and properties altered with or without subsequent thermal treatment.

Phase diagrams may play an important role in predicting the structure of such coatings under equilibrium or near-equilibrium conditions. Actually such diffusion coatings are composed of layers corresponding in order and kind to the single phase regions shown in the phase diagram at the diffusing temperature in a binary system. Figure 20, for example, illustrates the situation in the Fe–Zn (galvanizing) system. It is apparent that the "galvanized" surface contains layers corresponding to the single phase regions in the diagram when processing is done in a liquid zinc bath at 500°C. Subsequent cooling from the diffusing temperature may, of course, involve other reactions in accordance with the phase diagram, for example, a eutectic reaction in a liquid layer (Fe–Zn) or separation from solid solution to accommodate the solubility relationships. Thus, two phase layers may develop in a binary system. In any event, some idea of the structure of the coating and therefore its properties and behavior may be obtained from the phase diagram.

The reason for the formation of layers, separated by a sharp interface, is directly related to the phase diagram. It is apparent that while the composition may vary continuously in a single-phase region, the concentration must remain fixed at a given temperature for each phase in a two-phase region, regardless of the average composition of the alloy. The thickness of the layers depends primarily upon the rates of diffusion of the diffusing species in the various layers, which may vary significantly.

Post Treatment of Diffusion Coatings

Since the crystal structure, composition and therefore properties of the different layers vary markedly (intermediate phases often being brittle) heat treatment of the diffusion coating can often be beneficial. The phase diagram may again be helpful in predicting the changes. Thus reheating the galvanized surface, i.e., annealing, after the galvanizing causes the zinc to diffuse inward with consequent disappearance of the layers requiring high zinc concentration and corresponding improvement in the ductility of the coating to facilitate forming.

Post heat treatment is particularly interesting in the case of carburizing. The carburizing temperature and carburizing gas, i.e., CO or hydrocarbons in a carrier gas are generally controlled in such a way as to supply active carbon to the surface of a low-carbon steel (0.10–0.20% C) while the steel

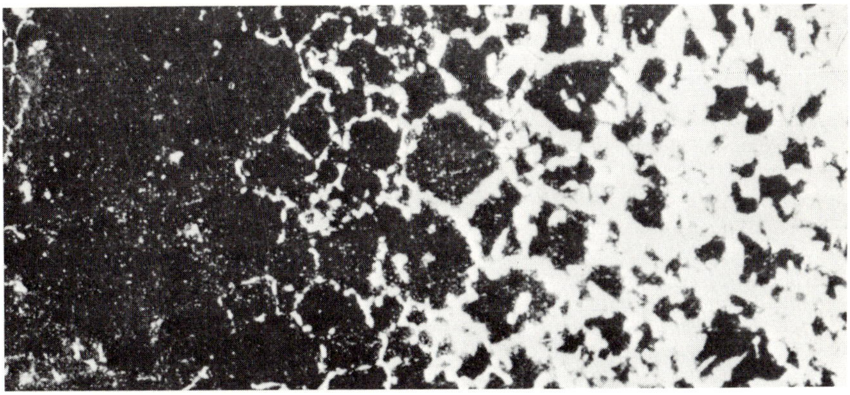

Fig. 21. Microstructure of carburized steel after slow cooling from the carburizing temperature. The carbon concentration varies continuously from about 1 wt % at the surface (left) to 0.20 wt % in the core (right). Microstructures vary from mixtures of Fe_3C (white network) and pearlite (dark) at the surface, through fully pearlitic zone (solid dark area), to mixtures of ferrite (white) and pearlite (dark) in the core. 75 ×.

is in the austenite range. The result is a variation in carbon content from the surface (~1%) to the core (original carbon content), the precise gradient and depth of case depending upon the carburizing time and temperature, with the possibility of some cementite at the immediate surface. Since carbon has such a pronounced influence on the transformation characteristics of austenite upon cooling, tending to increase both hardness and hardenability, control of the cooling rate after carburizing can produce a substantial variation in properties from the surface to the interior. Quenching at a suitable rate, for example, can produce high-carbon (1%) martensite on the surface with hardness R_c 63/65 and high strength, while producing low-carbon martensite or high-temperature transformation products having resistance to brittle failure in core. The structures and the properties of the case and core may be reasonably predicted by a knowledge of the likely response of iron–carbon alloys of corresponding compositions to cooling rates approaching equilibrium and nonequilibrium conditions using the phase diagram. Figure 21 shows the variations in structure produced by slow cooling a carburized case to room temperature.

References

American Society for Metals (1948). "Metals Handbook," pp. 1145-1268. American Society for Metals, Cleveland, Ohio.

American Society for metals (1959). "Precipitation from Solid Solution." American Society for Metals, Cleveland, Ohio.

AMERICAN SOCIETY FOR METALS (1964). "Heat Treating, Cleaning and Finishing," "Metals Handbook," Vol. 2, 8th ed. American Society for Metals, Metals Park, Ohio.

BAIN, E. C., and PAXTON, H. W. (1961). "Alloying Elements in Steel." American Society for Metals, Metals Park, Ohio.

BARRETT, C. S., and MASSALSKI, T. B. (1966). In "Structure of Metals," Chapt 11, pp. 270-305. McGraw-Hill, New York.

BARTLO, J. L. (1967). The Metallography of Titanium and Titanium Alloys, Res. Rep. R483, Reactive Metals, Inc., Niles, Ohio.

BRICK, R. M., GORDON, R. B., and PHILLIPS, A. (1965). "Structure and Properties of Alloys." McGraw-Hill, New York.

CLIMAX MOLYBDENUM COMPANY (1967). "Transformation and Hardenability in Steels." Climax Molybdenum Co., Ann Arbor, Michigan.

ELLIOT, R. P. (1965). "First Supplement to Constitution of Binary Alloys." McGraw-Hill, New York.

GROSSMAN, M. A. (1953). "Principles of Heat Treatment." American Society for Metals, Cleveland, Ohio.

GUTTMAN, L. *Solid State Phys.* **3**, 146–223.

GUY, A. G. (1959). "Elements of Physical Metallurgy." Addison-Wesley, Reading, Massachusetts.

HANSEN, M., and ANDERKO, K. (1958). "Constitution of Binary Alloys." McGraw-Hill, New York.

MARDER, A. R., and KRAUSS, G. (1967). *Trans. ASM* **60**, 651–660.

NEWKIRK, J. B., GEISLER, A. H., MARTIN, D. L., and SMOLUCHOWSKI, R. (1950). *Trans. AIME* **188**, 1249–1260.

NICHOLSON, R. B., THOMAS, G., and NUTTING, J. (1958–1959). *J. Inst. Metals* **87**, 429–438.

SEAGLE, S. R. (1965). "Titanium Metallurgy Course." Reactive Metals, Inc., Niles, Ohio.

UNITED STATES STEEL CORPORATION (1951). "Atlas of Isothermal Transformation Diagrams." U.S. Steel Corp., Pittsburgh, Pennsylvania.

UNITED STATES STEEL CORPORATION (1953). "Supplement to Atlas of Isothermal Transformation Diagrams." U.S. Steel Corp., Pittsburgh, Pennsylvania.

WERNER, F. E., AVERBACH, B. L., and COHEN, M. (1957.) *Trans. ASM* **49**, 823–841.

X

The Use of Phase Diagrams in the Joining of Metals

A. PRINCE

THE GENERAL ELECTRIC COMPANY LIMITED,
CENTRAL RESEARCH LABORATORIES,
HIRST RESEARCH CENTRE, WEMBLEY, ENGLAND

I. INTRODUCTION

Many techniques are used to join metals, but most processes can be categorized as variants of either fusion welding, solid-state bonding, brazing, or soldering. Fusion welds involve rapid heating and cooling cycles applied to similar metals or dissimilar metals, with or without filler metal. Solid-state bonding techniques may involve diffusion of atom species and their interaction to form new phases. Many brazed and soldered joints represent complex metallurgical interactions between the metals to be joined and the braze or solder. In each process a knowledge of the constitution of the alloys to be joined and of their likely phase reactions is of value in controlling the joining process and ensuring the reliability of the joints in service.

Naturally, phase diagrams do not stand by themselves. Of equal importance are transformation characteristics and diffusion kinetics. In this subject, as in many others, one may echo the words of Hume-Rothery: "The equilibrium diagram is the beginning of wisdom but not the end of it." It is not the purpose of this paper to review *in extensio* the literature but rather to make the point by selection, in the main from cases known personally to the author.

II. FUSION WELDING

One of the best-documented topics in welding science is that of weld-metal hot cracking. The foundryman meets it as hot tearing of castings. Verö (1935, 1936) was one of the first to realize that hot cracking was related to alloy constitution. The theory of hot cracking has since been developed by Bochvar (1939, 1947, 1952), Singer (1946, 1947), Pumphrey (1948), Medovar (1955), and Borland (1960). Referring to Fig. 1, we can distinguish three stages in solidification:

Stage 1. Formation of primary dendrites surrounded by the liquid phase.
Stage 2. Growth of the dendrites until interlocking results but in which there is still an interconnected network of the liquid phase.
Stage 3. Further growth of the solid phase has produced isolated pockets of liquid phase prior to final solidification.

Alloys in this system are considered to be susceptible to cracking when they have reached stage 2. However, any cracks formed can be healed by a flow of the liquid phase into the cracks. In stage 3 no relative movement of the liquid and solid phases is possible and any cracks formed cannot be healed. As will be seen from Fig. 1, the temperature range for stage 3 of the solidification process (the critical solidification range) rises to a maximum at the limit of solid solubility, point b. The amount of cracking experienced is related to the strain imposed in the critical solidification range. The strain is a function of the temperature interval between a, e and a, b, e (Fig. 1). The longer the critical solidification range the higher the contraction strain and the greater the cracking tendency. Experimentally, it is observed that the maximum susceptibility to cracking occurs with alloys whose composition is close to the limit of solid solubility in eutectic systems.

A weld pool does not freeze according to true equilibrium conditions. Nonequilibrium conditions depress the solidus by undercooling and limitation of diffusion. In this respect the critical solidification range is increased under practical conditions and the cracking tendency heightened. Borland (1960) has pointed out that it is not only the relative quantity of liquid that is important but also its distribution. If the liquid wets the solid phase such that a thin liquid film covers almost all grain faces, only a small contraction strain will be necessary to produce high stresses at the few solid bridges

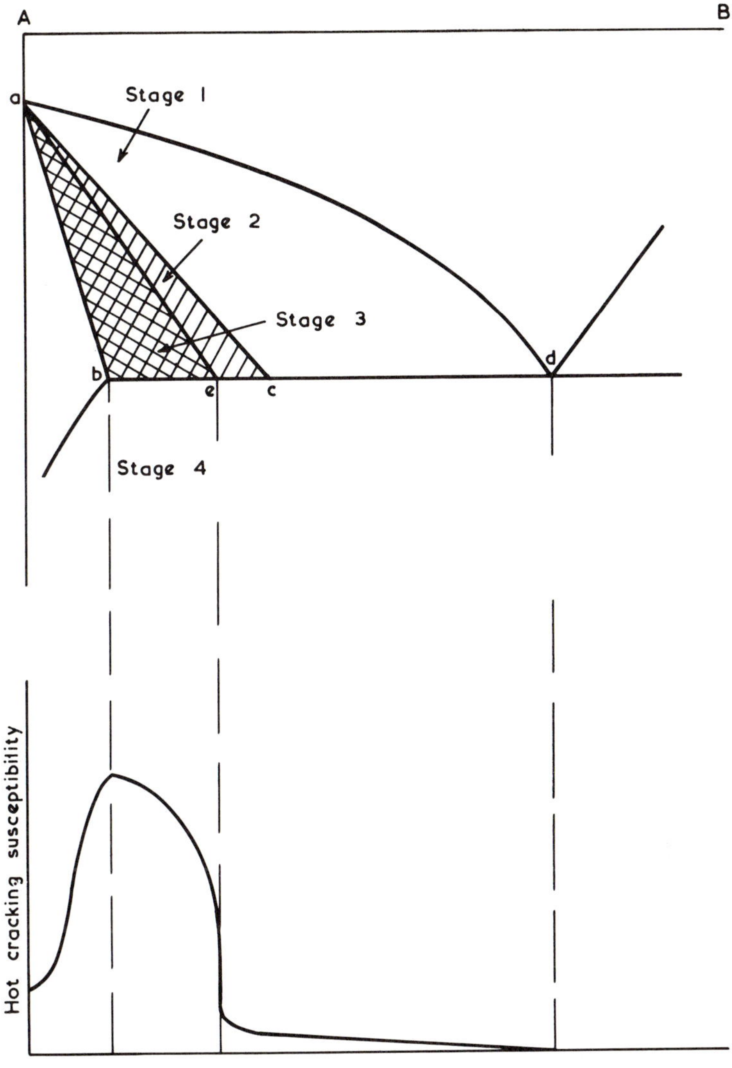

Fig. 1. Effect of constitution on cracking susceptibility in binary eutectic systems. (Courtesy The Welding Institute.)

between neighboring grains. If the liquid is confined to grain edges, the stresses imposed on the solid bridges will be lower and the alloy far less susceptible to cracking.

Many other examples could be given of the value of constitutional knowledge in fusion welding. Borland and Younger (1960) have published a detailed survey of the literature on cracking in welded Cr–Ni austenitic steels, linking many of the effects to variations in alloy constitution. Hull (1967) has recently used the generalized theory of Borland (1960) to explain the beneficial influence of delta ferrite on hot cracking of stainless steels.

The parent metal adjacent to the fusion line undergoes a high-temperature thermal cycle during welding. Heat-treatable alloys will therefore be subjected to a variety of heat treatments and structural changes in this heat-affected zone. As an example in this field, the paper by Hinde and Thorneycroft (1960) on the welding metallurgy of the Nimonic* alloys, is worth studying.

III. BRAZING

The real worth of alloy-constitution work in unraveling reactions in joining is to be seen when dealing with processes other than fusion welding. This is especially true in soldering and brazing. Gold-plated components are often used in the electronics and telecommunications industries. The reasons advanced for the use of gold are that it provides an inert surface that is tarnish and corrosion resistant (easing storage problems), it improves solderability, and provides a surface with low contact resistance. One interesting case of the use of a gold coating as a part of a brazed joint occurs in semiconductor-device manufacture. A silicon chip, the size of a pinhead across, is bonded to gold plated Nilo K (56% Fe, 19% Co, 25% Ni). The Au–Si binary is a simple eutectic system. The most reliable information (Heath, 1961; Gerlach and Goel, 1967) places the eutectic composition at 3.1 wt % Si as compared to the previously accepted value of 6-wt % Si (Hansen, 1958). When the silicon chip is placed in contact with the gold-plated Nilo K heated to a temperature above 370°C, reaction will occur between the gold and the silicon. This reaction produces liquid eutectic composition and from this is derived the term *eutectic brazing*. An illustration of this type of joint, Fig. 2, shows a 250-μ-diam gold wire bonded to silicon after surviving a life test of 14,000 h with a junction temperature of 175°C. The presence of primary gold dendrites at the silicon interface and primary silicon at the gold interface is associated with compositional differences in the molten bead when the bond was made, aggravated by gravity segregation (Au, 19.3 g/cc; Si, 2.4 g/cc; eutectic, 14 g/cc).

Eutectic brazing also finds application in many other circumstances.

† Registered trade name, International Nickel Limited.

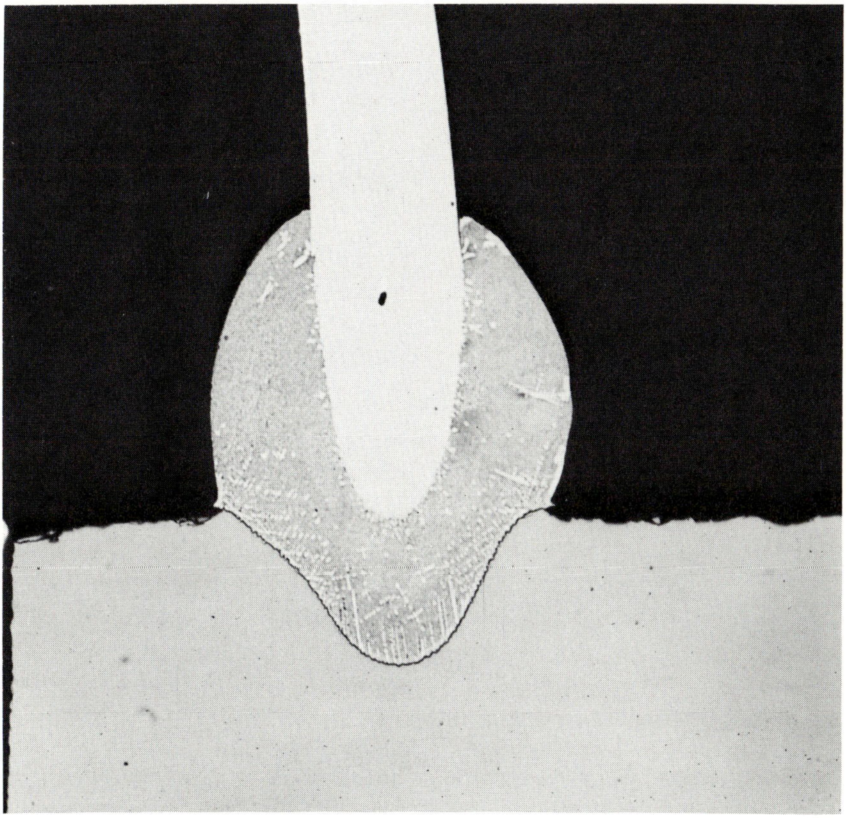

Fig. 2. Eutectic brazed joint between gold and silicon. 100 ×.

Owczarski (1962) utilized the low-melting-point eutectic found between $ZrFe_2$ and βZr, to join Zircaloy 2 to 18/8 stainless steel.

IV. SOLID STATE BONDING

A more complex situation was met with certain silicon transistors in which gold wires had been thermocompression bonded to aluminum evaporated onto silicon. Under conditions of severe overstress, in experiments aimed at identifying possible failure mechanisms, some of these bonds failed. It was observed that in these devices a dark purple coloration had grown out from the gold ball bond over the aluminum areas, and the observers linked the coloration with the failure. A colorful list of descriptive terms have been applied to this type of failure—purple plague, black death, and white plague. Cunningham (1965) quotes the purple deposit as having a growth rate along the aluminum of 10 mils/24 h at a temperature of 300°C.

The resolution of the mechanism of failure relies almost completely on a knowledge of the Au–Al phase diagram (Fig. 3). It is worth noting that Heycock and Neville's original work (1900) on the liquidus is still taken as authoritative and was essentially confirmed by Charquet (1967). Roberts-Austen described the purple brittle intermetallic phase $AuAl_2$ in the early 1890s (Roberts-Austen 1891, 1892). As can be seen from Fig. 3 the Au–Al system contains five intermediate phases: $AuAl_2$ (congruent), AuAl, Au_2Al (congruent), Au_5Al_2, and Au_4Al. When a bond is made between Au and Al some interdiffusion occurs and a thin purple phase can be seen even though

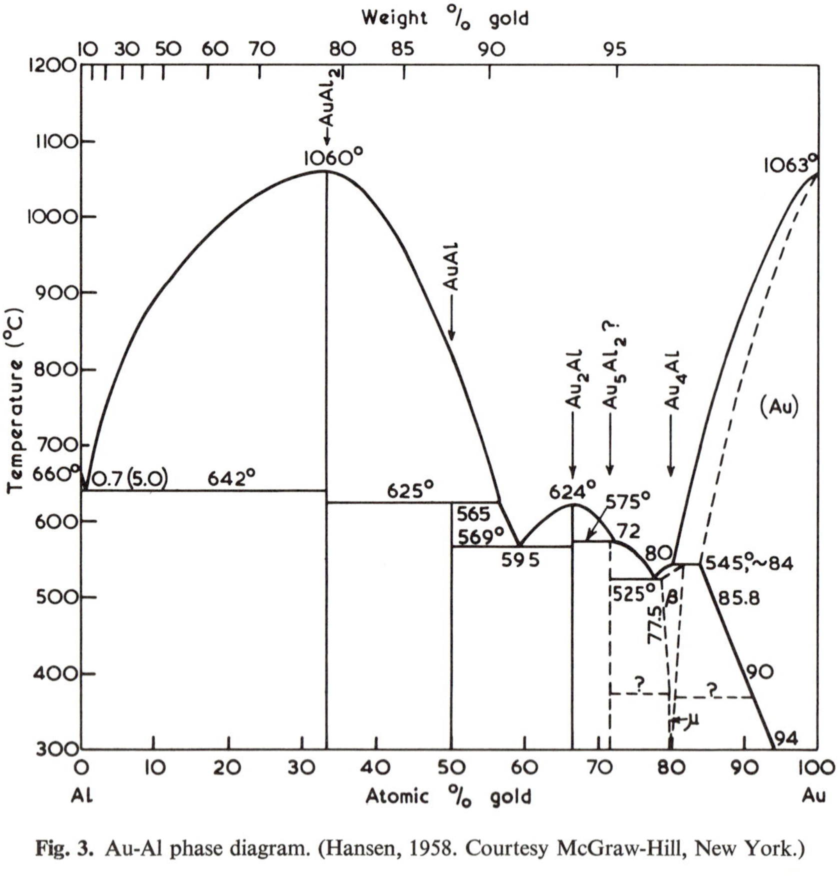

Fig. 3. Au-Al phase diagram. (Hansen, 1958. Courtesy McGraw-Hill, New York.)

the intermetallic layer is only of the order of 10,000Å thick. To accelerate diffusion bonds were heated at 300°C. Failure occurred at the interface between the Au wire and Au-rich intermetallic phase Au_4Al. The purple $AuAl_2$ phase was certainly not associated with the failure (Fig. 4). It will be noticed

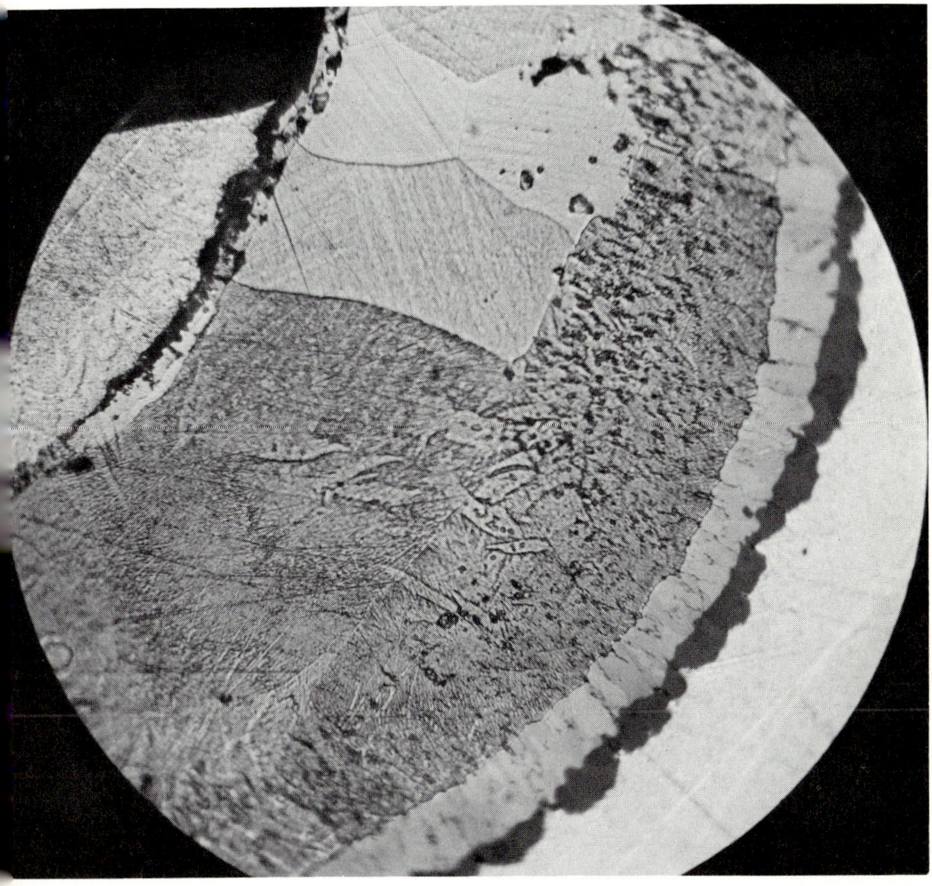

Fig. 4. Structure of Au-Al thermocompression bond after heating at 300°C. 740 ×.

that a good deal of porosity is present and Blech and Sello (1966) have postulated that this porosity, formed as a result of vacancy condensation produced by the higher diffusion rate of Au (Kirkendall phenomenon), is responsible for bond failure. The weakening effect of the porosity is also likely to be accentuated by the stresses set up at the Au interface by the volume change when Au_4Al forms (22% increase in volume).

V. SOLDERING

Soft soldering using Pb–Sn alloys is a means of joining often used in the electronics and telecommunications industries. When gold-plated surfaces are involved, as they often are, difficulties can arise in soft soldering. There is no real wetting problem. Indeed the reverse is true. Lead–tin alloys react extremely rapidly with gold. Herein lies the difficulty, since not only do brittle intermetallic phases form in the solder, but as the liquidus in the Au–Pb–Sn ternary system is of a relatively shallow slope near the Pb–Sn binary, a lot of gold can dissolve in the solder at normal soldering temperatures. For example a 40% Pb–60% Sn solder dissolves 11 wt % Au at 225°C, 17 wt % Au at 250°C, and 24 wt % Au at 275°C. In soft soldering to gold, it is recommended that the thickness of the gold deposit be kept below 5–μ to avoid excessive gold content in the finished joint, since the brittle $AuSn_4$ phase forms precipitates as platelets that provide easy fracture paths in the joint.

To arrive at the foregoing conclusion depended on a knowledge of the interaction between gold and lead–tin alloys. Berry and Johnson (1964) were the first to show that in normal circumstances lead, tin, and $AuSn_4$ were the equilibrium phases and they also showed that the addition of gold to a eutectic Pb–Sn alloy produced a flat liquidus. Subsequently the author established the eutectic nature of the pseudobinary section AuSn–Pb (Prince, 1966) and determined the equilibria, in the AuSn–Pb–Sn partial ternary system (Prince, 1967). Referring to Fig. 5, we see that a ternary quasi-peritectic reaction P_1 occurs at 275°C:

$$\text{Liq.} + \text{AuSn} \rightleftharpoons \text{Pb} + \text{AuSn}_2$$

followed by a second ternary quasi-peritectic reaction P_2 at 208°C:

$$\text{Liq.} + \text{AuSn}_2 \rightleftharpoons \text{Pb} + \text{AuSn}_4$$

and solidification is completed with a ternary eutectic reaction E at 177°C:

$$\text{Liq.} \rightleftharpoons \text{Pb} + \text{Sn} + \text{AuSn}_4$$

Even with slow cooling rates of 1°–3°C/min, the peritectic reaction does not go to completion and secondary separation, particularly of Pb + $AuSn_4$, is frequently degenerate. These are the conditions to be expected in the cooling of soldered joints and give confidence in applying the results of this type of alloy study to soldering conditions. The degeneracy of secondary separations has misled other workers (Whitfield and Cubbin, 1965) to postulate the presence of $AuPb_2$ in dendritic form in soldered joints.

To reduce the embrittlement of soldered joints the thickness of the gold coating should be kept to 5–μ maximum or soldering should be carried out on areas from which the coating has been removed by prior application of

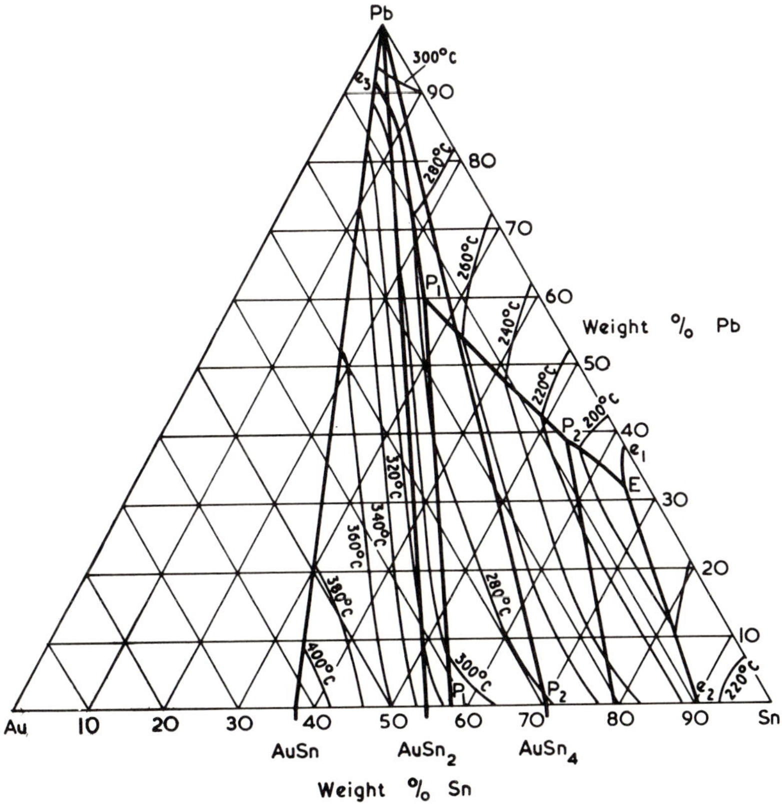

Fig. 5. Liquidus surface of AuSn–Pb–Sn system. (Courtesy Elsevier, Amsterdam.)

solder and wiping clean. The ternary diagram also suggests that a reduction in the quantity of $AuSn_4$ could be obtained by increasing the lead content of the solder. As tin is the active component in soft solder this would also have the effect of decreasing the reaction rate with the gold. Thwaites (1965) has shown that a 1-μ layer of gold over 5–7-μ of a tin–nickel alloy has low contact resistance, retains solderability after storage, and involves no risk of joint embrittlement.

Some of the most common soldering operations have not yet received the study they deserve. The soft soldering of brass is widely practiced and yet there is still a great deal of conjecture with regard to the reactions that occur (Saperstein and Howes, 1968; Griffiths and Charles, 1968). Discussion is normally based on the reaction of the tin with copper in terms of the formation of the binary Cu–Sn phases, Cu_6Sn_5 and Cu_3Sn. X-ray microprobe analysis shows however a significant concentration of zinc in the intermetallic phases formed, and it would seem that the report by Hartley (1927) of the formation of a Cu–Sn–Zn compound on exposure of brasses to molten tin should be investigated further. In the final analysis all such investigations will rely on a knowledge of the pertinent alloy constitution.

When the phase diagram is available, it is relatively easy to sort out the reactions that produced a given structure. For example, some recently unpublished work on soldered contacts for silicon transistors required an examination of the soldering of silver to copper using various SnAg solders. Figure 6 illustrates the microstructural features of joints made at 400°C

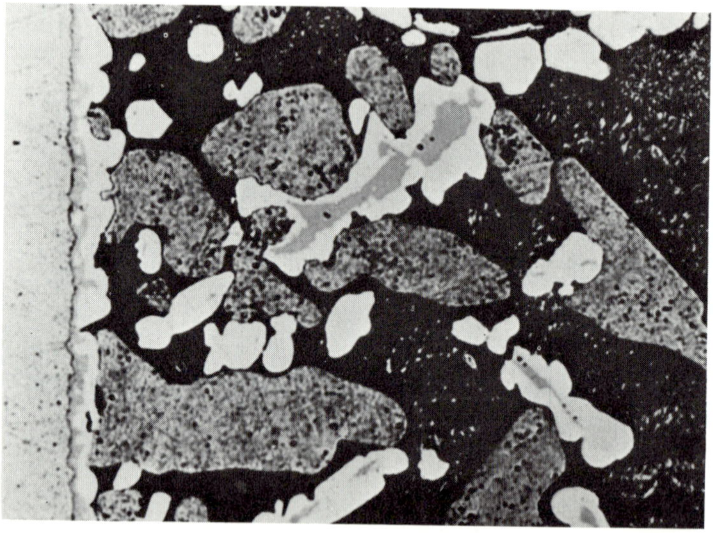

Fig. 6. Structure of joint produced by soldering copper to silver with a 97.5% Sn, 2.5% Ag solder. 500×. *Extreme left*: Cu, layer of Cu_3Sn: gray, layer of Cu_6Sn_5: white, solder: Cu_3Sn (gray), Cu_6Sn_5 (white), Ag_3Sn (mottled), eutectic matrix (black).

using a 97.5 wt % Sn–2.5 wt % Ag solder. Reaction layers of Cu_3Sn (gray) and Cu_6Sn_5 (white) formed at the interface with the copper surface on the left. Primary Cu_3Sn surrounded by a peritectic rim of Cu_6Sn_5 together with Ag_3Sn (mottled) set in a eutectic matrix constitute the solder microstructure. This interpretation is based on the constitutional work of Gebhardt and Petzow (1959) on the Cu–Ag–Sn ternary system (Fig. 7). At 400°C solution of copper is such as to bring the solder composition into the primary Cu_3Sn phase field. On cooling some primary Cu_3Sn separates, followed by the monovariant reaction, liquid + $Cu_3Sn \rightleftharpoons Cu_6Sn_5$, since the solder composition is in the area $u_4U_6u_3$ and not within the four-phase triangle for the reaction, liq + $Cu_3Sn \rightleftharpoons Cu_6Sn_5 + Ag_3Sn$ (350°C). The liquid will move

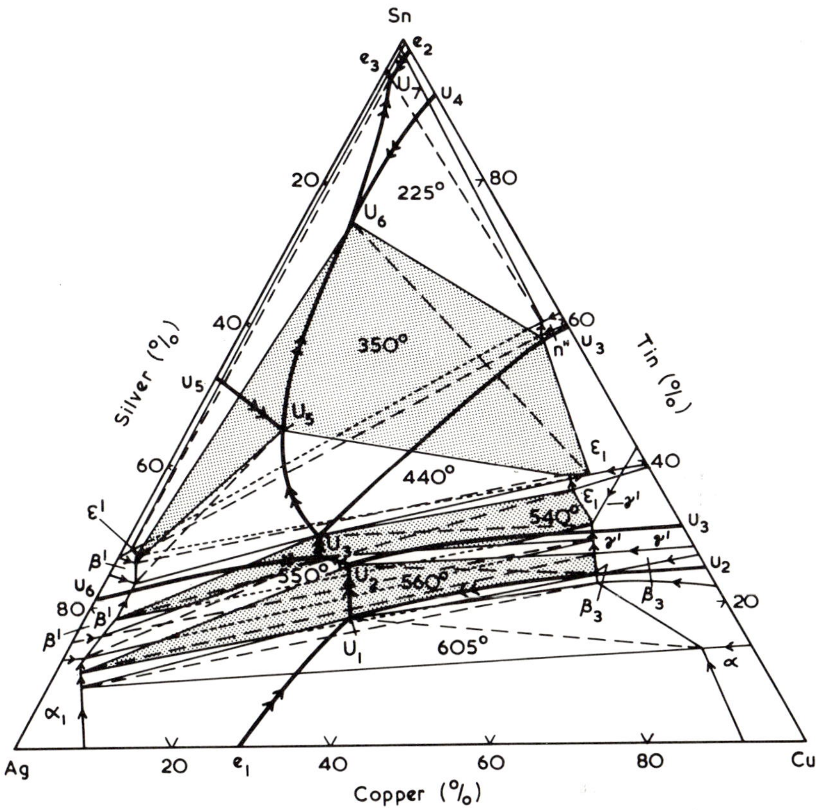

Fig. 7. Equilibria in the Ag–Cu–Sn system. (Courtesy Dr. Riederer-Verlag GMBH, Stuttgart.)

down the curve u_4U_6 until the peritectic rim of Cu_6Sn_5 surrounding the Cu_3Sn stops further reaction, when the liquid composition will move over the surface of primary crystallization of Cu_6Sn_5 to meet the monovariant curve U_6U_7 below point U_6. Separation of $Ag_3Sn + Cu_6Sn_5$ occurs to U_7 followed by the four-phase reaction, liquid + $Cu_6Sn_5 \rightleftharpoons Ag_3Sn$ + Sn (225°C). Solidification ends with the eutectic separation of Ag_3Sn + Sn.

VI. METAL–CERAMIC SEALS

The joining of metals to ceramics is an equally fascinating study which relies to a large extent on knowledge of equilibrium relationships. Such joints are of prime importance in the electronic valve and lamp industries. In general there are two main methods for sealing metals to oxide ceramics using alloys:

(1) refractory-metal techniques
(2) active-alloy brazing techniques.

In the refractory-metal technique seal formation depends on the glassy phase in an impure alumina ceramic wetting and entering the voids in a sintered molybdenum layer applied to the ceramic. The ceramic is bonded via a braze metal to an alloy with similar expansion coefficients [e.g., Nilo K (Kovar)]. The braze metal commonly used for vacuum devices is copper since it is readily available, has a low vapor pressure, and wets molybdenum without detectable alloying.

When a molybdenum-metallized ceramic is copper brazed to Nilo K or ferrous alloys, a layer-type deposit frequently forms at the interface between the molybdenum and the braze metal. Appreciable attack of the ferrous metal takes place, and a precipitate appears in the braze metal (Fig. 8).

Seebold and Birks (1961) detected the presence of one intermetallic compound, Fe_3Mo_2, in Fe–Mo diffusion couples held at 1100°C. In brazing the molten copper dissolves iron (see the Fe–Cu phase diagram for the solubility of iron in molten copper). The iron is deposited at the molybdenum surface by a process of isothermal mass transfer. It reacts with the molybdenum to form Fe_3Mo_2 in accordance with the phase diagram (Fig. 9).

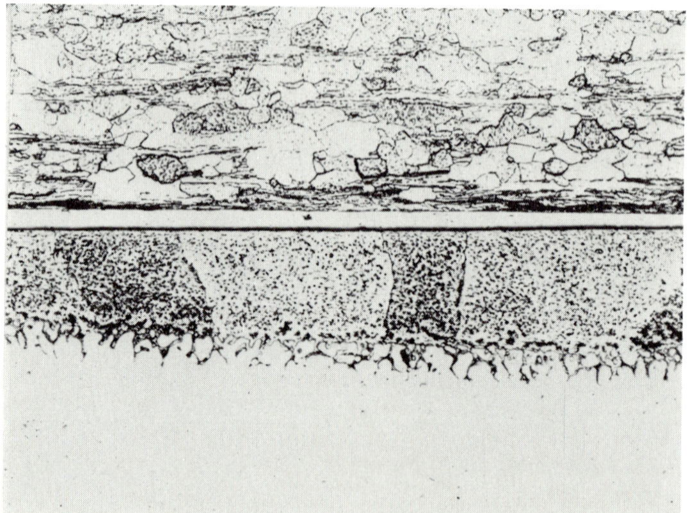

Fig. 8. Structure of molybdenum brazed to chromium-nickel stainless steel using copper braze metal. 200×.

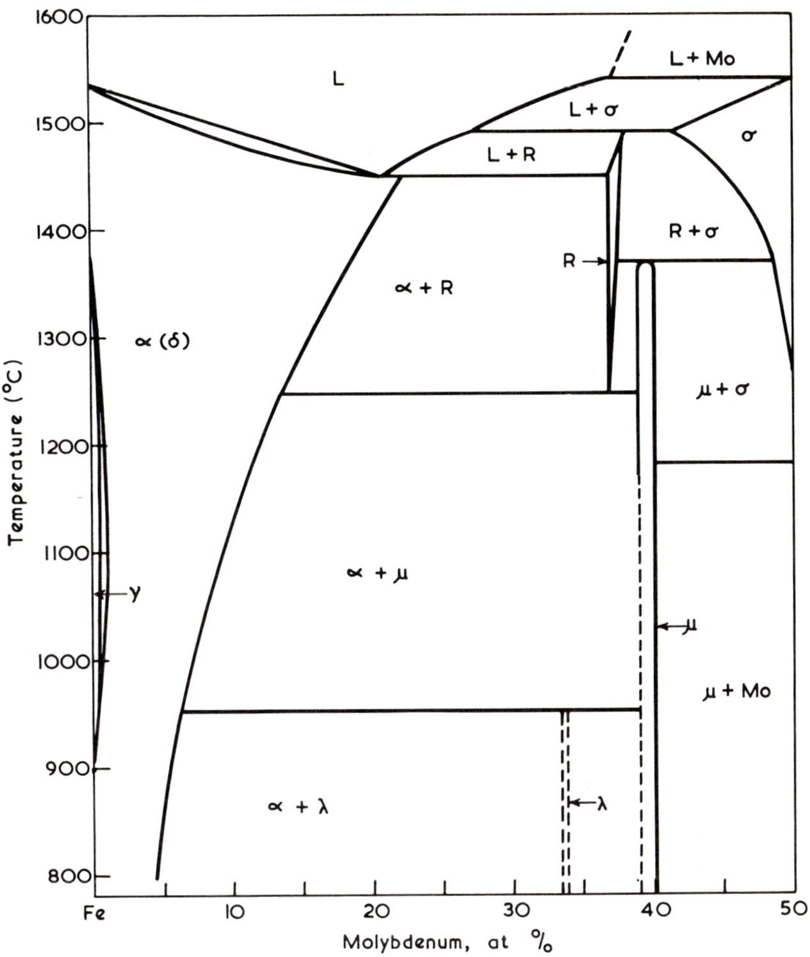

Fig. 9. Fe–Mo phase diagram. (Courtesy Iron and Steel Institute.)

The reaction occurs at the molybdenum interface since Mo is virtually insoluble in molten Cu. Since the braze metal is being depleted of dissolved iron by Fe_3Mo_2 formation more iron is dissolved at the iron–copper interface. The formation of Fe_3Mo_2 is the rate-controlling process.

When molybdenum is copper brazed to Nilo K, a hard layer with composition 57.5% Mo, 26% Fe, 8.5% Ni, and 8% Co, and of μ Fe_3Mo_2 structure, is formed. It should be noted that Fe_3Mo_2 contains 53.4 wt % Mo, 46.6 wt % Fe.

Samples of the quaternary alloy were synthesized and the thermal expansion coefficient of the alloy measured as twice that of the molybdenum. This expansion mismatch probably gives rise to strain sufficient to cause fracture and hence leak paths in valve assemblies. Since no data has been published on the Co–Fe–Mo–Ni system, the structure of the synthetic alloy was interpreted on the basis of the Co–Fe–Mo system, it being assumed that the cobalt content could be represented by the combined cobalt + nickel contents to give a ternary alloy containing 57.5% Mo, 26% Fe, 16.5% Co. In the Co–Fe–Mo ternary system (Das *et al.*, 1952) the μ phase extends across the diagram at 1200°C from the Fe–Mo binary to the Co–Mo binary. An alloy of the above composition would be entirely μ phase or μ with small amounts of molybdenum at 1200°C. Metallographic and x-ray analysis of the synthetic alloy heat treated at 1150°C showed its structure to consist of μ phase with some free molybdenum. Heat treatment at higher temperatures did not reveal the presence of the *R* phase (Sinha *et al.*, 1967), which may indicate limited compositional stability of this phase in the ternary system.

The active-alloy brazing technique for joining metals to ceramics uses braze metals containing alloy additions that have the ability to react chemically with the alumina ceramic. There is therefore no need to rely on the presence of the weak glassy phase in the ceramic as part of the bond. The active additions also promote wetting to the metallic part of the seal. Conventional active-alloy braze metals incorporate titanium or zirconium, which give the required wetting and reactivity to ceramics.

As an example of the active-alloy technique, mention may be made of the brazing of niobium. Fox *et al.* (1963) comment that niobium is in use, or will be in the future, for high-temperature nuclear reactors, heat exchangers, reentry vehicles, and other space applications. Any brazing alloy for niobium should be serviceable at high temperatures (i.e., have a melting point in the range 1000°–1350°C), resist liquid and vapor attack by alkali metals since this is usually the environment of most concern, and have little tendency to react with the niobium although wetting it easily and have the requisite mechanical properties. Alloy development was based on the high-melting-point metals titanium and zirconium (mp 1668° and 1850°C, respectively). The Ti–Zr system itself is attractive since there is a minimum in the liquidus at 50% Zr and 1610°C. It is necessary to find a ternary addition that lowers the liquidus further. Beryllium was found to be effective; 5 wt % Be lowers the melting point of Zr to 965°C (Fig. 10), and 10 wt % Be lowers the melting point of Ti to 1030°C (Fig. 11). It is not surprising that the alloy quoted by Fox *et al.*, 48 Ti–48 Zr–4 Be, had a solidus of about 1050°C. No work has been published on the Be–Ti–Zr ternary system but a glance at the constituent binary systems is sufficient to see that 4% Be added to the equiatomic Ti–Zr binary alloy will significantly depress the minimum in the Ti-Zr liquidus.

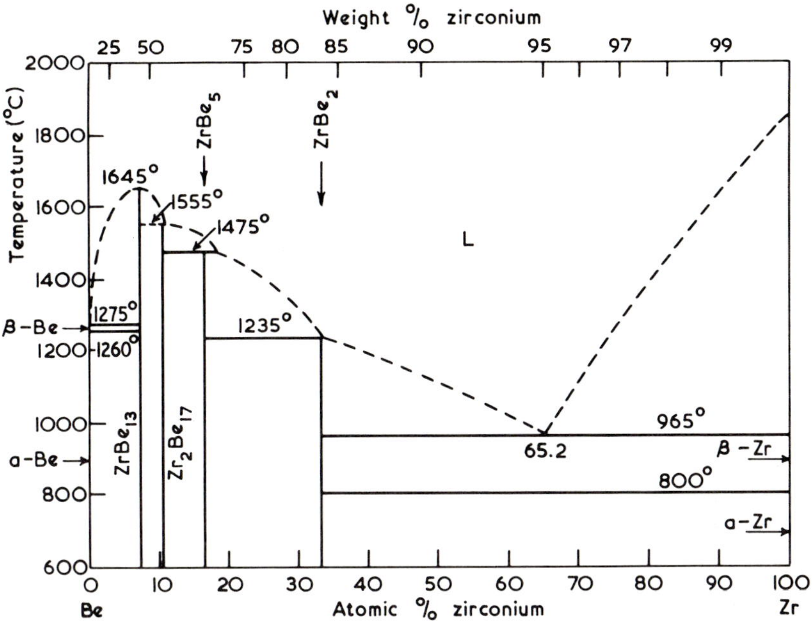

Fig. 10. Be–Zr phase diagram. (Hansen, 1958. Courtesy McGraw-Hill, New York.)

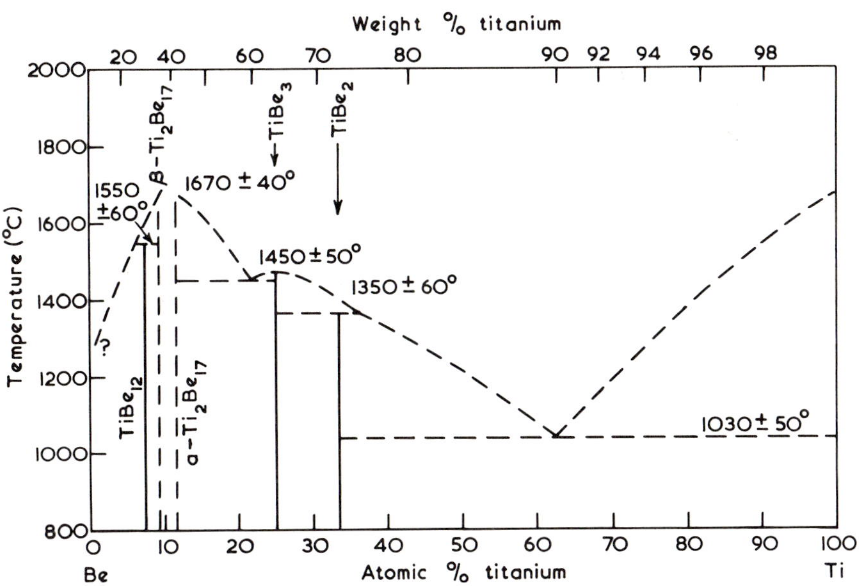

Fig. 11. Be–Ti phase diagram. (Hansen, 1958. Courtesy McGraw-Hill, New York.)

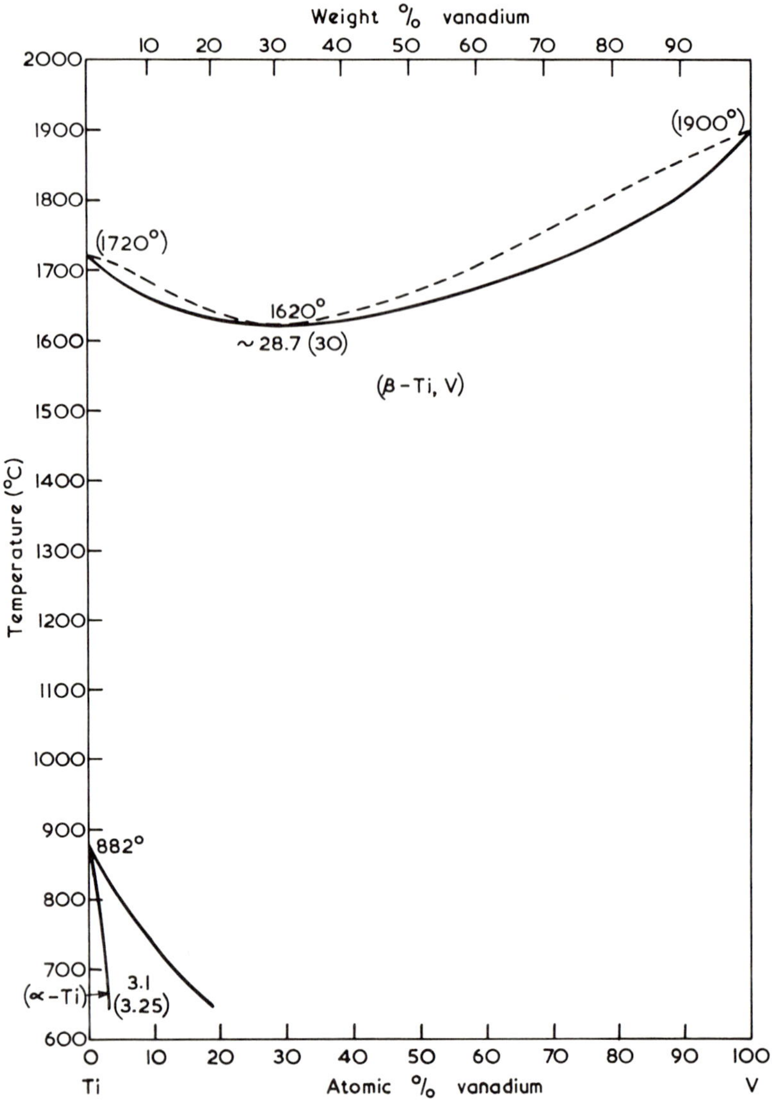

Fig. 12. Ti–V phase diagram. (Hansen, 1958. Courtesy McGraw-Hill, New York.)

Niobium also features as one of the components to be joined to high-purity alumina in high-pressure sodium vapor lamps. In our work (British Patent, 1964) we again took the Ti–Zr minimum melting point alloy as the base for study. The Nb–Ti–Zr system was initially considered since Nb and Zr also show a minimum in the liquidus at 22% Nb and 1740°C. The third binary system, Nb–Ti, exhibits complete solid-solution behavior. The work of

Mikheev and Belousov (1961) demonstrated that there is a continuous series of ternary solid solutions at high temperatures in the Nb–Ti–Zr system but there is a fold in the liquidus surface that runs from the Nb–Zr minimum at 1740°C down to the Ti–Zr minimum at 1610°C. All alloys in the ternary system possess excessive liquidus temperatures, making them somewhat impractical for use in lamp seals. Vanadium was considered a likely depressant of the liquidus temperature of Ti–Zr alloys since the Ti–V system (Fig. 12) contains a minimum in the liquidus at 30% V and 1620°C and the V–Zr system (Fig. 13) shows a eutectic reaction at 1230°C between βZr and Zr V_2.

The work of Nowikov and Baer (1958) shows a minimum in the liquidus at about 55% Zr, 26% V, 19% Ti, and 1220°C. Alloys with this composition have good wetting characteristics on alumina and form excellent seals (Fig. 14).

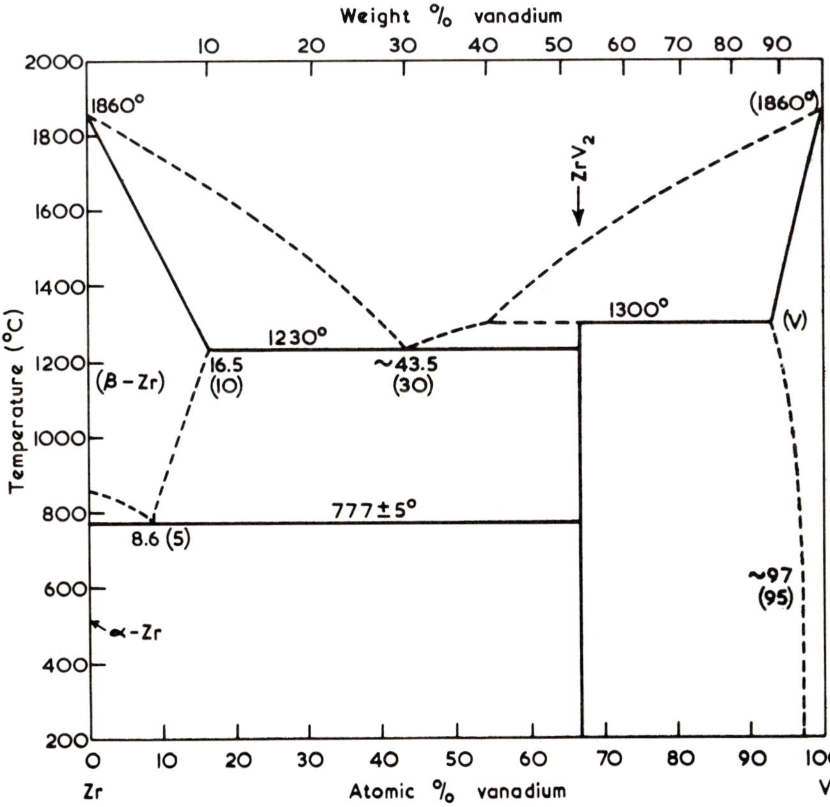

Fig. 13. V–Zr phase diagram. (Hansen, 1958. Courtesy McGraw-Hill, New York.)

Fig. 14. Structure of seal between niobium and pure alumina. 100×.

VII. CONCLUSION

It is rare to deal with simple binary systems in joining. More often than not multicomponent systems are involved, and the paucity of equilibrium data is such that intelligent simplification of the system is necessary to reduce the task of interpreting the reactions to manageable proportions. Often observations made in the course of work on the joining of metals provides information on the alloy phase equilibria concerned. Those who are interested in joining are generally not willing to spend time determining phase diagrams, but many pertinent observations on unknown equilibria must be buried in their files. An illustrative example will suffice. The grid of an experimental valve was made from Pt-clad Mo wire over which was fitted a Zr cap. During manufacture the assembly was heated to 1300°C. Afterwards the Zr cap was found to have fused to the wire. At this time there was no definitive work published on the Pt–Zr phase diagram. We heated the fusion zone and detected initial melting at 1130°C. Figure 15 shows the structure of the fused zone to consist of "βZr" grains surrounded by grain boundary eutectic. The later

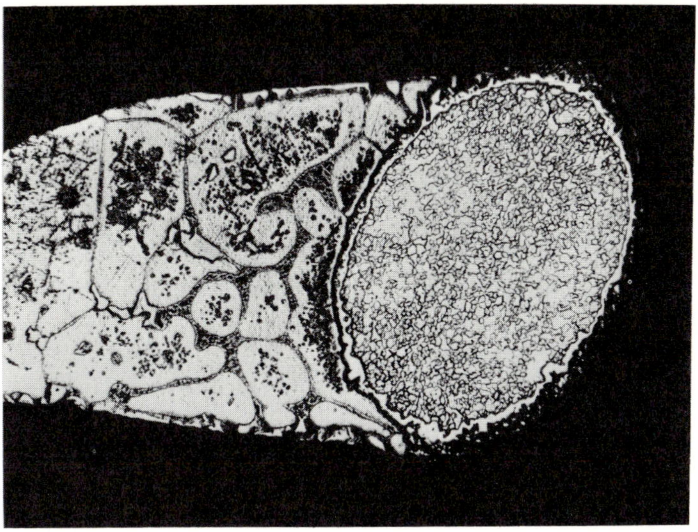

Fig. 15. Structure of zirconium strip fused to platinized molybdenum wire from a valve grid. 90×.

work of Kendall *et al.* (1961) indicated a eutectic reaction between βZr and $PtZr_2$ at 36 wt % Pt and 1185°C.

Industrial problems and research throw up many similar scraps of valuable data, most of which are never destined to enter the literature although they would constitute valuable additions to our knowledge. Unfortunately, metallurgy is subject to fashions and currently both phase equilibria and the science of joining are underrated. It is hoped that this contribution will show that the combination of both sciences is an exciting and rewarding field in which to work.

References

Berry, R. D., and Johnson, R. W. (1964). *Inst. Welding Conf. Brazing and Soldering*, Paper 9.

Blech, I. A., and Sello, H. (1966). *J. Electrochem. Soc.* **113**, 1052.

Bochvar, A. A., and Khakimdzhanova, M. K. (1939). *Metallurgy* **2**, 75.

Bochvar, A. A., and Sviderskaya, Z. A. (1947). *Izv. Akad. Nauk SSSR., Otd. Tekhn. Nauk.* **3**, 349.

Bochvar, A. A., and Novikov, I. I. (1952). *Izv. Akad. Nauk SSSR., Otd. Tekhn. Nauk.* **2**, 217.

Borland, J. C. (1960). *Brit. Welding J.* **7**, 508.

Borland, J. C., and Younger, R. N. (1960). *Brit. Welding J.* **7**, 22.

Charquet, D., Desre, P., and Bonnier, E. (1967). *Compt. rend.* **264**, 1637.
Cunningham, J. A. (1965). *Solid State Electron.* **8**, 735.
Das, D. K., Rideout, S. P., and Beck, P. A. (1952). *Trans. Am. Inst. Min. Met. Eng.* **194**, 1071
Fox, C. W., Gilliland, R. G., and Slaughter, G. M. (1963). *Welding J.* **42**, 535s.
Gebhardt, E., and Petzow, G. (1959). *Z. Metallk.* **50**, 597.
Gerlach, W., and Goel, B. (1967). *Solid State Electron.* **10**, 589.
Grififths, J. R., and Charles, J. A. (1968). *Metal Sci. J.* **2**, 89.
Hansen, M. (1958). "Constitution of Binary Alloys." McGraw-Hill, New York.
Hartley, H. J. (1927). *J. Inst. Metals* **37**, 193.
Heath, E. G. (1961). *J. Electron. Control* **11**, 13.
Heycock, C. T., and Neville, F. H. (1900). *Phil Trans. Roy. Soc.* A **194**, 201.
Hinde, J., and Thorneycroft, D. R. (1960). *Brit. Welding J.* **7**, 605.
Hull, F. C. (1967). *Welding J.* **46**, 399s.
Kendall, E. G., Hayes, C., and Swift, R. E. (1961). *Trans. Met. Soc. AIME* **221**, 445.
Medovar, B. (1955). *Avto. Svarka.* **8**, 79.
Mikheev, V. S., and Belousov, O. K. (1961). *Russian J. Inorg. Chem.* **6**, 972.
Nowikov, A., and Baer, H. G. (1958). *Z. Metallk.* **49**, 195.
Owczarski, W. A. (1962). *Welding J.* **41**, 78s.
Prince, A. (1966). *J. Less Common Metals* **10**, 365.
Prince, A. (1967). *J. Less Common Metals* **12**, 107.
Pumphrey, W. I., and Jennings, P. H. (1948). *J. Inst. Metals* **75**, 235.
Rigden, S. A. R., British Patent 1,065,023 (1964).
Roberts-Austen, W. C. (1891). *Proc. Roy. Soc.* (*London*) **49**, 347.
Roberts-Austen, W. C. (1892). *Proc. Roy. Soc.* (*London*) **50**, 367.
Saperstein, Z. P., and Howes, M. A. H. (1968). *Welding J.* **47**, 162s.
Seebold, R. E., and Birks, L. S. (1961). *J. Nucl. Mater.* **3**, 260.
Singer, A. R. E., and Cottrell, S. A. (1946). *J. Inst. Metals* **73**, 33.
Singer, A. R. E., and Jennings, P. H. (1947). *J. Inst. Metals* **73**, 197.
Sinha, A. K., Buckley, R. A., and Hume-Rothery, W. (1967). *J. Iron Steel Inst.* **205**, 191.
Thwaites, C. J. (1965). *Trans. Inst. Metal Finishing* **43**, 143.
Verö, J. (1953). *Mitt. berg-u. Lütt.* Abt. *Kg. ung. Palatin-Joseph Univ. Sopron.* **7**, 138.
Verö, J. (1936). *Metal Ind.* **48**, 431.
Whitfield, J., and Cubbin, A. J. (1965). *ATE J.* **21**, 2.

Author Index

Numbers in italics refer to the pages on which the complete references are listed.

K

L

Subject Index

T

U

V

W

X

Z